Applied ethology 2016

Standing on the shoulders of giants

ISAE2016

Proceedings of the 50th Congress of the International Society for Applied Ethology

12-15th July, 2016, Edinburgh, United Kingdom

Standing on the shoulders of giants

edited by:

Cathy Dwyer

Marie Haskell

Victoria Sandilands

OASES

Online Academic Submission and Evaluation System

This work is subject to copyright. All rights are reserved, whether the whole or part of the material is concerned. Nothing from this publication may be translated, reproduced, stored in a computerised system or published in any form or in any manner, including electronic, mechanical, reprographic or photographic, without prior written permission from the publisher:
Wageningen Academic Publishers
P.O. Box 220
6700 AE Wageningen
The Netherlands
www.WageningenAcademic.com
copyright@WageningenAcademic.com

EAN: 9789086862870
e-EAN: 9789086868339
ISBN: 978-90-8686-287-0
e-ISBN: 978-90-8686-833-9
DOI: 10.3920/978-90-8686-833-9

First published, 2016

Welcome to the 50th Congress of ISAE

In 1966, following the 'game-changing' publication of Animal Machines in 1964 and the Brambell Report in 1965, a small group of veterinarians began a Society for Veterinary Ethology in Edinburgh. They hosted their first meeting in the University of Edinburgh's Hume Tower, George Square on June 4th 1966, with a symposium where four scientific presentations were delivered. Fifty years on we are delighted to welcome the Society back to Edinburgh for the International Congress. Edinburgh last played host to the Congress in 1991, for the 25th Anniversary, where the name was changed to reflect the increase in scientists other than veterinarians who belonged to the society and so it became the International Society for Applied Ethology. This was also the first year that the organisers took the brave step of moving from a single session to two parallel sessions to accommodate the number of submissions. By 2016 the number of papers submitted to the meeting has increased a hundred-fold from that initial symposium, and we are honoured to be hosting the largest ISAE Congress ever.

In this our 50th year, we want to both look back and reflect on how far we have come, and reach forward to embrace new opportunities, new disciplines and new challenges that face applied ethology. We have borrowed from the words of Isaac Newton in setting the theme of this Congress: 'if I have seen further than others, it is by standing upon the shoulders of giants'. This perfectly encompasses our acknowledgement of the work of 'giants' in the field and our attempts to 'see further' from the vantage point of their scientific insights. We have chosen, as one of our most influential giants, the work of the great ethologist Nico Tinbergen, and have framed three Congress sessions around his seminal questions in animal behaviour. However, the application part of our name is also very important, as is our ability to respond and move with the times to provide relevant support to animal welfare and other grand challenges facing our world. So we have also developed sessions on trade-offs between animal welfare, applied ethology and other issues such as sustainability and environmental management; and the emerging field of positive welfare where applied ethology has much to offer. For those of us who have spent years of a research career watching videos, or measuring things by hand, the development of automation or novel technology represents an exciting advance in the types of research questions we can now ask, and so we also have a session, 'Novel Techniques', full of new developments to move our field forward.

We are really pleased that so many of you want to come to Edinburgh this summer to visit our beautiful city and enjoy the Congress. However, this has meant, with 50% more abstracts submitted than to any other Congress, that the number of people wanting oral presentations has vastly exceeded the number of talks we can include in the programme. To accommodate so much excellent research in applied ethology, we have borrowed from successful trials at other Congresses: each morning of the programme we will have a period when there will be three parallel sessions of talks; we have included in the programme short oral presentations of 5 minutes to allow more opportunities for oral presentations; and we have included again the possibility for poster presenters to make an additional video poster presentation to increase the visibility of their work. We hope that all these developments will mean that all presenters will get the opportunity to talk about their work at this great annual meeting of scientists, students and professionals in applied ethology.

A very warm welcome to Edinburgh this summer, and we look forward to the enduring ISAE traditions of scientific excellence in presentations, passion and new insights in discussions, and of course great dancing at the congress banquet.

Fàilte gu Alba

Cathy Dwyer, Marie Haskell and Victoria Sandilands

Acknowledgements

ISAE was jointly organised by:
- SRUC (Scotland's Rural College)
- University of Edinburgh
- World Animal Protection

Scientific Committee

Marie Haskell (Chair), Alistair Lawrence, Cathy Dwyer, Kenny Rutherford, Tamsin Coombs, Laura Dixon, Rick D'Eath, Fritha Langford, Pol Llonch, Jill Mackay, Malcolm Mitchell, Carol Thompson, Victoria Sandilands, Simon Turner, Francoise Wemelsfelder

Organising Committee

Cathy Dwyer (Chair), Natalie Waran, Mike Appleby, Marie Haskell, Fritha Langford, Victoria Sandilands, Simon Turner, Laura Dixon

Social Committee

Cathy Dwyer, Kenny Rutherford, Fritha Langford, Jill Mackay, Susan Jarvis, Tamsin Coombs, Leonor Valente, Irene Camerlink

Social Media

Lauren Robinson, Fritha Langford

Ethics committee

Anna Olsson (Chair), Portugal
Marie Jose Hotzel, Brazil
Alexandra Whittaker, Australia
Francois Martin, USA
Franck Peron, France
Francesco De Giorgio, The Netherlands

Technical support

University of Edinburgh, SRUC

Professional conference organisers:

Zibrant http://www.zibrant.com/

Referees

Michael Appleby
Gareth Arnott
Lucy Asher
Emma Baxter
Ngaio Beausoleil
Harry Blokhuis
Alain Boissy
Xavier Boivin
Liesbeth Bolhuis
Laura Boyle
Bjaarne Braastad
Stephanie Buijs
Oliver Burman
Andy Butterworth
Irene Camerlink
Sylvie Cloutier
Michael Cockram
Melanie Connor
Jonathon Cooper
Valerie Courboulay
Rick D'Eath
Antoni Dalmau
Ingrid De Jong
Suzanne Desire
Laura Dixon
Rebecca Doyle
Cathy Dwyer
Bernadette Earley
Sandra Edwards
Marisa Erasmus
Hans Erhard
Inma Estevez
Emma Fabrega
Mark Farnworth
Bjorn Forkman
Francisco Galindo

Jenny Gibbons
Derek Haley
Moira Harris
Marie Haskell
Suzanne Held
Paul Hemsworth
Mette Herskin
Sarah Ison
Susan Jarvis
Per Jensen
Margit Jensen
Linda Keeling
Jenna Kiddie
Ute Knierim
Paul Koene
Fritha Langford
Alistair Lawrence
Caroline Lee
Tina Leeb
Pol Llonch
Jill Mackay
Maja Makagon
Xavier Manteca
Jeremy Marchant Forde
Jessica Martin
Lindsay Matthews
Dorothy McKeegan
Rebecca Meagher
Michael Mendl
Stefano Messori
Susanne Millman
Daniel Mills
Michela Minero
Malcolm Mitchell
David Morton
Lene Munksgaard

Ruth Newberry
Christine Nicol
Birte Nielsen
Tomas Norton
Cheryl O'Connor
Ed Pajor
Elizabeth Paul
Carol Petherick
Fiona Rioja-Lang
Bas Rodenburg
Jeffrey Rushen
Kenny Rutherford
Mark Rutter
Victoria Sandilands
Lars Schrader
Karin Schütz
Janice Siegford
Lynne Sneddon
Mairi Stewart
Joe Stookey
Mhairi Sutherland
Janet Talling
Carol Thompson
Cassandra Tucker
Simon Turner
Anna Valros
Heleen Van De Weerd
Antonio Velarde
Morris Villarroel
Eberhard Von Borell
Marina Von Keyserlingk
Susanne Waiblinger
Natalie Waran
Daniel Weary
Francoise Wemelsfelder
Christoph Winckler
Hanno Wuerbel

Note of support following Japanese earthquake

In April 2016 a series of strong earthquakes occurred in Kumamoto in the Kyushu Region of Japan. It is estimated that at least 49 people were killed and more than 3000 people injured. Tokai University was also severely affected, and three students lost their lives.

As a result, members of the International Society of Applied Ethology from the region have been unable to attend ISAE 2016. Some authors have chosen to withdraw their abstracts, whereas others have retained their abstracts within the proceedings but cannot accompany their poster. The ISAE General Council and the ISAE 2016 Organisation Committee ask you to join us in extending our deepest sympathies to those affected by the tragedy, and wish the region a speedy and safe recovery.

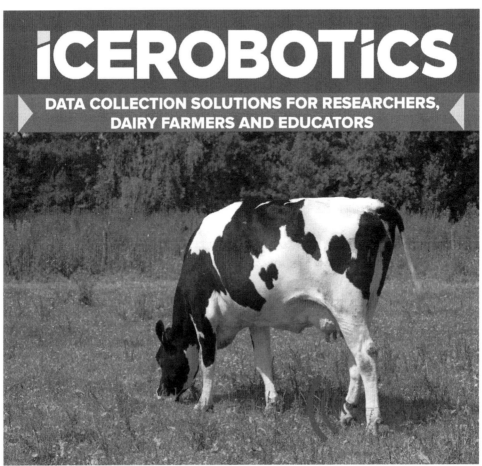

Applied ethology 2016

Dehorning is acutely painful. That's why a local anaesthetic is often given – but a few hours later its effect wears off and pain erupts. Co-administration of Metacam – newly licensed for dehorning pain – provides time-appropriate pain relief. So now, at last, you can make dehorning a metacomfortable experience for everybody.

PAIN ERUPTING AFTER THE LOCAL
YET ANOTHER THING METACAM TAKES CARE OF

THE DONKEY
SANCTUARY

HUMANE SOCIETY
INTERNATIONAL

Maps

University of Edinburgh's Pollock Halls and surrounds.

JMCC, South Hall and accommodation administered by the University (Chancellor's Court, Masson House and Salisbury Green) are shown.

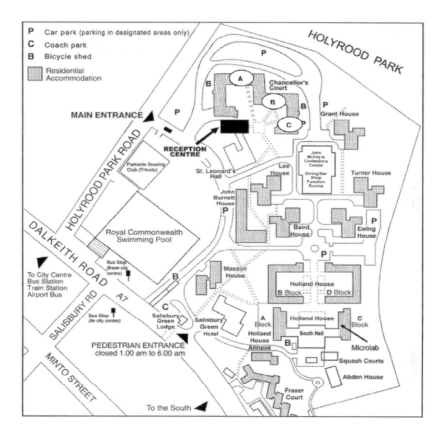

John McIntyre Conference Centre

South Hall

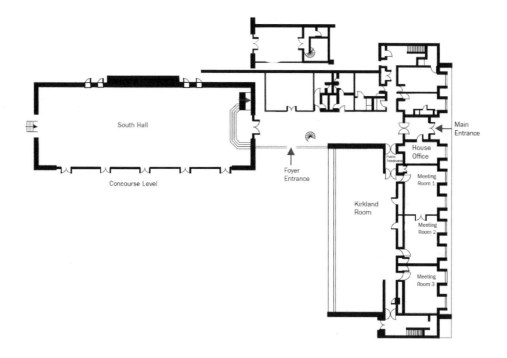

South Hall

Concourse Level

Foyer
Entrance

Kirkland
Room

House
Office

Public
Telephone

Meeting
Room 1

Meeting
Room 2

Meeting
Room 3

Main
Entrance

General information Congress

Venues
The first day of the Congress will be held at the Assembly Rooms, 54 George Street, Edinburgh EH2 2LR. Thereafter the congress will be located at the John McIntyre Conference Centre (JMCC), at the University of Edinburgh Pollock Halls site (Pollock Halls, 18 Holyrood Park Rd, Edinburgh EH16 5AY).

Official language
English is the official language of the ISAE2016 Congress

Registration and information desk
On Tuesday, 12th July, the registration desk will be at the Assembly Rooms, and open from 9:00am. Thereafter, it will be at JMCC and open from 7:30am on Wednesday and from 8:00am on subsequent days.
For further information please contact the congress organisers: ISAE2016@zibrant.com

Name badges
Name badges are required to be worn to allow admittance to the congress sessions, coffee breaks and lunches. Badges will be issued at registration.

Poster and Exhibition area
Posters will be shown in the Prestonfield Room at the JMCC and in the Kirkland Room, South Hall. Presenters of posters will be informed or where their poster will be located. Sponsors and other exhibitions will be shown in the Terrace at JMCC.

Internet Access codes
Assembly Rooms: free WiFi is available with no password required.
JMCC: free Wifi is available with the KEYSURE provider. Guests will need to register and this will give access for 24h.

Coffee breaks and lunches
Tea, coffee and refreshments will be located at the JMCC and South Hall sites. Lunches will be served in the 'Restaurant @ JMCC', downstairs at the JMCC.

Social programme
All bookings can be made via the Congress website (www.isae2016.co.uk)

Welcome Reception: Tuesday, 12th July, 18:15-21:00
The Welcome Reception will be held at the Assembly Rooms, 54 George Street, Edinburgh EH2 2LR (18:15-21:00h). Tickets are included in the Congress Registration fee and entry is open to all registered delegates and accompanying persons.

Wine and Cheese

A wine and cheese evening, with poster viewing, will be held on Wednesday, 13th July, 18:00-20:00h. JMCC/South Hall, Pollock Halls

Conference Dinner and Ceilidh: Friday, 15th July, (19:30-23:00)

The Conference Dinner and ceilidh will be held at the Assembly Rooms, 54 George Street, Edinburgh EH2 2LR. Tickets are purchased via the on-line booking system.

Excursions: Saturday 16th July

Booking can be made via the Conference website. Places on each tour may be limited.

- **The Hill and Mountain Research Centre and the Deanstoun Distillery** (09:00-17:30h, departs from Pollock Halls). Visit to SRUC's Hill and Mountain Research Centre near Crianlarich, Scottish Highlands and the Deanstoun Whisky Distillery.
- **SRUC Crichton and Cream O'Galloway** (09:00-17:30h, departs from Pollock Halls). Visit to SRUC's Dairy Research Centre at Crichton near Dumfries, and Cream O'Galloway dairy farm centre.
- **Sea Birds Catamaran Cruise** (09:30-1700h, departs from Pollock Halls). Visit the Scottish Seabird Centre at North Berwick and a boat trip to Bass Rock.
- **Research and Rosslyn Chapel** (09:00-13:00h, departs from Pollock Halls). Visit the research facilities at Royal (Dick) School of Veterinary Studies and Rosslyn chapel.
- **Edinburgh Walking Tour** (09:45-13:00; Leaves from Edinburgh Farmer's Market, which itself opens at 09:00h, Castle Terrace, Edinburgh). http://www.edinburghfarmersmarket.co.uk/location/. It takes approximately 30 min to walk from JMCC to Castle Terrace.

Tickets for all excursions can be purchased from the on-line booking system.

Farewell Party: Saturday 16th July (20:00-23:00)

The Farewell Party will be held at Teviot Row House, 13 Bristo Square, Edinburgh EH8 9AJ Teviot Row House, 13 Bristo Square, Edinburgh EH8 9AJ

Information about Edinburgh

Banking service, currency

There is an ATM at the JMCC, and others located nearby in Newington. The local currency is pound sterling (GBP).

Shopping and other facilities

There are plenty of shops, restaurants, museums, art galleries and other facilities located between the Assembly Rooms on George Street (where the Welcome Reception takes place) and the JMCC (where the rest of the conference takes place) . George Street, the Grassmarket, and the Royal Mile have a particularly high concentration of restaurants, pubs and artisan shopping. The Commonwealth Pool complex is next to Pollock Halls and has a large recreational pool and gym facilities. Arthur's Seat and Holyrood Park are immediately behind the JMCC and offer the opportunity for good walking and running. Edinburgh Castle is located at the west end of the Royal Mile, with Holyrood House at the south end. More information about places to eat, drink and visit as recommended by ISAE2016 can be found on the website www.isae2016.co.uk under the 'Plan your trip' tab.

Emergency Calls
999 or 122: emergency calls to ambulance, police and fire services

Local Conference Secretariat
Zibrant
Unit 2, 3 Jubilee Way
Faversham, Kent
ME13 8GD
www.zibrant.com

Transport to and from Edinburgh

By Air
Edinburgh Airport is located 8 miles (12 km) west of the capital and there are a range of transport options for travelling to and from the city centre. Edinburgh Airport offers direct and connecting flights to and from more than 100 UK, European and international destinations. For direct flight information, please visit the **Visit Scotland website** (https://www.visitscotland.com/)

By Car
Edinburgh is only 3 hours from Inverness and just over 2 hours from Aberdeen. Travel times from England are just as good, Birmingham is approximately 5.5 hours and Manchester and York 4 hours. To help plan your route: **AA Route Planner** (http://www.theaa.com/route-planner).
Scotland has three **National Tourist Routes** that you can take to drive to Edinburgh. These are scenic alternatives to the main roads and motorways that are well signposted and easy to follow, with visitor attractions along the way. (https://www.visitscotland.com/see-do/tours/driving-road-trips/)

By Train
All trains arrive at Waverley Station in the centre of the city, although some trains also stop at Haymarket.
The East Coast mainline linking Edinburgh and London King's Cross: you can travel between the cities in around 4 hours. Alternatively, you can travel overnight; a sleeper service runs between London Euston and Edinburgh Waverley 6 nights a week. Visit the **National Rail Enquires** or **ScotRail**. (http://www.nationalrail.co.uk and https://www.scotrail.co.uk)

By Bus
Edinburgh is well placed on the Scottish motorway network so arriving by bus is easy, and can be a lot cheaper. Buses and coaches arrive at St Andrews Square Bus Station in Edinburgh's City Centre. Visit **Lothian Buses**. (http://lothianbuses.com/)

Scientific and social programme

Tuesday 12 July

16:00	**Welcome and Presidential address**	
	SESSION 1: Opening Ceremony	
16:30	Mike Appleby	Finding Gold: 50 years of SVE and ISAE
16:50	Maja Makagon	Walking on the wild side
17:10	Christine Nicol	Wood-Gush Lecture: Decisions, decisions: Animals, scientists and the time that is given us
18:15 - 21:00	**Welcome Reception. Venue: Assembly Rooms**	

Wednesday 13 July

8:30	**PLENARY: Niko Tinbergen and W.H. Thorpe: very different answers to the question of animal awareness. *Marian Stamp Dawkins*. Venue: JMCC, Holyrood Park Road, South Hall**	
	SESSION 2 at South Hall: Tinbergen's questions: function and evolution	SESSION 3 at Pentland East: Cattle
9:15	Large homerange sizes and hunting styles involving prolonged chase predict stereotypic route-tracing in the Carnivora. *Georgia Mason*	The relationship between standing behaviour in transported calves and subsequent blood glucose and CK concentrations. *Andrew Fisher*
9:30	It's mine! Factors associated with different types of resource guarding behaviour patterns in companion dogs. *Jacquelyn Jacobs*	Effects of short and no dry period on behaviour of dairy cows. *Akke Kok*
9:45	Function of maternal behaviour towards the end of the lactation period: The self-sacrificing mink mother? *Jens Malmkvist*	Effect of feed allowance at pasture on lying behaviour of dairy cows. *Keelin O'Driscoll*
10:00	Improving gait in ducks and chickens – the importance of the ancestral phenotype. *Brendan Duggan*	Is maternal behaviour affected by peri-conception diet? *Giuliana Miguel-Pacheco*
10:05	Effects of genetic background and selection for productivity on behavioral traits in laying hens. *Anissa Dudde*	Interaction of dairy cattle and robot simulating objects of varying designs. *Renate Doerfler*
10:10	Genetic loci associated with dog-human communication and dog sociality. *Mia Perrson*	The association between the welfare of culled dairy cows and price achieved at auction market. *Kerin Rosen*

10:15	**Coffee**

Wednesday 13 July

	SESSION 2 at South Hall: Tinbergen's questions: function and evolution	SESSION 4 at Pentland East: Diverse species, new challenges and reflections	SESSION 5 at Pentland West: Pain/Health
10:45	Evolution of domestication in chickens – is tameness the key? *Per Jensen*	Religion and Animal Welfare – An Islamic Perspective. *Abdul Rahman Sira*	Effect of topical healing agents on scrotal lesions after surgical castration in 4-5 month old beef calves. *Sonia Marti*
11:00			Assessment of time of administration of subcutaneous Meloxicam on indicators of pain after knife castration of weaned calves. *Daniela Melendez*
11:05			Lameness, productivity, and cow behaviour in dairy herds with automated milking systems. *Meagan King*
11:10	The social environment affects maternal endocrine profiles but not offspring behaviour in Japanese quail. *Vivian Goerlich-Jansson*	Rehoming of research animals – analyzing animal welfare pitfalls and presenting possibilities. *Christel Moons*	Cow-level risk factors for lameness in transition dairy cows – A study on grazing dairy herds in southern Brazil. *Rolnei Daros*
11:25	Benefits of social information in cattle management using social network analyses. *Paul Koene*	"A man of science should attend only to the opinion of men of science" – animal research regulation 150 years after C Bernard. *Anna Olsson*	Passive immunity of suckling calves and associations with colostrum management routines on organic dairy farms. *Julie Føske Johnsen*
11:40	Calling by domestic piglets during simulated crushing and isolation: a signal of need? *Gudrun Illmann*	Cold-blooded Care: identifying novel methods of welfare assessment for reptiles. *Sophie Moszuti*	Pain in primates: Using behaviour and facial expressions as a welfare tool. *Kris Descovich*
11:55	The defensive behavioral patterns of white-lipped and collared peccary: an approach for peccary's conservation. *Selene Nogueira*	Out like a light? The effects on sleep of being a nocturnal mouse in a diurnal lab. *Amy Robinson-Junker*	Effects of transport on the clinical condition of sows destined for slaughter. *Katrine Fogsgaard*
12:10	Malfunctional male mink: effects of enrichment and basal ganglial traits on stereotypic behaviour and mating success. *Maria Díez-León*	The importance of burrowing, climbing and standing upright for laboratory rats. *Joanna Makowska*	Systematic review of on-farm factors affecting leg health in broilers. *Ida Just Pedersen*
12:15	Effects of UV light provision on broiler chickens' fear response and lighting preferences. *Antonia Patt*	Use of an outdoor run by dairy goats is reduced under adverse weather conditions. *Nina Keil*	Assessing the physical well-being of dogs in commercial breeding facilities. *Judi Stella*

Wednesday 13 July

| 12:20 | Personality in Labrador Retriever dogs: underlying genetics and effects of lifestyle. *Marie Haskell* | Effects of three different methods of restraint for shearing on behaviour and heart rate variability (HRV) in alpacas. *Susanne Waiblinger* | Effect of prepartum physical activity on behaviour and immune competence of non-lactating dairy cows. *Randi Black* |
| 12:25 | Early-life behavioural development of lines divergently selected on feather pecking behaviour. *Jerine Van der Eijk* | Animals and us: history, exciting developments, cross-pollination and new directions for applied ethology research. *Jennifer Brown* | Can gastric health issues in veal calves be linked to poor welfare? *Laura Webb* |

12:30   LUNCH: SEMINAR: Ice Robotics 'Cows in the Cloud'

14:00   PLENARY: A Comparative Science of Emotion. *Elizabeth Paul & Mike Mendl*.
Venue: Pentland East, Holyrood Park Road

	SESSION 6 at South Hall: Individual Differences	SESSION 7 at Pentland East: Emotion
14:45	Animal personality: what does it mean for our understanding of animal welfare? *Lisa Collins*	In your face: Indications of emotional facial expressions in pigs. *Inonge Reimert*
15:00	Fearfulness and access to the outdoor range in commercial free-range laying hens. *Hannah Larsen*	Face-based perception of emotions in dairy goats using 2-D images. *Lucille Bellegarde*
15:15	The relationship between heat stress tolerance and temperament of dairy cows. *Maria Jorquera Chavez*	Facial indicators of positive emotions in rats. *Kathryn Finlayson*
15:20	Stereotypic behaviour in standard cages is an alternative to a depression-like response, being awake but motionless, in mice. *Carole Fureix*	Emotion recognition in dogs, orangutans, and human infants. *Min Hooi Yong*
15:25	Is leaner meaner? The effect of leanness on pig-pig aggression and pig-human interactions in finisher gilts. *Lorri Jensen*	Perception of emotional valence in goats. *Luigi Baciadonna*

15:30   Coffee

16:00	Personality, plasticity, and predictability of shelter dogs during interactions with unfamiliar people. *Conor Goold*	Vocal expression of emotions in pigs. *Elodie Briefer*
16:15	Individual-based variation in grazing behavior of rangeland-raised beef cows. *Laura Goodman*	Exploring whether ear postures are a reliable indicator of emotional state in dairy cows. *Helen Proctor*
16:30	Sociability in grazing dairy cows is related to individual social network attributes and behavioural synchrony. *Kees Van Reenen*	Characterising fear in dogs. *Bjorn Forkman*
16:45	Fear and its relationship to social aggression in group-housed swine. *Carly O'malley*	Behavioural assessment of the habituation of captured feral goats to an intensive farming system. *David Miller*

17:00	Breed, gender and age variation in behavioural traits in domestic cats (Felis silvestris catus). *Bjarne Braastad*	Welfare indicators for sheep: relationship between QBA and behavioural measures. *Susan Richmond*
17:15	Patterns of range access of individual broiler chickens in commercial free-range flocks. *Peta Taylor*	Qualitative Behaviour Assessment as a tool for monitoring cat welfare in vet nurse practice. *Francoise Wemelsfelder*
17:20	Links between owner adult attachment style and the behaviour of the dog during challenging situations. *Therese Rehn*	Lateralization during agonistic behaviour in pigs. *Sophie Menneson*
17:25	Changes in autonomic balance of pigs in different behavioral contexts with special focus on coping type. *Annika Krause*	What is the emotional state of dairy calves and young stock? *Marta Brscic*
17:30	Responses in a social isolation test are related to spontaneous home pen activity in dairy cattle. *Jill Mackay*	Use of conditioned place preference to investigate emotion transfer in domestic hens. *Joanne Edgar*

18:00 - 20:00	**Wine and Cheese at the Posters. Venue: JMCC/South Hall, Holyrood Park Road**

Thursday 14 July

8:30	**PLENARY: The long reach of early life: How developmental experience shapes adult behaviour in the European starling. *Daniel Nettle*. Venue: South Hall, Holyrood Park Road**
	SESSION 8 at South Hall: Tinbergen's questions: behavioural development	SESSION 9 at Pentland East: Management and Improving Environments
9:15	Effects of prenatal stress and postnatal enrichment on exploration and social behaviour of suckling piglets. *Sophie Brajon*	A nest by any other name: Pre-laying behaviour of hens in furnished cages. *Michelle Hunniford*
9:30	Cognition of dairy calves exposed to nutritional enrichments during the milk-feeding stage. *Kelsey Horvath*	The behaviour of beef cattle unloaded for feed, water, and rest en route during long distance transportation. *Derek Haley*
9:45	Impact of perinatal nutrition on spatial cognitive performance of pigs later in life. *Caroline Clouard*	The effect of alternative feeding strategies on feather condition and corticosterone levels in broiler breeder pullets. *Elyse Mosco*
10:00	Effects of prenatal stress (PNS) on lamb behaviour, perinatal temperature and weight change. *Leonor Valente*	Both tail docking and straw provision reduce the risk of tail damage outbreaks. *Mona Larsen*
10:05	Development of fearfulness in horses: modulated by maternal care? *Janne Christensen*	Management factors associated with mortality of dairy calves. *Leena Seppä-Lassila*
10:10	Rearing complexity affects fearfulness and use of vertical space in adult laying hens. *Margrethe Brantsaeter*	Influence of milking-parlour size on the behaviour of dairy cows. *Yamenah Gómez*

10:15	**Coffee**

Thursday 14 July

	SESSION 8 at South Hall: Tinbergen's questions: behavioural development	SESSION 9 at Pentland East: Management and Improving Environments	SESSION 10 at Pentland West: Pigs and Poultry
10:45	Prenatal learning: a forgotten but essential driver for behavioural development. *Peter Hepper*	Effects of long-term exposure to an electronic containment system on the behaviour and welfare of domestic cats. *Naima Kasbaoui*	Perching behavior in broiler breeders. *Sabine Gebhardt-Henrich*
11:00		Feeder space relates to aggression, jostling, and feeder sharing in laying hens. *Janja Sirovnik Koscica*	Predation of free range laying hens. *Monique Bestman*
11:05		Dairy cow preference for different types of outdoor access during the night. *Anne-Marieke Smid*	Effects of keel bone fractures on individual laying hen productivity. *Christina Rufener*
11:10	Ontogeny of coping styles in commercial sows: from birth to adulthood. *Kristina Horback*	The effects of nesting material on the toxicological assessment of cyclophosphamide. *Brianna Gaskill*	Effect of different laying hen strains on daily egg laying patterns and egg damage in an aviary system. *Silvia Villanueva*
11:25	Early life in a cage environment adversely affects spatial cognition in laying hens. *Fernanda Tahamtani*	Enrichment and management influence the emotional state of farm mink. *Anne Sandgrav Bak*	Do domestic birds avoid the fecal odours of conspecifics? *Bishwo Pokharel*
11:40	Maternal age affects leghorn chick response to the Anxiety Depression model. *Leanne Cooley*	Using animal protein and extra enrichments to reduce injurious pecking in furnished cage housed laying hens. *Krysta Morrissey*	Effect of switching sows positions during lactation on performance and suckling behaviour of early socialized piglets. *Jaime Figueroa*
11:55	Does an active fetus become an active lamb? *Tamsin Coombs*	Shelter 'runways' to improve bird distribution in the outdoor area of free-range laying farms. *Isabelle Pettersson*	Behavioural responses of commercially housed pigs to an arena and novel object test. *Amy Haigh*
12:10	Calf-cow contact during rearing affects social competence and stress reactivity in dairy calves. *Edna Hillmann*	Cow comfort, evaluated by lameness, leg injuries, and lying time, after facility and management changes in dairy barns. *Emily Morabito*	Nursing behaviour and teat order in pigs reared by their dam or nurse sow. *Oceane Schmitt*
12:15		Using elevated platforms to improve broiler leg health on commercial broiler farms. *Eija Kaukonen*	Does split marketing affect animal welfare in organic fattening boars? *Jeannette Lange*

Thursday 14 July

Time	SESSION 11	SESSION 12	
12:20		Behaviour of Danish weaner and grower pigs is affected by the type and quantity of enrichment material provided. *Franziska Hakansson*	Application of Social Network Analysis3 in the study of post-mixing agsgression in pigs. *Simone Foister*
12:25	Locomotor style of laying hens on inclinded surfaces. *Chantal Leblanc*	Measures of cow welfare in a hybrid pasture dairy system. *Cheryl O'Connor*	Weaner piglet activity measured with commercial motion sensors in groups with and without tail biting. *Sabine Dippel*

12:30   LUNCH: SEMINAR Donkey Sanctuary

14:00   PLENARY: More than Refinement – improving the validity and reproducibility of animal research. *Hanno Würbel*. Venue: Pentland East, Holyrood Park Road

Time	SESSION 11 at South Hall: Humans & Animals	SESSION 12 at Pentland East: Positive Welfare
14:45	Human attitudes and their relevance to animal welfare. *Grahame Coleman*	Validating play behavior of cattle as a positive welfare indicator: a review of research. *Jeff Rushen*
15:00	Student perceptions of animal welfare in intensive and extensive animal production environments. *Janice Siegford*	Behind those eyes: Associating eye wrinkle expression and emotional states in horses. *Sara Hintze*
15:15	Perceptions of current and future farmers and veterinarians regarding lameness and pain in sheep. *Carol Thompson*	Playful pigs:Evidence of consistency and change in play depending on litter and developmental stage. *Sarah Brown*
15:20	A prospective exploration of farm, farmer and animal characteristics in human-cow relationships: an epidemiological survey. *Alice De Boyer Des Roches*	Inducing positive emotions: cardiac reactivity in sheep regularly brushed by a human. *Priscilla Tamioso*
15:25	The burden of domestication – a representative study of behaviour and health problems of privately owned cats in Denmark. *Peter Sandoe*	Spreading happiness: Induced social contagion of positive affective states and behaviours in monkeys via audio/video playback. *Claire Watson*

15:30   Coffee

Time	SESSION 11	SESSION 12
16:00	Predicting human directed aggressive behaviour in dogs. *Rachel Orritt*	Investigating positive emotional contagion in rats. *Jessica Lampe*
16:15	Tickling pet store rats: impacts on human-animal interactions. *Megan Lafollette*	Does the anticipatory behaviour of chickens communicate reward quality? *Nicky Mcgrath*
16:20	Positive human-bird interactions in farmed ostriches: a way forward for improved welfare and production? *Maud Bonato*	Pharmacological manipulation to validate indicators of positive emotional states in dogs. *Linda Keeling*

16:40   AGM . Venue: Pentland East, Holyrood Park Road

Friday 15 July

8:30	**PLENARY: From food-hoarding titmice to stressed poultry: integration of functional and mechanistic approaches to behaviour.** *Tom Smulders.* **Venue: South Hall, Holyrood Park Road**	
	SESSION 13 at South Hall: Tinbergen's questions: causation	SESSION 14 at Pentland East: Novel Techniques
9:15	Feather pecking genotype shows increased levels of impulsive action in a delayed reward task. *Patrick Birkl*	The potential of transects to reveal the effect of the environment during broiler welfare assessment. *Xavier Averos*
9:30	Using epidemiology to understand stereotypic behavior of African and Asian elephants in North American zoos. *Brian Greco*	How to assess cumulative experience in laboratory animals? *Colline Poirier*
9:45	Neighbour Effects Confirm that Stereotypic Behaviours in Mink are Heterogeneous. *Andrea Polanco*	Characterisation of short and long-term changes in mechanical nociceptive thresholds in pigs following tail injury. *Pierpaolo Di Giminiani*
10:00	Behavioural activity of dairy cows on the day of oestrus v. mid luteal phase. *Gemma Charlton*	Noise around hatching hampers chick communication and reduces hatching synchronisation. *Bas Rodenburg*
10:05	Do laying hens have a motivation to grasp while night time roosting? *Lars Schrader*	Effects of a water spraying system on lying and excreting behaviours of fattening pigs in heat stress. *Hieu Nguyen Ba*
10:10	On-farm risk factors for non-nutritive sucking in group-housed organic Simmental dairy calves. *Verena Größbacher*	Age differences in exploratory and social behaviour in dairy cows. *Alexander Thompson*

10:15 Coffee —————————————————————————————————

	SESSION 13 at South Hall: Tinbergen's questions: causation	SESSION 14 at Pentland East: Novel Techniques	SESSION 15 at Pentland West: Cognition
10:45	Affective states and proximate behavioural control mechanisms. *Lorenz Gygax*	Citizen Science: a next step for applied animal behaviour. *Julie Hecht*	Attention bias: a practical measure of affective state in farm animals. *Caroline Lee*
11:10	Effects of dietary protein and amino acid supply on damaging behaviours in pigs kept under diverging sanitary conditions. *Yvonne Van Der Meer*	Can oral sucrose solution alleviate castration pain in piglets? Results of an evaluation using a behavioural test. *Yolande Seddon*	Serotonin depletion in a cognitive bias paradigm in pigs. *Jenny Stracke*
11:25	Explaining daily feeding patterns in pigs: modelling interaction between metabolic processes and circadian rhythms. *Iris Boumans*	External validity of single and multi-lab studies: considering reaction norms. *Bernhard Voelkl*	On how impulsivity can affect responses in cognitive tests of laying hens. *Elske de Haas*
11:40	Effect of docking length on tail directed behaviour and aggression in finisher pigs. *Karen Thodberg*	Relationship between sow conformation, farrowing floor type and posture change characteristics using accelerometer data. *Stephanie Matheson*	Correlation between cognitive bias and other measures of stress in rats. *Timothy Barker*

Friday 15 July

11:55	Pigs' fighting ability in the application of game-theory models to address aggression. *Irene Camerlink*	Porcine Facial Inference: Facial Biometrics as noninvasive proxy measures for innate aggression in piglets. *Catherine McVey*	Dogs with abnormal repetitive behaviours do not show differences in relative cognitive bias than matched unaffected dogs. *Bethany Loftus*
12:10	Different behavioural responses of stabled horses to delayed feeding and polymorphisms of the dopamine D4 receptor gene. *Moyuto Terashima*	Welfare assessment of Low Atmospheric Pressure Stunning in chickens. *Jessica Martin*	Crib-biting in horses: stress and learning capacity. *Sabrina Breifer*
12:15	Feather pecking in laying hens – do control and case flocks differ regarding compliance with recommendations? *Lisa Jung*	Accelerometer to quantify inactivity in laying hens with or without keel bone fractures. *Teresa Casy-Trott*	Observing calves' behaviour during learning in a colour discrimination task throws light on their cognitive processes. *Alison Vaughan*
12:20	Confinement before farrowing affects performance of nest building behaviours but not progress of parturition in prolific sows. *Christian Fink Hansen*	The effect of pressure vest on the behaviour, salivary cortisol and urine oxytocin of noise phobic dogs. Anne-*Maria Pekkin*	Pavlovian decision-making in a counter-balanced go/ no-go judgement bias task. *Samantha Jones*
12:25	Behavioural, physiological and scientific impact of different fluid control protocols in the rhesus macaque (Macaca mulatta). *Helen Gray*	Thermal imaging to monitor development and welfare in broilers. *Katherine Herborn*	Social learning in ungulates: effect of a human demonstrator on goats in a spatial learning task. *Christian Nawroth*

12:30   LUNCH: Student Event: Meet the Professors in the JMCC restaurant

14:00   PLENARY: Breed-specific stress-coping characteristics and welfare concerns on pig production in China. *Ruqian Zhao*. Venue: Pentland East, Holyrood Park Road

	SESSION 16 at South Hall: Social Behaviour	SESSION 17 at Pentland East: Trade-offs and Synergies
14:45	A measure of fear of humans in commercial group-housed sows. *Lauren Hemsworth*	Can animal welfare science have a role in creating a sustainable future for animal agriculture? *Janice Swanson*
15:00	Effect of group size and health status on social and feeding behaviour of transition dairy cows. *Margit Bak Jensen*	How to make sure cows age well: incorporating economics, environment and welfare for truly sustainable animal farming. *Elsa Vasseur*
15:15	Mixing sows into groups during lactation: presence of piglets reduces aggression. *Emma Greenwood*	An assessment method for the management of farm animal welfare in global food companies. *Heleen Van de Weerd*
15:20	Sows with low piglet mortality are not more careful towards their piglets than sows with high piglet mortality. *Janni Hales*	A decade of progress in the United States working to eliminate intensive confinement of hens and gestating sows. *Sara Shields*

Friday 15 July

15:25	Behavioural reactions of horned and dehorned dairy cows to herd mates whose horn status was manipulated. *Janika Lutz*	Comfort among crisis: is comfort of working animals a luxury within the international development context? *Rebecca Sommerville*

15:30 Coffee ───────────────────────

16:00	Potential influencing factors on social behaviour and animal welfare state in large Danish dairy cow herds. *Marlene Kirchner*	The big picture: synergies and trade-offs between animal welfare and other sustainability issues. *Eddie Bokkers*
16:15	Milker behaviour affects the social behaviour of dairy cows. *Stephanie Lürzel*	Organic pig husbandry in Europe: Do welfare and environmental impact go hand in hand? *Christine Leeb*
16:30	Social structure stability in farmed female capybaras (Mammalia, Rodentia). *Sergio Nogueira-Filho*	Know thy neighbour - does loose housing at farrowing and lactation positively influence sow welfare post-weaning? *Sarah Ison*
16:35	Spontaneous, intensive shoaling in laboratory zebrafish: quantifying a previously undescribed social behaviour. *Courtney Graham*	Dog and human behaviour in a mass street dog sterilization project in Jamshedpur, India. *Tamara Kartal*

16:45   Closing Ceremony. Venue: Southall and Pentland East

19:30 - 00:00   Congress Banquet. Venue: Assembly Rooms, George Street

Saturday 16 July

	Excursions
	Excursions
09:00-17:00	The Hill and Mountain Research Centre and the Deanstoun Distillery
	Buses depart from Pollock Halls (JMCC)
09:00-17:30	SRUC Crichton and Cream O' Galloway
	Buses depart from Pollock Halls (JMCC)
09:30-17:00	Sea Birds Catamaran Cruise
	Buses depart from Pollock Halls (JMCC)
09:00-13:00	Research and Rosslyn Chapel
	Buses depart from Pollock Halls (JMCC)
09:45-13:00	Edinburgh Walking Tour
	Departs from Edinburgh Farmer's Market, Castle Terrace (you must make your own way to the market)

20:00-late   Farewell Party. Venue: Teviot Row House Student Union, Charles St Lane

Table of contents

Session 03. Cattle

Oral presentations

Poster presentations

Session 04. Diverse species, new challenges and reflections

Oral presentations

Session 05. Pain/health

Oral presentations

Poster presentations

Session 06. Individual differences

Oral presentations

Poster presentations

Session 07. Emotion

Oral presentations

Session 08. Applied Ethology and Tinbergen's questions: Behavioural development

Oral presentations

Poster presentations

Session 09. Management and improving environments

Oral presentations

Session 10. Pigs and poultry

Oral presentations

Poster presentations

Session 11. Humans and animals

Oral presentations

Session 12. Positive welfare

Oral presentations

Poster presentations

Session 13. Applied Ethology and Tinbergen's questions: Causation

Oral presentations

Session 14. Novel techniques

Oral presentations

Session 15. Cognition

Oral presentations

Session 16. Social behaviour

Oral presentations

Session 17. Trade-offs and synergies between welfare and other major global issues

Oral presentations

Poster presentations

Finding gold: 50 years of SVE and ISAE

Michael Appleby[1], Ian J.H. Duncan[2] and J. Carol Petherick[3]
[1]*World Animal Protection, 5th floor, 222 Gray's Inn Road, London WC1X 8HB, United Kingdom,*
[2]*University of Guelph, Animal & Poultry Science, Stone Road, Guelph, ON, N1G 2W1, Canada,*
[3]*The University of Queensland, QAAFI, 25 Yeeppoon Road, Rockhampton, QLD 4701, Australia;*
michaelappleby@worldanimalprotection.org

As animals are an integral part of our world, it can be argued that applied ethology is essential for a sustainable future. Given this, our wonderful and indispensable society is surprisingly modest. We were founded as the Society for Veterinary Ethology in Edinburgh in 1966, following publication of the Brambell Report in the UK in 1965. We changed our name to the International Society for Applied Ethology in 1991 and celebrate our Golden Jubilee in Edinburgh in 2016. From 37 founders, we now have over 500 members in about 35 countries, although that still suggests a need for further expansion. The society is famously friendly, with traditions at meetings including support for students, informality and dancing. That character obviously depends on our members, including officials, and this presentation will give some glimpses of our people and our activities over the years. Indeed, applied ethology is obviously good for you: three of our founders are still active in the field. Our logo is emblazoned with 'Ethos,' meaning the characteristic spirit of a people or community – which we interpret to include animals. Let us agree that in summarising the ethos of the ISAE we can find gold in the past, and great potential in the future.

Walking on the wild side: increasing interdisciplinary approaches in the next 50 years

Maja Makagon
Department of Animal Science, Univeristy of California, Davies, One Shields Avenue, Davis, CA 95616, USA; mmakagon@ucdavis.edu

A quick glance through the proceedings of any Congress of the International Society for Applied Ethology serves to highlight the multidisciplinary nature of this field. By taking multiple research approaches to address common issues, important strides have been made toward understanding how factors such as resource availability, environmental design, social grouping, temperament, nutrition, genetics and early life experiences affect the ways in which animals perceive and respond to their housing environments. While the multidisciplinary approach has served our field well, a more integrative interdisciplinary approach will further elevate and expand the broader impacts of our work. Integration of information from across scientific disciplines can and should be done at a number of levels. We should continue to integrate research conducted across the sub-disciplines of applied ethology, for example by connecting behavioral responses of individuals to underlying physiological processes and welfare outputs. At the same time, we should strive to strengthen and build new connections with disciplines outside of applied ethology. I will discuss recent and current research that has benefited from broadly interdisciplinary collaborative efforts and will suggest areas where increased interdisciplinary approaches are likely to help spring the field of applied ethology forward in the next 50 years.

Decisions, decisions: animals, scientists and the time that is given us
Christine Nicol
University of Bristol, School of Veterinary Science, Langford House, BS40 5DU, United Kingdom;
c.j.nicol@bris.ac.uk

Decision making is the process of forming preferences based on the appraisal of simple or complex stimuli, followed by the selection and execution of a choice. The decisions made by animals have long played a central role in applied ethology. Such decisions are considered by many to be the 'gold standard', and by some to be the 'only', means of establishing a meaningful animal-centred concept of animal welfare. Interest in how animal decisions might be related to animal welfare can be traced back to the earliest days of applied ethology, when lively debate took place regarding the interpretation of preference tests. In more recent years the cognitive and emotional influences on stimulus appraisal and decision-making have been studied. Compared to the early days, far more information is now available about animals' capacity to utilise private and public information, encode reward, plan ahead, and exhibit self-control. The influence of previous experience is also better understood in both absolute and relative terms. In the context of animal welfare, studies that assess the impact of high levels of arousal on decision-making are pertinent. These lines of work help us to understand both the constraints and the capacity of domestic animals to make decisions that can inform us about their welfare. The study of animal decision making is now a complex and multi-disciplinary area but this should be seen as an opportunity and not a threat. The underlying structure, consistency and rationality of animal decisions can be analysed and used to provide a measure of confidence that true preferences are being expressed. As we look towards the future we can see that work on animal decision-making will be a vital adjunct of studies of animal emotion, since the only way of defining the valence of a stimulus or a situation in operational terms is via an animal's decision to approach or avoid. New studies of animal preferences are also still needed to inform practical improvements to animal housing and management, and to validate other indicators of welfare. Examples will be given in each context. We scientists also have decisions to make; notably what to do with the time that is given us. The talk will conclude with the argument that a renewed focus on the acquisition of meaningful animal-centred information would be time well spent.

Niko Tinbergen and W.H. Thorpe: very different answers to the question of animal awareness

Marian Stamp Dawkins

University of Oxford, Zoology, South Parks Road, Oxford OX1 3PS, United Kingdom; marian.dawkins@zoo.ox.ac.uk

Niko Tinbergen's 4 questions formed – and still form – the backbone of ethology, but curiously, during the 1970s and 1980s it was applied ethology that was responsible for much of the most innovative work on two of the questions – development and causation. This was because animal welfare science investigates how animals respond to a variety of situations (causation) and how these responses are shaped by early experience (ontogeny), even when sociobiological thinking was making mainstream ethology neglect these questions in favour of questions about function and adaption. Tinbergen was, however, very insistent that asking about the subjective experiences of animals was not properly part of ethology. This was not because he denied the existence of such subjective experiences but because he thought that they could not be studied scientifically and were therefore no more than guesswork. An early voice against Tinbergen's behaviorist views was W.H. Thorpe of Cambridge University. In a remarkable Appendix to a UK Government report on farm animal welfare published in 1965, Thorpe set out an agenda for what he saw as the future study of animal welfare that included, as a top priority, the study of animal subjective feelings. Thorpe's contribution to the subsequent development of animal emotions and animal welfare science has been relatively neglected and I hope to use this talk to set the record straight.

Large home range sizes & hunting styles involving prolonged chase predict stereotypic route-tracing in the Carnivora

Georgia Mason[1], Jeanette Kroshko[1], Ros Clubb[2], Laura Harper[1], Emma Mellor[3] and Axel Moehrenschlager[4]
[1]University of Guelph, Animal Biosciences, Guelph, Ontario N1G 2M7, Canada, [2]Royal Society for the Prevention of Cruelty to Animals, Southwater, West Sussex, RH13 9RS, United Kingdom, [3]University of Bristol, School of Veterinary Sciences, Langford, Bristol, BS40 5DU, United Kingdom, [4]Calgary Zoological Society, Centre for Conservation Research, Zoo Road, Calgary, Alberta, T2E 7V6, Canada; gmason@uoguelph.ca

Evolved aspects of behavioural biology pre-adapt some species to cope well with captivity, but put others at risk of stress. Correspondingly, in zoos, some Carnivora tend to show little or no stereotypic behaviour (SB), while other species in that order are very prone to SB. Phylogenetic comparative methods can reveal largescale patterns across taxa and so identify the fundamental origins of welfare problems. We therefore used these (via contrasts generated in Mesquite) to identify species-level risk factors for SB. Candidate predictor variables were natural ranging behaviour, territoriality, specific aspects of natural foraging, wild activity levels, cranial volume, and IUCN Red List status. Previous research had already identified naturally long daily travel distances, and being large-bodied and wide-ranging, as SB risk factors. Via extensive literature search, we nearly doubled the size of this original SB database, and then imposed stricter quality controls (e.g. on minimum sample sizes for inclusion). Patterns in the resulting 23-species dataset (whose medians for stereotypic time-budgets ranged from <1% to c.55%) were investigated. These confirmed naturally large ranges ($T_{1,13}$=3.42, P=0.002) and long daily travel distances ($T_{1,10}$=2.00, P=0.037) as risk factors for route-tracing SBs. They also revealed that the home range size effect is independent of body mass (although body mass & range size together predicted SB most strongly: P<0.0001), and that it explains the apparent daily travel distance effect (which vanished if range size was controlled for). Furthermore, a new finding emerged: that naturally long chase distances during hunts also predicted more severe route-tracing ($T_{1,3}$=4.21, P=0.012). These two risk factors appeared to act independently. Overall, naturally wide-ranging Carnivora with long chase distances are thus most prone to extensive route-tracing in captivity. These results suggest likely strategies for the most effective enrichment of enclosures to help zoos reduce or eliminate SB and replace it with more naturalistic activity. They also suggest that aspects of being wide-ranging and a pursuit predator have evolved to be inflexible 'behavioural needs' that must be met even when a Carnivore's homeostatic requirements have been accomodated.

It's mine! Factors associated with different types of resource guarding behaviour patterns in companion dogs

Jacquelyn Jacobs[1], Jason Coe[1], David Pearl[1], Tina Widowski[2] and Lee Niel[1]
[1]University of Guelph, Population Medicine, Ontario Veterinary College, 50 Stone Rd. East, Guelph, Ontario, N1G 2W1, Canada, [2]University of Guelph, Animal Biosciences, 50 Stone Rd. East, Guelph, Ontario, N1G 2W1, Canada; jjacob01@uoguelph.ca

Resource guarding (RG) involves the use of specific behaviour patterns to control access to items of potential 'value' to the dog. Of particular concern are aggressive patterns, due to safety concerns, but other patterns of RG behaviour are prevalent and include avoidance (i.e. via positioning of head or body, or location change) and rapid ingestion (i.e. rapid ingestion of consumable item). Previous research has not investigated the etiology of RG aggression in depth, nor have the additional RG patterns been considered. Dog owners (n=3,068) were recruited through social media to answer questions regarding dog- and household-related factors, as well as their dog's current and past RG patterns around people. Participants were screened for their ability to identify different RG patterns from video, and were removed from the study if they incorrectly identified any pattern of RG. This resulted in a final sample of 2,207 participants representing information for 3,589 dogs. Multiple multi-level logistic regression models (SAS 9.4) were developed to determine the association between independent variables of interest and each RG pattern, and key results (odds ratios and 95% confidence intervals) are presented here. Dogs with increasing levels of impulsivity were more likely to display RG rapid ingestion (OR: 1.15 [1.08-1.23]), and aggression (OR: 1.15 [1.08-1.24]; P<0.001 for each). Dogs with increasing levels of fearfulness were also more likely to display RG aggression (OR: 1.20 [1.12-1.28]; P<0.001). Neutered males (P<0.01) and mixed breeds (P<0.05) were more likely to be RG aggressive compared to dogs of other sexes, neuter statuses, and breeds. Teaching dogs to reliably 'drop' items when requested was associated with a reduced likelihood of RG aggression (OR: 0.75 [0.61-0.92]; P<0.01) and avoidance (OR: 0.75 [0.61-0.92]; P<0.001). Furthermore, the addition of palatable food during mealtime was associated with an increased likelihood of less severe RG behaviour (OR: 1.77 [1.27-2.46]; P<0.01), whereas removal of the food dish during mealtime was associated with an increased likelihood of expressing more severe or frequent RG behaviours (OR: 1.88 [1.16-3.06]; P<0.05). Relationships between the three types of RG patterns were varied, suggesting that RG behaviour patterns are flexible when humans are involved. The results highlight various factors that might predispose dogs to RG behaviour and potential methods for prevention of RG aggression, and can serve as a basis for future longitudinal RG research to establish causation.

Function of maternal behaviour towards the end of the lactation period: the self-sacrificing mink mother?

Jens Malmkvist[1], Dennis Dam Sørensen[1], Torben Larsen[1], Rupert Palme[2] and Steffen Werner Hansen[1]
[1]Aarhus University, Animal Science, Blichers Alle 20, P.O. Box 50, 8830 Tjele, Denmark, [2]University of Veterinary Medicine, Biomedical Sciences, Veterinärplatz 1, 1210 Vienna, Austria; jens.malmkvist@anis.au.dk

The optimal timing of separating the mink dam from the litter at farms is suggested to be a balance between the partly conflicting needs of the mother and the kits. Early removal of the dam or partial removal of the litter may protect the dam against exhaustion. Little is known, however, about dam stress and maternal motivation around the time of separation. Therefore, we investigated the effects of separating the dam from the litter, using brown first-parity farm mink dams (Neovison vison; n=374) balanced for birth date but otherwise randomly assigned to two treatment groups: The dam was taken away from the litter either day 49±1 (7 w, n=185) or day 56±1 (8 w, n=189) after birth. The aim was to investigate whether the dams experienced stress/had a different motivation to take care of the litter after 7 and 8 weeks, estimated by non-invasive determination of cortisol (FCM: Faecal Cortisol Metabolites) and dam behaviour including calls the first week after separation (D0, D1, D7). The two treatment groups had an equal litter size at time of separation (age 7 w: 5.5±0.17; 8 w: 5.5±0.17 kits; range 1-11; P=0.76). Likewise, there was no significant difference in dam body weight (7 w: 1,420±15.0 g, 8 w: 1,404±14.7 g; range 930-1,680 g, P=0.43). However, the litter size negatively influenced both the dam weight and her BCS (P<0.001), regardless of the litter age. Cortisol peaked D1 and dams separated at litter age 7 weeks had higher concentrations of cortisol during the first week after removal (day of separation, D0: 18.8%, D1: 34.5%, D7: 36.9% higher FCM) than dams separated at litter age 8 weeks (P=0.014). Likewise, the dam calls increased on the separation day, peaking on the first day after separation (D1). The proportion of dams with calls decreased with litter age at separation (P=0.024; e.g. D1, 7 w: 18.5%, 8 w: 11.7%). We interpret these results as a higher maternal motivation in dams at 7 weeks than at 8 weeks after birth. Additionally, the amount of dam calls after separation decreased with the litter size (P=0.022; for lower/median/upper quartiles; 1-4 kits: 20.7%, 5-6 kits: 19.0%, 7-11 kits: 5.8%). Thus in addition to litter age, the size of the litter is important for the maternal motivation. These factors should therefore be taken into account in the optimal separation time on mink farms.

Improving gait in ducks and chickens – the importance of the ancestral phenotype

Brendan M. Duggan, Paul M. Hocking and Dylan N. Clements
The Roslin Institute and R(D)SVS, University of Edinburgh, EH25 9RG, United Kingdom;
brendan.duggan@roslin.ed.ac.uk

After the broiler chicken, the Pekin duck is the most numerous animal grown for meat consumption globally. Since the broiler chicken has experienced extensive gait problems associated with its high growth rate, there are concerns that similar gait abnormalities may occur in the future in the Pekin duck. In order to understand how selection for improved meat yields has affected walking behaviour in chickens and ducks, bone morphology and gait were assessed objectively in divergent lines. The Pekin duck was compared to its slower growing ancestor, the mallard, and the broiler chicken was compared to the slower growing layer chicken. For each line, at three, five and seven weeks, 12 birds were allowed to walk over a pressure sensing walkway. After culling, the leg bones of these birds were scanned by computed tomography to assess how leg morphology has changed through selection for rapid growth and high meat yield. Results were analysed by ANOVA, accounting for age and sex. Each line walked at its own comfortable velocity range, which differed from other lines (P<0.005). Duck lines walked faster than chickens and, within each species, lighter lines moved at a faster pace than their heavier conspecifics. Heavier lines had a wider stance than their lighter conspecifics when walking (mean (sd) for heavy duck and chicken lines respectively: 11.3 (1.2), 12.1 (1.9) cm; mean (sd) for lighter duck and chicken lines respectively: 6.0 (0.7), 1.8 (0.4) cm; P<0.001) which theoretically improves stability during standing but may reduce balance when walking. Heavy lines also spent more time than lighter lines supporting their mass over two feet while walking (mean (sd) for heavy duck and chicken lines respectively: 0.38 (0.07), 0.38 (0.09) sec.; mean (sd) for lighter duck and chicken lines respectively: 0.03 (0.02),0.04 (0.01) sec.; P<0.001), further suggesting that these heavy meat lines are less balanced when walking. There were pronounced differences in leg morphology between the species. These differences may be an adaptation for swimming which may hinder locomotion on land. Duck legs reached adult size by five weeks, allowing them greater opportunity to remodel their bones to handle loads imposed on them. Selection for improved meat yield and rapid growth has led to changes in morphology and gait in both chickens and ducks. The growth rates in both birds are similar but it is not yet known why fewer gait problems are reported in Pekin ducks compared to broiler chickens. This may be due to earlier leg development or the different leg morphology observed in ducks, allowing these birds to maintain balanced gait after a shift in the body's centre of mass due to muscle growth.

Effects of genetic background and selection for productivity on behavioral traits in laying hens

Anissa Dudde[1], Lars Schrader[1], Steffen Weigend[2], Lindsay Matthews[3] and E. Tobias Krause[1]
[1]Federal Research Institute for Animal Health, Friedrich-Loeffler-Institut, Dörnbergstr. 25/27, 29223 Celle, Germany, [2]Institute of Farm Animal Genetics, Friedrich-Loeffler-Institut, Höltystraße 10, 31535 Neustadt, Germany, [3]University of Auckland, 22 Princes Street, 1010 Auckland, New Zealand; anissa.dudde@fli.bund.de

Resource allocation theory (RAT) suggests that highly productive layer hens sacrifice activity to direct energy to maintain egg output. Productivity and genotype have been implicated in differences in fearfulness and stress reactivity in layers but their relative contributions have not been fully explored. We aimed to determine the separate effects of selection for high productivity and genetic strain on fearfulness and sociality. Therefore, we used 80 hens from a 2×2 group design, with two white and two brown layer lines which differed in laying performance (200 vs 300 egg/y) within colours, respectively, to measure two key behavioral dimensions, fearfulness and sociality, using six different test paradigms (visual cliff, emergence from a box, social approach, novel-object, mirror approach and tonic immobility). The data from the tests were analysed using General Linear Models with genetic background (white/brown) and productivity level (high/low) and their interaction as explanatory variables. The highly productive lines were consistently and significantly less social than the less productive lines, and the white were more social than brown layers (e.g. in the social approach test: mean times to reach conspecifics 11 vs 37 s for white vs brown layers (GLM factor $F_{1,73}$=29.17, P<0.0001); 36 vs 14 s for high vs low productivity (GLM factor $F_{1,73}$=25.03, P<0.0001)). Genetic background but not productivity had a significant effect on fearfulness, with the white layers being more fearful (e.g. mean time to end tonic immobility: 321 vs 182 s for white vs brown layers (GLM: $F_{1,73}$=14.47, P<0.001)). Our results show that genetic background and selection for production efficiency can affect different behavioural dimensions differently i.e. sociality but not fearfulness has been affected strongly by selection for productivity. This needs to be taken into account when characterising the effects of genotype on behavioural traits. Further, RAT can explain the effects of selection for production on sociality but not fearfulness, and the impact of unintentional changes in behavior that accompany intense selection pressure on productivity deserves further study effects on welfare or function.

Genetic loci associated with dog-human communication and dog sociality

Mia Persson, Dominic Wright, Lina Roth, Petros Batakis and Per Jensen
AVIAN Behaviour Genomics and Physiology Group, IFM Biology, Linköping University, 58183 Linköping, Sweden; mia.persson@liu.se

Unlike their ancestors, dogs have evolved unique social skills allowing them to communicate and cooperate with humans. Previously, we found a significant heritability for variation in human-directed social behaviors in laboratory beagles. Therefore, the aim of the current study was to identify candidate genes associated with this variation. To do this, a Genome-Wide Association Study was performed on human-directed social behaviors filmed during the 'unsolvable problem' paradigm. We recorded the propensity for dogs to initiate physical interactions with an unknown human while attempting to solve an unsolvable problem. Genotyping was done with a HD Canine SNP-chip of 190 beagles from a unique laboratory population, bred, kept and handled under highly standardized conditions. The genotyped sample consists of the individuals with the highest and lowest scores in human-directed social behaviors from theprevious study. After quality control, 85,172 SNPs remained and were analyzed using the GEMMA software to correct for population stratification, inbreeding and sex. This revealed a genetic marker on chromosome 26 within the SEZ6L gene, significantly associated with time spent close to the human ($P=4.87\times10^{-8}$; $R^2=0.15$) and in physical contact with the human ($P=5.25\times10^{-7}$; $R^2=0.13$). Another suggestive locus on chromosome 26 within the ARVCF gene was also identified ($P=7.94\times10^{-7}$; $R^2=0.14$) and three additional genes, including the COMT gene, were present in the same linkage block. In humans, the SEZ6L gene has previously been associated with autism and the COMT gene has been associated with aggression in adolescents with ADHD, all of which affects social abilities. These results contribute to a better understanding of the genetics of dog sociality, which can lead to a better understanding of dog domestication and potentially improve selective breeding. This is, to our knowledge, the first genome-wide study to present candidate genes for dog sociality and inter-species communication.

Evolution of domestication in chickens – is tameness the key?

Per Jensen, Beatrix Agnvall and Johan Bélteky
Linköping University, IFM Biology, Linköping University, 58183 Linköping, Sweden;
perje@ifm.liu.se

Domestication is a powerful proof-of-principle for evolution. Understanding the processes driving it may help explaining how evolution can modify large complexes of phenotypic traits in response to changes in selection pressures. Furthermore, it is highly relevant from a welfare point of view, since it has been adapting populations of animals to a life under human auspice for thousands of years. Domesticated animals tend to evolve complexes of phenotypic traits, which are similar across species, including changes in size, reproduction and social behaviour. Previous research has suggested that this may be a result of correlated selection responses, perhaps reflecting similar underlying genetic architectures. Starting from an outbred population of 100 Red Junglefowl, ancestors of all domesticated chickens, we selected strains for high or low fear of humans during five generations (each generation maintained at about 100 birds), based on a standardized behaviour test. We then assessed effects on a range of phenotypes, and on brain gene expression in birds from the different selection lines. In generation four and five, Low Fear birds were heavier at hatch (22.9 vs 22.3 g) and grew faster to 200 days of age (200 days weight: 975 vs 880 g; both weights: P<0.001). Low Fear birds were also more socially dominant in a pair-wise food competition test (8 vs 3 aggressive pecks/60 min), showed less fear of novel objects (80 vs 150 s latency to approach), had higher basal metabolic rate as chicks (27 vs 26 ml O_2/min/kg and higher feed conversion as adults (5.5 vs 3.9% growth/feed intake) (all effects P<0.05). Brain gene expression profiles, measured in 24 hypothalamic samples differed consistently between the selection groups, and the main pathways enriched were related to reproduction and immunology (GO-analysis, P<0.05). The results support the idea that tameness may have driven the evolution of domesticated phenotypic complexes, and may have a bearing for our understanding of rapid evolutionary processes in general.

The social environment affects maternal endocrine profiles but not offspring behaviour in Japanese quail

Vivian C Goerlich-Jansson[1], Esther MA Langen[1,2] and Nikolaus Von Engelhardt[3]
[1]Utrecht University, Animal in Science and Society, Yalelaan 2, 3508 TD Utrecht, the Netherlands,
[2]University of Bielefeld, Animal Behaviour, Morgenbreede 45, 33615 Bielefeld, Germany,
[3]Plymouth University, School of Biological Sciences, Drake Circus Plymouth, Devon PL4 8AA,
United Kingdom; v.c.goerlich-jansson@uu.nl

An individual's behaviour and physiology, and ultimately welfare status, is strongly affected by its social environment. In birds, independent studies have shown that social density affects both circulating hormones in breeding females and deposition of hormones to the yolk, thereby inducing non-genetic grand-/parental effects on offspring development. In this project we combined the above aspects to further explore in which way the social environment influences hormone mediated maternal effects and offspring phenotype in Japanese quail (*Coturnix c. japonica*). The experiment followed four generations to test for grand-/maternal effects. Females in the parental generation (P0) were housed either in a pair (one female, one male), or in a group (three females, one male). The F1 offspring was housed under social conditions that either matched or differed from the maternal treatment, while birds in the F2 and F3 generation were all kept in groups. In each generation we performed behavioural and physiological tests to correlate phenotypes between mothers and daughters. P0 pair females had higher baseline plasma corticosterone concentrations (ng/ml; pair: 2.9 ± 0.6, group: 2.1 ± 0.7, $P=0.01$), but did not respond differently to a standardized restraint stress protocol (ng/ml; pair: 13.6 ± 6.1, group: 11.6 ± 7.9; $P=0.7$). In both treatments, however, plasma corticosterone levels negatively correlated with number ($P=0.006$) and hatch mass of F1 offspring ($P=0.03$). Although P0 pair females also had higher baseline plasma testosterone concentrations (ng/ml; pair: 0.7 ± 0.2, group: 0.5 ± 0.2; $P<0.01$), the average yolk testosterone concentrations did not differ between treatments (pg/mg; pair: 12.4 ± 0.4, group: 12.3 ± 1.0, $P=0.9$). We did not find correlations between maternal hormone levels and offspring behaviour during tonic immobility or in an emergence test, nor did (grand-) maternal treatment predict offspring behaviour in these tests (F1, F2, F3, $P>0.08$). We show that the social environment does affect a female's hormonal state, though this does not seem to result in strong maternal effects on offspring behaviour. Nevertheless, we found the general pattern that high plasma levels of corticosterone, the main avian stress hormone, negatively affect reproductive output. Japanese quail have been domesticated for decades, therefore might be coping well with different social conditions, while fundamental biological trade-offs still exist.

Benefits of social information in cattle management using social network analyses

Paul Koene
Wageningen University & Research, Animal Welfare, De Elst 1, 6708 WD Wageningen, the
Netherlands; paul.koene@wur.nl

Many domestic animals tend to be large, non-selective feeders occupying open habitats. They are socially organized non-territorial species, typically occurring in relatively large groups in their natural environments. Such grouping of animals has costs and benefits. A strong social network is functional in maintaining the group and coping with many challenges. In captivity such networks may or may not exist, dependent on actual group size and living conditions. Social relations of individuals determine social stress, social support and even disease transmission and are important for health and welfare. In dairy cows the focus is mainly on individual performance and measurements despite the fact that social information is shown to be important in managing introductions and removals, estrus, lameness, cow traffic with respect to feeding (fence) and milking (robot). Social influence can be both positive and negative. Relevant behavioural information are interactions, rank order, nearest neighbors, furthest neighbors, friends, approaches, avoidances, time spend together and many social behavioural elements, such as vocalizations. Many of the interactions and associations can be automatically measured using sensor technology. In this review, the social organization of cattle in different contexts is analyzed using published and new data of cattle. Interactions and/or associations are analyzed using Social Network Analysis (SNA) using UciNet and NetDraw visualization. Natural social organizations – using SNA – are shown in Heck cattle (fenced and free-range), Galloways and Scottish Highland cattle (free-range), while social organizations of dairy cows, beef bulls and calves are shown in a farm setting (old data re-analyzed). Also, practical applications of the use of SNA are shown in calves and in the effect of fencing on heifer and dairy cow social behaviour (new data). The presented information shows a functional and evolutionary interpretation of the social organization of cattle. Using the social information in the barn will improve actual management by more precise interpretation of performance measurements and the opportunity to guide the behavior in cows, bulls and calves. Furthermore, assessments of individual and group welfare are possible, facilitating benchmarking. The conclusion is that social networks are important for the health and welfare management of cattle. Furthermore, the automatic measurement of location, movements and/or nearest neighbors for management purposes is feasible and automatic recording and follow-up management actions will probably become an essential part of cattle management in the near future.

Calling by domestic piglets during simulated crushing and isolation: a signal of need?

Gudrun Illmann[1], Kurt Hammerschmidt[2], Marek Špinka[1] and Céline Tallet[3]
[1]Institute of Animal Science, Prague, Czech Republic, Department of Ethology, Přatelství 815, 10400 Prague, Czech Republic, [2]German Primate Center, Cognitive Ethology, Kellnerweg 4, 37077 Göttingen, Germany, [3]INRA, Agrocampus Rennes, UMR1348 Rennes, France; gudrun.illmann@vuzv.cz

This study examined whether piglet distress vocalizations vary with age, body weight and health status, according to the predictions of the honest signalling of need evolutionary model. Vocalizations were recorded during manual squeezing (a simulation of being crushed by mother sow) and during isolation on Days 1 and 7 after birth in piglets from 15 litters. We predicted that during squeezing, younger, lighter and sick piglets would call more intensely because they are in higher risk of dying during crushing and therefore they benefit more from the sow's reaction to intensive vocalization. For isolation, we predicted that lighter and younger piglets would call more because they are more vulnerable to adverse effects of the separation. Calls were analyzed in their time and frequency domain. The rate of calling, call duration, proportion of high-pitched calls and eight acoustic parameters characterizing frequency distribution and tonality were used as indicators of acoustic signalling intensity. To assess which factors affected the acoustic quality of calls we used linear mixed models, with age, weight and the interaction between age and weight as between-subject fixed factors. Piglets that experienced 'squeezing' on Day 1 produced more intense acoustic distress signalling than on Day 7. On Day 1, piglets called more often ($F_{1, 55}=7.1$, $P=0.01$, Day 1: 28.7±1.7 vs Day 7: 20.9±2.6, mean±SEM), and had a higher proportion of high-pitched calls ($F_{1, 55}=14.8$, $P=0.0001$, Day 1: 71.6%±4.9, Day 7: 45.3%±4.9) than Day 7. Lighter piglets called more during squeezing than heavier piglets ($F_{1,55}=7.1$, $P=0.011$). Health status did not significantly affect any of the indicators of intensity of vocalization during squeezing. In isolation, none of the parameters of vocalization intensity were affected either by the age or by the weight of the piglets. In summary, the model of honest signalling of need was confirmed in the squeezed situation, but not in the isolation situation.

The defensive behavioral patterns of white-lipped and collared peccary: an approach for peccary's conservation

Selene Nogueira, Aline Reis, Stella Lima, Thaise Costa and Sérgio Nogueira-Filho
Universidade Estadual de Santa Cruz, Applied Ethology Laboratory, Rod. Jorge Amado km 16, Salobrinho, 45662900, Brazil; seleneuesc@gmail.com

Defensive behavioral patterns can affect an animal's fitness and may play a role in species conservation status. To test this hypothesis, we compared the risk assessment and defensive behavioral responses of the white-lipped peccary (Tayassu pecari, WLP) and the collared peccary (Pecari tajacu, CP), which hold different conservation statuses (WLP, vulnerable and CP, least concern). We used an adapted mouse defense test battery (MDTB) paradigm, consisting of four consecutive tests. Two of the tests simulated a novel environment, while the other two simulated a predator attack. In addition to differences on risk assessment and defensive threat/attack behavioral patterns between species, we compared flight initiation distance and flight speed. We analyzed the data in each test by the Principal Component Analysis. We used the factor scores received from each animal in the first principal component (PC1) in each test to evaluate the discrimination between species using t-tests. It was possible to differentiate the species in the PC1 scores in all tests: The novel environment test (CP:0.94±1.15 and WLP:-0.94±0.97; t=3.96; P=0.0009), the foraging/eating in a novel environment test (CP:1.32±0.58 and WLP:-1.32±0.76; t=8.72; P<0.0001), the chase test (CP:-0.90±1.40 and WLP:0.90±0.59; t=-3.73; P=0.001), and the forced contact test (CP:-1.52±0.52 and WLP:1.52±1.07; t=-8.07; P<0.00001). When facing novel environment and predator risk challenges, the WLP showed more exploratory and defensive threat/attack behavioral patterns, shorter flight initiation distances, and lower flight speeds, while the CP showed more cautious and retreat patterns, longer flight initiation distances, and higher flight speeds. The results confirmed our hypothesis that the collared peccary's awareness may help to prevent the decrease of its population, whilst the white-lipped peccary's exploratory and confrontation behavioral patterns in overhunted areas might contribute to include this species in the vulnerable category.

Malfunctional male mink: effects of enrichment and basal ganglial traits on stereotypic behaviour and mating success

María Díez-León[1], Sylvia Villegas[1], Rob Duprey[2], Mark Lewis[2] and Georgia Mason[1]
[1]University of Guelph, 50 Stone Rd E, N1G2W1 Guelph, Canada, [2]University of Florida, 1149 Newell Dr, 32611 Gainesville, USA; mdiez@uoguelph.ca

Neurophysiological and behavioural data implicate aspects of basal ganglia (BG) function, even malfunction, in caged animals' stereotypic behaviour (SB). For example, SB often co-occurs with perseveration (inappropriate repetition under test), likely reflecting the BG's behavioural sequencing role. But how BG function, SB and perseveration interrelate to affect competence in fitness-related behaviours (e.g. mating) is unknown, and this could provide an objective way to identify whether SB is pathological. We tested the hypothesis that enrichment-induced BG changes reduce both SB and perseveration, and enhance mating success in mink. We raised 16 males in non-enriched (NE) and 16 in enriched (E) cages from birth until 2 years old, and assessed their SB, perseveration and mating success (copulations won in a mate choice task) over two breeding seasons. Mink were then humanely killed for analysis of neuronal metabolic activity via cytochrome oxidase staining of the following BG areas: dorsal striatum (caudate, putamen), globus pallidus (externus, GPe; internus, GPi) and subthalamic nuclei. GPe:GPi ratios were calculated to assess relative activation of inhibitory vs excitatory pathways. As expected, NE mink stereotyped more (E:2.5±0.7; NE:33.2±7.1; $F_{1,30}$=13.6, P<0.001), and tended to perseverate more (E:0.4±0.08; NE:0.6±0.07; $F_{1,12}$=2.85, P=0.06). They gained fewer copulations too (E:3.2±0.6; NE:1.9±0.4; $F_{1,14}$=3.2, P<0.05), their low mating success being predicted by high SB (r^2:0.4; $F_{1,24}$=4.25, P<0.05) and perseveration (r^2:0.3; $F_{1,11}$=4.96, P<0.05). However, our measures of BG function were only partially explanatory: higher caudate activity predicted greater perseveration (r^2:0.3; $F_{1,10}$=4.18; P<0.05); in E mink only, lower putamen activity predicted mating success (r^2:0.3 $F_{1,13}$=4.62; P=0.05); and high GPi activity predicted more copulations in NE mink only (r^2:0.4; $F_{1,10}$=5.16; P<0.05). Correcting for multiple testing rendered all three results as trends. Also, though E mink had higher GPe:GPi ratios (E:0.1±0.05; NE:-0.1±0.05; $F_{1,24}$=4.7, P<0.05), enrichment did not affect neural activity in a single BG area. Thus enrichment effects on these structures did not explain how housing affects SB and mating success, suggesting instead that individual differences in the neural bases of routine-formation interacted with the presence/absence of enrichment to affect males' success with females. Overall, mink SB is thus malfunctional, but its precise neurological bases remain unclear. We are therefore reanalysing stained brain sections to assess whether changes in another BG region, the nucleus accumbens, are implicated instead.

Effects of UV light provision on broiler chickens' fear response and lighting preferences

Antonia Patt and Rachel Dennis
University of Maryland, Animal and Avian Sciences, 8127 Regents Drive, College Park, 20742, MD, USA; apatt1@umd.edu

Chickens are tetrachromatic, meaning that they perceive both ultraviolet (UV) and visible spectrum of colors. Birds use the UV spectrum in identifying social cues and discriminating between foodstuffs. It has been shown that the presence of UV light can alter behavioral displays. In the present study we investigate the impact of UV light on broiler behavior, including fearfulness, and lighting preference. Birds were randomly assigned to one of 24 9-bird pens with *ad libitum* feed and water. Half of the pens (n=12) were exposed to UV light for 4 hours/day, while controls had only white light. At 5 wk, 3 birds/pen were moved to pens of either the same or opposite treatment, resulting in 4 treatments for the remaining 2 wks: continuous UV, early UV exposure, late UV exposure, white light only (controls). In wks 4 and 6, the birds' fear responses were tested in a social isolation and flight distance (fear of human) test. Additionally, the birds' lighting preferences were tested in spectral and intensity preference tests. Data was analyzed with SAS 9.3 software using mixed models analysis. During the social isolation test (wk 4) latency to vocalize tended to be shorter in UV birds (1.7±1.4 s) compared to controls (5.1±1.4 s; P=0.09). Differences in 1st (P=0.01) and 3rd (P<0.0001) quartile frequencies of vocalizations indicate a higher energy distribution in calls of UV exposed birds compared to controls. The flight distance of birds to an unfamiliar human (wk 6) was shorter in UV birds than in controls from 0-10 s (69±2.3 and 88±2.4 cm, respectively; P=0.04). In a 4 arm maze (spectral preference test consisting of UV light, white light, both lights, no light arms), birds in the UV treatment spent more time in the arm providing UV light (56±14 and 14±14 s, respectively; P=0.04) and tended to do so in the arm providing both UV and white light (23±8 and 4±8 s, respectively; P=0.07) compared to animals in the control treatment. No differences were found between treatments in time spent in the center of the maze (P=0.98) or the arms that provided white light only (P=0.38) or no light (P=0.15). When tested for their preferred intensity of both UV and white light, birds did not differ in their preference for UV light intensity. However, early UV exposure birds showed a preference for lower white light intensity (P=0.02), whereas late UV exposure birds showed a preference for higher white light intensity (P=0.01) compared to controls. Our results show that birds exposed to UV light are more fearful during social isolation, but less fearful of humans than control birds, and have altered relative lighting preferences. These findings show the need for more research into the impact of UV lighting on broiler well-being.

Personality in Labrador Retriever dogs: underlying genetics and effects of lifestyle

Marie J. Haskell[1], Sarah E. Lofgren[2], Joanna Ilska-Warner[3], Sarah Blott[4], Enrique Sánchez-Molano[3], Dylan N. Clements[3] and Pamela Wiener[3]
[1]SRUC, West Mains Rd, Edinburgh, EH9 3JG, United Kingdom, [2]University of Edinburgh, R(D) SVS, Midlothian, EH25 9RG, United Kingdom, [3]The Roslin Institute, Midlothian, EH25 9RG, United Kingdom, [4]University of Nottingham, Sutton Bonnington, LE12 5RD, United Kingdom; marie.haskell@sruc.ac.uk

Canine personality is of interest to dog owners, and also offers researchers the possibility to investigate the effects of genetics and environment on behavioural traits. The aim of our study was to assess the effects of housing and work/exercise on personality, and to investigate the interactions with genetic background. Owners of Labrador Retrievers were identified from the UK Kennel Club records and sent two questionnaires: the C-BARQ canine personality questionnaire, which consists of 102 questions on behavioural characteristics such as aggression, anxiety and trainability and a management questionnaire on physical characteristics, care and housing of the dogs. A sub-set of owners were also asked to provide a DNA sample (saliva swab) from their dogs. Pedigree data was retrieved from Kennel Club records. There were 2020 respondents to the C-BARQ survey, 1978 dogs with both the management questionnaire and C-BARQ and 885 with all data. PCA was used to identify 12 personality traits from the questions in C-BARQ. Logistic regression was used to investigate the associations between personality and management, and linear mixed models were fitted in ASReml to estimate heritability. The 'role' of the dog had the largest effect. Gundogs had better scores for 'trainability' than showdogs or pets (scores: 4.3, 4.0, 3.6; P<0.05). Showdogs showed more favourable scores for 'human and object fear' than gun or pet dogs (0.22, 0.28, 0.29; P<0.05). Dogs exercised up to 2 h/day showed more 'non-owner aggression' than those exercised for longer (0.43, 0.43, 0.37, 0.33 for <1, 1-2, 2-3,>4 h/day; P<0.05). Dogs housed entirely outdoors showed less 'human and object fear' than those housed entirely or partially indoors (0.28, 0.29, 0.22 for indoors, mixed and outdoors; P<0.05). Heritability estimates ranged from 0.03±0.04 for 'owner-directed aggression' to 0.38±0.08 for 'fetching', which is similar in range to other studies. Heritabilities greater than 0.20 were found for six traits including 'fear of noises', 'non-owner directed aggression' and 'trainability'. The spread of genotypes indicated a separation between gun dogs and dogs with other roles. Sub-populations appear to exist in this breed that are evident in the behaviour and the genetic background. Aspects of management, such as exercise, appear to modify the expression of personality traits, often in relevance to the role of the dog.

Early-life behavioural development of lines divergently selected on feather pecking behaviour

Jerine Van Der Eijk[1,2], Aart Lammers[1], Bas Kemp[1], Marc Naguib[2] and Bas Rodenburg[2]
[1]*Wageningen University, Adaptation Physiology Group, De Elst 1, 6708 WD Wageningen, the Netherlands,* [2]*Wageningen University, Behavioural Ecology Group, De Elst 1, 6708 WD Wageningen, the Netherlands; jerine.vandereijk@wur.nl*

The behavioural development of an animal is affected by the interplay between its genetic background and its environment. In laying hens, feather pecking is a damaging behaviour that involves pecking and pulling at the feathers or tissue of conspecifics, negatively affecting welfare. Feather pecking is a heritable trait, but its development can be affected by many factors, including environmental factors. In order to better understand the development of this damaging behaviour, we characterised the early-life behavioural development of lines selected both for and against feather pecking. We used genetic lines selected for high (HFP) and low (LFP) feather pecking and an unselected control line. Lines were housed separately in groups of 19 birds per pen, with 8 pens per line. Group size was reduced by 2-3 birds at 0, 5 and 10 weeks of age. There were two batches that differed two weeks in age. Birds were tested at 0 and 10 weeks of age in a novel object test, at 4 weeks of age in a novel environment test and at 15 weeks of age in an open field test. Data were analysed using mixed models, with selection line as fixed factor and pen nested within batch as random factor. When data were not normally distributed the non-parametric Kruskal Wallis test was used. HFP birds had a shorter latency to approach the novel object compared to LFP and control birds at both 0 weeks of age (HFP=36 s, LFP=120 s and control=117.5 s, $F_{2,20}$=36.52, P<0.0001) and 10 weeks of age (HFP=16.88 s, LFP=79.75 s and control=61 s, $F_{2,20}$=12.60, P=0.0003). Furthermore, HFP birds had a shorter latency to vocalize compared to LFP and control birds at both 4 weeks of age (HFP=5.48 s, LFP=15.16 s and control=16.2 s, X^2_2=42.23, P<0.0001) and 15 weeks of age (HFP=26.34 s, LFP=50.27 s and control=37.63 s, X^2_2=15.59, P=0.0004). Thus, based on these three tests, HFP birds were less fearful at all studied ages compared to control and LFP birds. In addition, HFP birds showed a more pro-active coping style than control and LFP birds. In conclusion, our results suggest that selection for feather pecking affects early-life behavioural characteristics. These results can help to better understand the development of feather pecking behaviour, and possibly to identify early-life behavioural characteristics as potential indicators of feather pecking.

Open or Under? Occupancy and dust bathing in different litter areas by 4 strains of laying hen

Nicholas Newsome[1], Ahmed Ali[1], Dana Campbell[2] and Janice Siegford[1]
[1]Michigan State University, Animal Science, 474 S Shaw Ln, East Lansing, 48824, USA, [2]University of New England and CSIRO, Armidale, NSW, Australia; newsomen@yahoo.com

Aviary housing systems for laying hens typically contain a litter area outside of, and sometimes under the housing units, providing hens the opportunity to perform a variety of behaviors such as dust bathing and foraging. How hens occupy and use covered vs uncovered litter areas is not fully understood, particularly with respect to strain differences. The objective of this study was to compare occupancy and dust bathing behavior of covered and uncovered litter areas among 4 strains of laying hens (Hy-Line Brown, Bovan Brown, DeKalb White and Hy-Line W36). The number of hens on litter was counted and compared with the number of hens dust bathing and analyzed in relation to time of day (midday or evening) and location on litter (open=OP; covered=CO). At 36 wk of age, two aviary units were observed for each strain (144 hens/unit). For 3 consecutive days, hens were observed every 30 min between (12:00 – 14:00 and 16:00 – 18:00). Synchronized live observations (CO) and video recordings (OP) were used to count the number of hens occupying and dust bathing in the litter with the data analyzed using ANOVA with post hoc Tukey test. Because both strains of brown hens (BR) behaved similarly to each other and both strains of white hens (WH) behaved similarly to each other, for analysis brown strains were grouped together and white strains were grouped together. More WH than BR were in both litter areas during midday (WH: 75.45±10.01, BR: 62.41±9.64) and evening periods (WH: 73.25±7.93, BR 63.28±3.27, $P<0.05$ for both). During the midday, BR occupied the CO litter area in higher numbers (BR: 39.94±10.93, WH: 27.42±9.22, $P<0.05$) while more WH were in the OP litter area (WH: 41.57±11.84, BR: 26.31±10.42, $P<0.05$). In the evening, BR utilized OP and CO litter areas to the same degree (OP: 34.67±4.45, CO: 33.70±10.09) while WH continue to use the OP litter more than the CO litter area (OP: 42.53±2.63, CO: 31.30±13.46, $P<0.05$). At midday, the pattern of hens dust bathing paralleled the use of litter areas for all strains, with more BR dust bathing in CO litter (CO: 8.30±5.63, OP: 2.19±0.62, $P<0.05$) and more WH dustbathing in OP litter (CO: 7.44±5.15, OP: 10.66±3.50, $P<0.05$). However, during evening observations, white and brown strains dust bathed at higher rates in CO litter compared to OP litter ($P<0.05$ for both). Hens occupied and dust bathed in all areas, but white and brown strains occupied litter areas differently and circadian differences were observed in use of areas for dust bathing. Further research is needed to understand the factors underlying these patterns to optimally design aviary litter areas and match strains of hens to them.

Are laying ducks stressed when they cannot access a nest?
Lorelle Barrett, Shane Maloney and Dominique Blache
The University of Western Australia, 35 Stirling Highway, Crawley, 6009, Perth, Australia;
lorelle.barrett@research.uwa.edu.au

In commercial duck farming, floor-laying of eggs is a common problem. Factors such as competition for nest boxes, the motivation to access a nest, or stress related to difficulties with accessing a nest have not been thoroughly examined. We investigated the impact that the inability to access a nest had on 2 indicators of stress; egg corticosterone levels and elevated core body temperature (stress hyperthermia). Twelve 28-week old laying Pekin ducks were trained to access a nest. Ducks could access a nest site by pushing a door that was loaded with increasing weight and eventually blocked to prevent nest access. Each weight load was in place for 4 consecutive nights. Ducks were habituated to the overnight housing for the 2 weeks immediately before the experiment commenced. Birds were housed individually overnight, with visual and auditory contact of peers. Prior to testing, temperature data loggers were surgically implanted in the abdomen. Eggs were collected daily and concentrations of corticosterone in egg albumin were measured using a radioimmunoassay. Data sets from 7 birds were deemed appropriate for use in the statistical analysis. Corticosterone concentrations were analysed using repeated measures ANOVA. The association between hyperthermia and failure to access the nest site was analysed using logistic regression. Corticosterone in egg albumin did not increase with inability to access the nest or increasing workload (mean 14.23±6.07, range 3.06-29.83 ng/ml). The likelihood of hyperthermia increased whe birds performed the maximum workload (odds ratio=240.1, $P<0.001$) or were unable to access the nest (odds ratio 182.7, $P<0.01$), compared with when they had free access to the nest. On the night of first failing to access the nest, the median number of attempts made by birds was 17 (range 0-71), but no attempts were made on the followng 3 nights. However, on these 3 nights, hyperthermia was associated with first look at the nest in 61% of events. Egg corticosterone does not seem to be a relevant indicator of acute stress. Although the cause of hyperthermia on the first night of failure to access the nest cannot be differentiated between intense muscular activity during repeated attempts and psychogenic stress, the results obtained on the 3 subsequent nights show that ducks experienced stress hyperthermia when unable to nest in their preferred site. In the context of commercial production systems, the measurement of stress hyperthermia can provide a useful index of psychogenic stress, and will be helpful to assess strategies to improve nest availability (e.g. by decreasing nest competition) that could improve duck welfare and production efficiency.

Vocal playback affects investigative and social behaviour in captive African lions (*Panthera leo*)

E Malachowski[1], J Cooper[1] and S Marsh[2]
[1]University of Lincoln, School of Life Sciences, Lincoln, LN6 7TS, United Kingdom, [2]Yorkshire Wildlife Park, Doncaster, DN4 6TB, United Kingdom; jcooper@lincoln.ac.uk

Captive lions often have unusual pride structures which may affect social stability. This study investigated lions' response to familiar and unfamiliar vocalisations, to assess if they react to intra-specific signals in same way as their wild counterparts and if the vocalisations promote investigation and affiliative behaviours such as allo-licking and head rubbing. The study population consisted of 10 lions in three closely related groups: an 8 year old female with two 4 year old male offspring; an 11 year old female with two 5 year old male offspring; and an 11 year old female with three 7 year old female offspring. All had been transferred to YWP in 2010 and housed in 3 one hectare enclosures, mainly laid to grass, with elevated mounds, rock shelters and a deep pond. Data collection was conducted in summer 2014. After 4 days of observation to record baseline data, three treatments were used twice over 6 days: familiar lion vocalisations ('roars' recorded from YWP), unfamiliar lion vocalisations (two recordings from wild lions) and unfamiliar cross species vocalisation control (dolphin). Orientation to speakers, duration and form of locomotion, affiliative social responses, vocalisation and yawning were recorded. Frequency and duration of behaviours were analysed using GLM ANOVA with group as random factor and lion nested within group with Tukey test to determine treatments effects. Roar playback resulted in increased locomotion (7.8% of time vs 0.6%, F=14.3, P<0.001), speaker orientation (12.2% vs 1.2%; F=25.1, P<0.001), yawning (6.0 vs 1.0; f=17.4, P<0.001), licking (3.0 vs 0.2; F=118, P<0.001) and head rubbing (2.6 vs 0.5; F=36.1, P<0.001) compared to baseline/control. Head-rubbing (3.6 vs 1.6; t=4.03, P=0.002) and licking (3.8 vs 2.2; t=3.95, P=0.003) were more frequent with unfamiliar roars than familiar roars, and lions showed shorter latency to orientate to speaker with unfamiliar roars (3.3 vs 6.5 s; t=2.68, P=0.039). Vocalisations were rarely recorded in response to the three treatments. The data indicate that captive lions respond to biologically meaningful vocalisations and can distinguish between familiar and unfamiliar signals. Locomotion after recordings was largely associated with investigation and patrolling the enclosure, and did not appear stereotypic. The increases in head rubbing and licking were consistent with a more inclusive social network within groups. These findings suggest that playback of vocalisations could provide environmental enrichment for captive lions at least in the short term, increasing both investigative behaviour and positive social interaction.

Estimation of heritability and environmental effects of number of scored lesions in group-housed pigs

Kaitlin Wurtz[1], Carly O'malley[1], Janice Siegford[1], Ronald Bates[1], Catherine Ernst[1], Nancy Raney[1] and Juan Steibel[1,2]
[1]*Michigan State University, Animal Science, 474 S. Shaw Lane, East Lansing, MI 48824, USA,*
[2]*Michigan State University, Fisheries and Wildlife, 480 Wilson Road, East Lansing, MI 48824, USA; wurtzkai@msu.edu*

A welfare concern for group-housed pigs is aggression, particularly in the first 24 hours following grouping with novel individuals. Use of enrichment or alternative pen designs have practical constraints and have shown limited success in reducing aggression. Our study aims to estimate genetic parameters and environmental effects influencing the number of skin lesions in group-housed pigs to provide alternative means to reduce aggressive interactions through genomic selection. Purebred Yorkshire pigs (n=527) were strategically mixed into new groups at 3 ages: weaning (~27 d of age), move to finisher pens (~72 d of age), and at approximately 100 kg (~150 d of age, gilts only). Total number of fresh lesions (<24 h old, bright red marks on skin) were counted prior to mixing (baseline), 24 h post mixing, and 3 wk post mixing. Each individual was genotyped using a 70 k SNP chip (Neogen Corporation – GeneSeek Operations). A genomic best linear unbiased prediction model was fit utilizing the gwaR package in R with fixed effects of sex (gilt or barrow), replicate (4 levels), observer (3 levels), weight as a covariate, and random effects of pen (38 levels) and genetic additive effect (527 levels). The response variable was either total number of skin lesions within the first 24 h following remixing (mix) or total number of skin lesions present 3 wk post mixing (stable). Analyses were performed assuming a genomic relationship matrix for the genetic additive effect. Proportion of total variance explained by additive effects (heritabilities) were 0.096 (P=0.023), 1.857E-05 (P=0.500), and 0.134 (P=0.071) for mix scores at weaning, grow-finish, and gilts at ~100 kg, respectively, and 0.030 (P=0.160), 2.279E-06 (P=0.500), and 0.167 (P=0.060) for stable scores, respectively. Percent of variance explained by pen effects ranged from 0.011 to 0.359, with the remainder of variance attributed to residual effects. Heritabilities for total number of lesions were only significantly different (P<0.05) from zero for mixing at weaning, with variation at the additional time points being largely explained by pen and residual effects. However, it is important to note that these results are preliminary and focus on one measure of aggression. The number of animals will ultimately be doubled as the remaining pigs in the study are genotyped, and we continue analyses using additional measures of aggression as well as conduct genome-wide association studies to identify regions within the genome responsible for the observed variation.

The relationship between standing behaviour in transported calves and subsequent blood glucose and CK concentrations

Andrew Fisher, Bronwyn Stevens, Melanie Conley, Ellen Jongman, Sue Hides and Peter Mansell
The University of Melbourne, Animal Welfare Science Centre and Faculty of Veterinary and Agricultural Sciences, 250 Princes Highway, Werribee Vic 3030, Australia; adfisher@unimelb.edu.au

Young dairy calves may be transported from their farm of birth to a grower property or meatworks. Suitable space and flooring are thought to be important for calf resting comfort and welfare during journeys under extensive conditions. Research indicates that some calves have low blood glucose concentrations and/or elevated blood creatine kinase (CK) after the transport and consequent feed withdrawal period, indicating welfare risks due to hypoglycaemia and muscle bruising. The study was approved by the regulatory Ethics Committee and the objective was to undertake an analysis of the standing and lying behaviour of these calves, to determine whether calves that spent a greater journey proportion standing also experienced reduced blood glucose and elevated CK. Fifty-nine male Holstein-Friesian dairy calves (5 to 9 d of age) that had received colostrum after birth were clinically assessed for suitability for transport and in three replicates either: (1) held *in situ* on farm (control; n=15); (2) transported for 6 h (n=14); (3) transported for 12 h (n=15); or (4) transported for 1 h to a holding facility where they were kept for 6 h and then transported for 5 h (n=15). The space allowance in transport was the industry standard of 0.3 m^2 per calf. IceTag™ loggers recorded standing and lying behaviour of each calf, and blood data from the end of the total transport and feed withdrawal period were used in regression analyses within treatment to identify the influence of the proportion of time spent standing during transport on the final glucose and CK concentrations. Calves were clinically assessed and all fed immediately following the study period. For final blood glucose concentrations, 12% of calves had values below the normal published reference range, however there was no overall effect of the regression model (F=0.163, df=4, P>0.05), and no effect within treatments, with calves in the 12-hour transport treatment having the closest result to a significant relationship (P=0.15). Similarly, although calves in the 12-hour transport group had elevated CK concentrations compared with other treatments, the regression analysis for standing time % against Log [CK] indicated a P value of 0.19, and for the overall model P=0.305. It is concluded within the conditions of this study, that although sufficient room and flooring conditions to encourage suitable lying behaviour may be important for calf comfort, standing behaviour per se was not an influence on the extent of hypoglycaemia and elevated CK, and that other factors such as subclinical calf health may have contributed.

Effects of short and no dry period on behaviour of dairy cows

A. Kok[1], R.J. Van Hoeij[1], B.J. Tolkamp[2], M.J. Haskell[2], A.T.M. Van Knegsel[1], I.J.M. De Boer[1] and E.A.M. Bokkers[1]
[1]Wageningen University, Animal Sciences, P.O. Box 338, 6700 AH Wageningen, the Netherlands, [2]SRUC, West Mains Road, Edinburgh EH9 3JG, United Kingdom; akke.kok@wur.nl

A non-lactating period or 'dry period' (DP) of 6-8 weeks before calving is common practice. It is often seen as a rest period for the cow, and it maximises milk yield in the next lactation. A higher milk yield, however, can result in a more severe negative energy balance (NEB) in early lactation, associated with impaired health and fertility. A short or no DP reduces milk yield and the NEB after calving, at the expense of the 'rest period' before calving. We studied effects of a short (30 days) and no DP on walking, lying, and feeding behaviour of dairy cows 4 weeks before and 4 weeks after calving. Moreover, relations between milk yield, NEB, and behaviour were analysed. 86 Holstein-Friesian cows were given a short (n=29) or no (n=57) DP. All cows were housed in a free stall system with cubicles. Lactating cows were milked twice daily and received a lactating cow diet; dry cows received a dry cow diet. Diets were provided in feeding boxes that recorded visits, time spent feeding, and feed intake. Walking and lying behaviour were measured with accelerometers. DP categories were compared with unpaired t-tests; periods before and after calving were compared with paired t-tests. Before calving, cows with no DP tended to spend less time lying than cows with a short DP (12.7 vs 13.5 h/d; P: 0.05), whereas they walked more (1,134 vs 661 steps/d; P<0.01). Also, cows with no DP spent 0.9 h/d less on feeding than cows with a short DP (P<0.01). After calving, cows had shorter lying times than before calving (P<0.01). Cows with no DP spent 1.0 h/d more time lying and had 4.1 kg/d higher feed intake than cows with a short DP. The number of steps and feeding duration did not differ between the DP categories. Cows with no DP had a 3.5 kg/d lower milk yield (P: 0.01) and less severe NEB (-26 vs -197 kJ kg$^{-0.75}$/d; P<0.01) 4 weeks after calving than cows with a short DP. There was no correlation between daily lying and feeding time, or between daily lying time and milk yield. In conclusion, cows with no DP spent less time lying and feeding and walked more during their extended lactation than cows with a short DP, which is in line with not having a 'rest period'. Although both groups spent similar times on feeding after calving, cows with no DP spent more time lying than cows with a short DP. Moreover, lying and feeding behaviour were more constant between 4 weeks before and 4 weeks after calving for cows with no DP than for cows with a short DP. A DP may not be a necessity for rest, and no DP increased daily lying time and feed intake in early lactation.

Effect of feed allowance at pasture on lying behaviour of dairy cows

Keelin O'driscoll, Eva Lewis and Emer Kennedy
Teagasc, Pig Development Department, Animal and Grassland Research and Innovation Centre,
Fermoy, Co. Cork, Ireland; keelin.odriscoll@teagasc.ie

Pasture based dairying systems are the most efficient in Ireland when herbage utilisation is maximised so that feed costs remain low. However, grass growth can be low in spring which could result in a challenge to dairy cow welfare. Lying behaviour is often used as a welfare indicator for cows, and the daily pattern varies with daily herbage allowance (DHA). This study compared lying behaviour of lactating dairy cows at a variety of daily herbage allowances; 120%, 100%, 80% and 60% of intake capacity (IC). Cows were managed at 100% IC until 28±8 DIM, assigned to treatment on 9 Mar (d0) for 42 days, then returned to 100% IC. Dataloggers that record stand/lie at 30 sec intervals were fitted for 3 days prior to the experiment (d-6 to d-3; PRE), for 5 days from d16 (MID), and d37 (LATE), and for 5 days in June, approx. 7 weeks after cows returned to a DHA of 100% IC (POST). Total daily lying time, lying bout duration and no. lying bouts per day were investigated using a mixed model approach, accounting for repeated measures, and using PRE recordings as covariates. There was no overall effect of DHA on daily lying time, but it was higher in LATE (09:48:28±00:09:56) and POST (09:58:44±00:07:30) than in MID (08:31:54±00:11:05; $P<0.001$). In LATE there was an effect of treatment, with 60% cows having numerically lower daily lying times than the other treatments. Lying time of 120% cows did not change over time, but in all other treatments was lower in MID than POST ($P<0.01$). Daily lying time of 60% cows did not increase between MID and LATE, but did for both 80% and 100% cows ($P<0.01$). The DHA affected bout duration ($P<0.01$); 120% cows had longer lying bouts (01:14:12±00:03:20) than 60% (01:02:35±00:01:55; $P=0.01$), 80% (01:02:37±00:01:56; $P=0.001$) and 100% (01:00:15±00:01:48; $P<0.01$). This treatment effect was significant in LATE ($P<0.01$) and POST ($P<0.05$). The no. bouts per day was also affected by DHA ($P<0.05$); 60% cows had fewer bouts than 120% cows ($P<0.05$), than cows fed at or above 100% ($P<0.05$), and than all other treatments combined ($P=0.01$), whereas 120% cows had more than all other treatments combined ($P<0.05$). Lying times normally increase as lactation progresses, and cows on higher DHA treatments appeared to achieve this sooner than the 60% cows, driven by both longer and/or more lying bouts. At pasture, cows generally lie when not grazing. Differences in duration of lying behaviour could thus be due to variation in the efficiency of grazing at different herbage allowances. However, 7 weeks after DHA restriction concluded, 60% cows had similar lying times to the other treatments. Thus a period of feed restriction in early lactation did not have a long lasting reductive effect on lying behaviour.

Is maternal behaviour affected by peri-conception diet?

Giuliana G Miguel-Pacheco[1], Juan H Hernandez-Medrano[1,2], Duane H Keisler[3], Joerg-Peter Voigt[1] and Viv E.A. Perry[1]
[1]*University of Nottingham, School of Veterinary Medicine and Science, College Road, Sutton Bonington, LE12 5RD Nottinghamshire, United Kingdom,* [2]*Universidad Nacional Autonoma de Mexico, Animal Reproduction, Circuito Interior s/n, 04510 Ciudad Universitaria, Mexico,* [3]*University of Missouri, Division of animal sciences, 920 East Campus Drive, Columbia, MO 65211, USA; giuliana.miguelp@gmail.com*

The aim of this study was to investigate the effect of a low protein diet during breeding on maternal behaviour in beef cattle. Twenty-eight first parity South Devon heifers located in two farms (n=14 each) were allocated to two isocaloric diet treatments during preconception (two months before AI): High (H, 18% crude protein (CP)) or Low (L, 10%CP). After calving, 26 cow-calf pairs (n=11 females and n=15 males) were penned individually, fitted with an accelerometer (within 6 hours post-calving) and fed hay ad-libitum and concentrate twice daily (Farm 1) or silage ad-libitum (Farm 2). In both farms, pens were 18-20 m^2 with concrete flooring covered with deep straw bedding and fitted with video cameras recording continuously. To evaluate cow behaviour, 24-hour recordings were divided in 6 hour periods with the first 3 hours of each block analysed for cow maternal behaviours: total time licking or sniffing calf (nozzle in contact with any body part of the calf, minutes/12 h), total nursing time (calf's head is under cow's belly in the udder area, minutes/12 h) and total standing time (minutes/24 h). Data was analysed using unbalanced ANOVA with diet, farm, sex, calf weight and their interactions used as fixed factors. Cows body weight at calving was not different between diet groups (P=0.08; H cows (n=14): 534.00±15.99 kg, L cows (n=12): 585.71±20.52). Total nursing time was significantly higher in L cows (P=0.04; 36.51 min/12 h) compared to H cows (25.48 min/12 h); consequently there was a significant difference in the total standing time (P=0.03) between treatments. Licking/sniffing calf duration was comparable between treatments (P=0.21). Sex and calf weight did not have a significant effect (P>0.10) on the behaviour variables; but farm had a significant effect (P=0.03) on total standing time with farm 1 (961.78±39.39 min/24 h) being higher than farm 2 (745.91±39.67 min/24 h). In conclusion, the results from the present study suggest that a low maternal protein before conception increase maternal nursing time that may be linked to calf's suckling behaviour.

Interaction of dairy cattle and robot simulating objects of varying designs

Renate Luise Doerfler, Lena Hierstetter and Heinz Bernhardt
Technical University of Munich, Agricultural Systems Engineering, Am Staudengarten 2, 85354
Freising, Germany; renate.doerfler@wzw.tum.de

Robotic cleaning systems in present-day dairy farms are autonomous mobile objects in the activity areas of dairy cattle. These robots can detect obstacles only after coming in contact with them. As a consequence, collisions between animals and the robotic system can cause injuries to the cows and can impair the functionality of the robot. In order to minimize such collisions, knowledge about the interaction between cows and mobile objects is vital. Therefore, this study aimed to test the influence of the size, the shape and other attributes of a mobile object on the animal-object interaction and the potential risk of injuries for dairy cattle. The experiment was conducted over a period of four weeks at the university experimental farm Veitshof. Cameras monitored the behaviour of 30 cows when five differently designed objects with integrated electric motors moved alternately up and down the walkway. A Latin square design was used to present dairy cattle with: a small object which was defined as the reference, a tall object, a tall object equipped with cow horns, a tall object equipped with a wooden stick and an object shaped as a calf. Continuous recording was applied in order to measure the exact durations of 15 different behaviour patterns. Statistical analysis was carried out using linear mixed models. Results indicated that the time animals spent in passive contact with the object significantly increased between the small object and the tall object equipped with horns (26.71 vs 40.25; P=0.033) and between the small object and the calf-shaped object (26.71 vs 38.07; P=0.008). Similarly, the time in which the cows actively came into contact with the object significantly rose between the small object and the tall object with the horns (14.48 vs 27.04; P=0.003) and between the small object and the tall one with the stick (14.48 vs 18.45; P=0.013). In a distance of more than 2.5 m from the object the time cows spent standing on the walking area significantly increased between the small object and the large object equipped with horns (234.84 vs 252.56; P=0.018) and between the small object and the calf-shaped object (234.84 vs 292.23; P=0.045). In summary, most contacts between cows and the robot simulating objects occurred in the tall objects with attributes and the calf-shaped object. The risk of injuries is higher for cows interacting with these objects. Contrary to our expectations, the outcomes of this study support the conclusion that a mobile object which is small in size and neutral in character is more appropriate as a robotic cleaning system.

The association between the welfare of culled dairy cows and price achieved at auction market

Kerin Rosen[1], Pol Llonch[2] and Fritha Langford[3]
[1]*University of Edinburgh, Royal (DIck) School of Veterinary Studies, Easter Bush, Midlothian, EH25 9RG, United Kingdom,* [2]*Universitat Autonoma de Barcelona, School of Veterinary Science, 08193 Bellaterra, Cerdanyola del Vallè, Spain,* [3]*SRUC, Animal and Veterinary Sciences, West Mains Road, Edinburgh, EH9 3JG, United Kingdom; fritha.langford@sruc.ac.uk*

Dairy cows that are no longer productive are often culled by sending them to market to go to slaughter. Buyers have limited time to see the cows in pens prior to the sale and the in the ring during the sale. Buyers could use weight, visual body condition, mobility, general demeanour and behaviour in the ring to judge how much to offer. We hypothesised that buyers would use weight alone to base the price to pay at the auction. As the first step of understanding buyers' decisions, the study aimed to (1) assess the welfare of culled dairy cattle sent to an auction in the UK, and (2) identify associations between welfare indicators and the price achieved at an auction market. Over a ten-week period of Spring 2015, 383 cull dairy cows were assessed at Lanark Auction (Scotland, UK) for welfare assessment measures including body condition score (BCS), mobility, hair loss, swelling, and cleanliness (using AHDB Dairy score cards). A 14-term, fixed-list Qualitative Behavioural Assessment of 2-mins duration was carried out when cows were in the pre-sale pen. Breed, age, weight, and price achieved at auction were obtained from market reports and cattle passports (with farmer consent). Farmers that had brought a cull cow to market were asked the number of lactations, recent health problems and the reason for culling the cow. Correlations were used to assess relationships between continuous variables. Univariate generalized linear mixed models (GLMM) were conducted for each continuous and categorical variable; significant variables were retained for multivariate GLMM analysis. 74% of the cattle were second lactation cows or older, 26% were heifers. Weight had the strongest positive relationship with price (r=0.86, P<0.001) Age, BCS, lactation number, and breed type positively related to price paid in the auction even after weight had been taken into account. Mobility, swelling, hair loss and cleanliness scores each had a weak negative correlation with price paid. There was no relationship between the QBA terms and price paid. Although the reason for culling did not relate to price paid, cows which farmers reported to have had more than one case of mastitis in the last year achieved a lower price at market (r^2=-0.22, P<0.05). The results suggest that older, bigger cows in good body condition achieve higher prices at market, which is not surprising. It also gives an indication that poor welfare such as lameness and repeated mastitis results in a lower price at market.

Short-term consistency of behavioural on-farm measures of animal welfare in dairy cattle

Christoph Winckler[1], Gerhard Koch[1], Angelika Wassermann[1] and Marlene Kirchner[2]
[1]University of Natural Resources and Life Sciences, Vienna, Department of Sustainable Agricultural Systems, Gregor-Mendel-Strasse 33, 1180 Vienna, Austria, [2]University of Copenhagen, Department of Large Animal Sciences, Grønnegårdsvej 8, 1870 Frederiksberg C, Denmark; christoph.winckler@boku.ac.at

The aim of this study was to evaluate the consistency over time for several behavioural on-farm measures of dairy cattle welfare. For this purpose, all behavioural measures of the Welfare Quality[*] (WQ) assessment protocol for dairy cattle (e.g. time taken to lie down, incidence of agonistic interactions, avoidance distance, Qualitative Behaviour Assessment (QBA)), as well as the rising behaviour and two indices of stall use were assessed. Data were gathered on 30 Austrian dairy farms with cubicle housing; each of two trained observers assessed 15 farms twice at an interval of four days. Three out of ten WQ measures (i.e. % animals that can be touched (mean prevalence at first visit: 43.1%), % animals with avoidance distance <50 cm (mean 45.4%), QBA score (mean WQ score at first visit: -1.1)) and two rising score categories (i.e. % animals with undisturbed rising behaviour (mean 57.5%) and % animals with short pause on carpal joints (mean 34.5%)) showed a Spearman rank correlation coefficient greater than 0.7 (all $P<0.001$). Correlation coefficients for the other WQ measures ranged from 0.34 (number of displacements; $P>0.05$) to 0.63 (% animals with avoidance distance >100 cm; $P<0.001$). Also the indices of stall use showed only low consistency ('cow comfort index': rs 0.47; 'stall use index': 0.42; all $P<0.05$). Reasons for the low consistency over time at single measure level presumably are: small sample size, low incidence in the sample, response to small changes in the environment and low intra-observer reliability. Consistency partly improved with aggregation of WQ parameters (rs 0.58 and 0.79 at the 'Expression of social behaviours' and 'Human-animal relationship' criterion level and 0.87 at the 'Appropriate Behaviour' principle level; all $P<0.001$). Our results show that even at short intervals of only a few days, intra-farm variation of behavioural measures of dairy cattle welfare may be substantial, which calls for repeated assessments in order to obtain reliable results.

Resting behaviour of steers after a low dose of lipopolysaccharide

Kosuke Momita[1], Masumi Yoshida[2], Masayoshi Kuwahara[2], Daisuke Kohari[1], Hiroshi Ishizaki[3], Yoshihiro Muneta[4] and Ken-ichi Yayou[5]
[1]Ibaraki University, Inashiki-gun, 300-0393, Japan, [2]The University of Tokyo, Bunkyo-ku, 113-8657, Japan, [3]Institute of Livestock and Grassland Science, Nasushiobara, 329-2793, Japan, [4]National Institute of Animal Health, Tsukuba, 305-0856, Japan, [5]National Institute of Agrobiological Sciences, Tsukuba, 305-8602, Japan; kosuke_m2@yahoo.co.jp

Early detection of illness is important for animal welfare and productivity. Injections of bacterial lipopolysaccharide (LPS) have often been used to study sickness behaviour. High dose of LPS (25µg/kg of body weight [BW]) decreased time spent lying in milking cows. On the other hand, low dose of LPS (25 ng/kg of BW) increased the time spent lying inactive in calves. In most of the studies using LPS, however, physiological responses were prominent. Thus we examined the effects of a lower dose of LPS on respiration rate, rectal temperature, and resting behaviour of steers to help to detect the early onset of illness. Six steers (6-8 months) were intravenously injected both with LPS (5 ng/kg of BW) and with saline as a control using a crossover design. Rectal temperature and respiration rate were recorded just before injection, and at 1 h intervals during 12 h after the injection. Behaviour was continually sampled for 12 h. Lying behaviour without rumination was classified into following resting behaviour categories: (1) lying vigilance (LV: head lifted up and moving actively); (2) Behavioural status of non-rapid-eye-movement sleep (NREM: head lifted up and still if the behaviour lasted at least 16 s); (3) Behavioural status of rapid-eye-movement sleep (REM: head resting against body or ground if the behaviour lasted at least 16 s). The effects of LPS on resting behaviors were analyzed using the Wilcoxon signed rank test. And the effects of LPS on respiratory frequency and rectal temperature were analyzed using the two-way repeated-measures ANOVA. Steers showed significant increase in respiration rate during 1-5 h and in rectal temperature during 2-5 h after the LPS injection (P<0.05). The body temperature never increased more than 39.5 °C maximum. LPS did not affect the total time spent lying without rumination, LV and REM. On the other hand, LPS significantly increased the time spent NREM (SAL vs LPS: 45.3±26.0 min vs 79.8±26.7 min; P<0.05). LPS did not influence the mean duration of NREM, but significantly increased the frequency of NREM (P<0.05). In the present study, we could mimic mild physiological responses during the beginning of acute gram-negative bacteria infection. Although overall time spent lying was not influenced in such a mild infection, more detailed analysis of resting behaviour, such as behavioural status of NREM sleep analyzed in the present study could be an indicator of early onset of illness.

Duration of sleep posture of milking cows in Japanese dairy farms

Michiru Fukasawa, Tokushi Komatsu and Yumi Higashiyama
NARO Tohoku Agricultural Research Center, 4 Akahira, Shimo-kuriyagawa, Morioka, Iwate,
020-0198, Japan; shakecat@affrc.go.jp

Sleep is important for the health and welfare of mammals. Sleep in a relaxing posture is innately dangerous for prey animals, such as cattle, therefore sleep posture may be used as an indicator of perceived safety and comfort. In this study, we investigated the duration of sleep posture (sleep time) of milking cows, which we defined as the period when their head is relaxed on their shoulder in a lying position. This study was conducted at 12 Japanese dairy farms located in northern Tohoku, Japan. All cows were kept in tie-stall sheds during whole day and were milked twice daily. Measurements were taken in summer and autumn. Five cows from each farm were selected at random each season. The sleep time was measured over 24 h using an accelerometer fixed on the middle occipital with a halter. Urinary cortisol concentration and bulk milk quality were also measured. Data were downloaded and analysed by accompanying software. In total, data were collected from 85 cows (summer 41, autumn 44) which were both primi- and multiparous in late lactation stage. Sleep time was significantly affected by farm. Cows in TMR feeding farm tended to sleep longer. On the other hand, lying time was not affected by farm. Sleep/lying ratios was significantly different among farms. Cows slept significantly less in summer (49.2±3.5 min) than in autumn (60.7±4.3 min). Lying time was also shorter in summer. However, sleep/lying ratios was not different between seasons. Sleep time was not correlated with urinary cortisol. Solid non-fat content in bulk milk was positively correlated with sleep time (r=0.72). These results shows that seasonal factors, such as sunshine duration and temperature, would affect both sleep and lying time. On the other hand, some management differences among farms would affected on sleep time only.

Milk allowance and hunger behaviour in dairy calves

Katrina Rosenberger[1], Joao H. C. Costa[2], Heather W. Neave[2], Marina A. G. Von Keyserlingk[2] and Daniel M. Weary[2]
[1]*Institute of Animal Husbandry and Welfare, Department for Farm Animals and Veterinary Public Health, Veterinärplatz 1, 1210 Wien, Austria,* [2]*Animal Welfare Program, Faculty of Land and Food Systems, 191, 2357 Main Mall, BC V6T 1Z4, Vancouver, Canada; katrinarosenberger91@gmail.com*

The number of attempts an animal makes to feed may be considered as evidence of hunger. Calves fed limited amounts of milk are known to visit automated feeders many times a day, even when these visits do not result in a milk meal (non-nutritive visits). The aim of this study was to assess how milk ration and weaning affect the number of such visits. Fifty-six Holstein were fed 6, 8, 10 or 12 l/d until d 41, when milk was abruptly reduced by 50%, and later gradually reduced (by 20% per day), starting at d 50, such that calves were completely weaned at d 55. The lowest ration (6 l/d) was higher than the average milk allowance on Canadian farms. The number of rewarded and unrewarded visits to the milk feeder (as well as milk intake) were recorded daily until 10 wk of age by the automated feeding systems. The effect of treatment was analyzed using the MIXED procedure in SAS. The study was approved by the UBC Animal Care Committee. Before weaning, calves on the 6 l treatment had a high number of unrewarded visits (averaging 11.1 ± 0.73 visits/d) in comparison to all other treatments (e.g. 0.4 ± 0.78 visits/d for the 12 l treatment). When the milk rations declined by 50% at d 41, non-nutritive visits increased for all calves, but were still highest for calves fed less milk (e.g. 19.6 ± 1.64 visits/d for the 6 l treatment vs 10.1 ± 1.74 visits/d for the 12 l treatment). Treatments did not differ after weaning. These results indicate that feeding more milk reduces signs of hunger in calves, both before and during the weaning process, and adds to the increasing evidence in support of feeding high volumes of milk to improve welfare and productivity of dairy calves.

Behavioural changes in dairy steer calves infected with *Mannheimia haemolytica*

Nicole L. Eberhart[1], Jennifer M. Storer[2], Marc Caldwell[2], Arnold Saxton[1] and Peter D. Krawczel[1]
[1]University of Tennessee, Knoxville, Animal Science, 2506 River Drive, Knoxville, TN 37996, USA,
[2]University of Tennessee, Knoxville, College of Veterinary Medicine, 2406 River Drive, Knoxville,
TN 37996, USA; neberhar@vols.utk.edu

Visual detection of bovine respiratory disease lacks accuracy. Improving detection methods could reduce unneeded treatment and improve welfare. The objective was to determine behavioural changes of Holstein dairy steer calves in response to a Mannheimia haemolytica infection. Behaviour data were collected over 7 d from 12 calves aged 4-5 month with body weight of 166±13 kg. On d 0, treatment calves (MH; n=6) were inoculated with 3-5×10^9 cfu of M. haemolytica suspended in 5 ml of phosphate buffered saline (PBS) followed by 60 ml wash PBS. Control calves (n=6) were inoculated with 65 ml of PBS. Control and MH calves were group housed in an open-sided covered barn with dirt floor pens. Pens were separated by 10 m and measured 20×20 m with 66.7 m^2 per calf. All calves were acclimated to pens 7 d prior to inoculation and monitored for disease 2× daily. Prior to d 0, treatments were assigned to create groups balanced by body weight. Accelerometers were attached 2 d prior to inoculation. Data, recorded at 1-min intervals, were summarized into daily lying times (h/d) and lying time laterality (h/d). Rectal temperatures (RT) and clinical illness scores (CIS; 1 = normal; 4 = severe illness) were recorded 2× daily following inoculation. RT were collected while calves were restrained in the headlocks. CIS were assessed from outside of the pen on unrestrained calves. All procedures described were approved by the University of Tennessee Institutional Animal Care and Use Committee. A mixed model was used to determine effect of treatment, date and their interaction on total lying time, RT and CIS. Calf within treatment was treated as a random variable and observations were repeated by day. Laterality was analyzed using the above model plus split-plot effects of lying side and its interactions. Compared to control calves, MH calves lay down longer on d 0 (16.4±0.5 vs 14.4±0.5 h/d, P<0.01). MH calves spent more time lying on their right side than their left side (7.8±0.3 vs 6.8±0.3 h/d, P=0.01). No lateral preference was evident from control calves (P=0.22). Mean RT of MH calves 12 hr post inoculation (41.3±0.3 °C) exceeded 40 °C and was greater than control calves (39.2±0.3 °C; P<0.001). Mean CIS of MH calves (2.2±0.1) was greater than control calves (1.0±0.1) 12 h post inoculation (P<0.001). Unbiased continuous behavioural monitoring, focused on laterality, may offer more accurate detection of respiratory disease in calves and improve welfare. Establishing thresholds for behavioural changes is necessary for on farm application.

Inter-observer reliability of animal based measures in the Welfare Quality® protocol for dairy cows

Mikaela Mughal[1], Lilli Frondelius[2], Juhani Sepponen[2], Erja Tuunainen[3], Pirjo Kortesniemi[3], Matti Pastell[2] and Jaakko Mononen[1,2]
[1]*University of Eastern Finland, Department of Environmental and Biological Sciences, P.O. Box 111, 80100 Joensuu, Finland,* [2]*Natural Resources Institute Finland (Luke), Green Technology, Viikinkaari 4, 00790 Helsinki, Finland,* [3]*Animal Health ETT, P.O. Box 221, 60101 Seinäjoki, Finland; mikaela.mughal@uef.fi*

When selecting the specific welfare measures for the Welfare Quality® (WQ) protocols, their validity, feasibility and reliability were taken into account, the latter including both intra- and inter-observer reliability. Animal based measures (ABM) were preferred to resource and management based measures in the protocol, as they were considered more valid in most cases. They present, however, much sources of variation in relation to different observers and times of observation. We tested the inter-observer reliability (IOR) of the ABM included in the WQ cattle protocol as a part of a project where the protocol is applied for the first time on dairy farms in Finland. The full WQ assessment was conducted on six voluntary dairy farms (loose housing, number of dairy cows 45-130) in November-December 2015 by two trained assessors. The ABM evaluations were conducted simultaneously but independently by the assessors from the same animals or segments of the animal or the barn, or from different animals, depending on the criterion. IOR of criterion scores was tested with Pearson correlation ($P<0.05$ if $r>0.90$ n=6) for all the criteria including ABMs (the number of individual measures indicated): Absence of prolonged hunger (C1; 1), Comfort around resting (C3;6), Absence of injuries (C6;2), Absence of disease (C7; 10), Expression of social behaviours (C9; 1), Good human-animal relationship (C11; 1) and Positive emotional state (C12; 1). For the behaviour based criteria correlation varied from low to very high: C3, 0.38; C9, 0.998; C11, 0.68 and C12, 0.93. For criteria based on clinical observations, i.e. C1 and C6, the correlations were 0.23 and 0.78, respectively. In C7, including both animal and management based measures, r between the two observers was 0.92. There was high consistency between the two observers in four (C6, C7, C9 and C12) out of the seven scores. For some of the measures within the criteria C11 and C3, the observers tested partly different animals; since the WQ is meant for welfare assessment on herd level, this was not accounted for in the analysis. The low correlation of scores in C1, in turn, is most likely due to one outlier in our small sample. A similar, although smaller, outlier can be seen in the scores for C3. The results show that animal based measures included in WQ are mostly reliable, but sources of error for C11 need further examination. For C1 and C3 the results are less promising.

Timing of overall animal welfare assessment of dairy cattle at the farm level in Japan

Ken-ichi Takeda[1] and Ai Morimoto[2]
[1]Shinshu University, Academic Assembly, Institute of Agriculture, 8304 Minamiminowa, 3994598 Nagano, Japan, [2]Shinshu University, Graduate school of Agricultre, 8304 Minamiminowa, 3994598 Nagano, Japan; ktakeda@shinshu-u.ac.jp

Assessments for animal welfare (AW) of dairy cattle at the farm level should take place in late winter, because although the rearing environment may differ depending on the season, winter is the least favourable season. However, the extent of seasonal changes and basal climatic condition are different in various parts of the world. Here, we aimed to clarify the optimal evaluation timing for AW during the summer weather in Japan with high temperature and humidity, and we compared our evaluation results of dairy cattle in summer with that in winter. Ten dairy farms employing one of two housing types (tie-stall system and free-stall system) were visited, and the welfare of all cattle reared on these farms was assessed during summer (Aug. to Sep.) and winter (Dec. to Jan.). The welfare levels were scored using the assessment methods of the Shinshu Comfort Livestock Farm Certification Standard. This model contains 65 measures based on the following 'Five Freedoms': (F1) freedom from hunger and thirst (BCS, etc.); (F2) freedom from discomfort (heat and cold stresses such as panting and shivering, etc.); (F3) freedom from injury, or disease (Hock score, etc.); (F4) freedom to express normal behaviour (presence or absence of abnormal behaviour, etc.); and (F5) freedom from fear and distress (flight distance, etc.). The overall score (maximum 500 points) was the total of each freedom score (each of 100 points). The overall scores for summer and winter were 387.9±30 and 396.2±34.5, respectively. No difference was noted in the freedom scores of F1, F2, F4, and F5 between summer (F1=78.1±8.5, F2=88.6±7.4, F4=62.7±14.4, F5=74.4±12.3) and winter (F1=83.4±7.1, F2=86.6±8.5, F4=61.3±18.3, F5=75.9±14.2). However, significant differences were noted for the freedom score of F3 between summer (83.9±8.0) and winter (88.9±5.4) (paired t-test, t=3.12, P=0.014). Further, a significant difference was noted in the ratio of number of cows with dirty hind legs and udders between summer (39.1±9.5% and 18.6±8.2%) and winter (18.4±13.8 and 1.0±1.3) (paired t-test, t=3.48, P=0.008; t=6.73, P<0.001). The quantity of liquid consumed and urine, tail swishing times, and number of pest flies that alight on the body increase during summer in east and southeast Asian areas such as Japan, because of the high temperature and humidity. Therefore, we consider that the best timing for overall AW assessment of dairy cattle in these regions is summer, unlike that in Europe, where the dry winter is suitable.

Assessment of dairy cattle welfare in different types of dairy farms in India

Chandan Kumar and Madal lal Kamboj
National Dairy Research Institute Karnal Haryana India, Livestock Production Management,
LPM Division, NDRI Karnal Haryana 132001 India, 132001, India; drchandan@outlook.com

Animal welfare issues have grown in importance in recent years not only in developed countries but also in developing countries where improvement of animal welfare practices can lead to not only improved production and health of the animals but also increased trade opportunities. The animal welfare issues and problems in different dairy farms may be different due to different farming practices and use of modern production technology. So far little effort has been made in India to understand dairy cattle welfare or to identify the indicators of welfare or to assess the level of welfare. Therefore the status of dairy cattle welfare under our different dairy farming systems needs to be studied so that the animal welfare issues and management practices which jeopardize dairy cattle welfare could be identified and a strategy could be developed for enhancing the dairy cattle welfare. So keeping these points in mind the present study will be carried out with following objectives: (1) to develop dairy cattle welfare measurement scale based on identified welfare indicators; (2) to assess the level of dairy cattle welfare in different types of dairy farm in Haryana and to suggest suitable measures for improving welfare of dairy animals. As per methodology suggested by Calamari and Bertoni based on 12 welfare criteria, 20 indicators were selected in which 10 was output and 10 was input indicators and welfare status was assessed by a validated scale on 50 dairy farms comprises of small, medium and large dairy farms in rural as well as urban areas. Welfare score of housing and other facilities was significantly higher in case of large dairy farms ac compared to small and medium dairy farms. Total welfare score of feeds and feeding facilities of larger dairy farms is significantly in larger dairy farms as compared to small and medium dairy. Among the animal health, performance and behaviour did not differ significantly between small, medium and large dairy farms. The overall total welfare score was significantly higher in large dairy farms as compared to small and medium. About a half (49%) of the total farms were ranked as poor on welfare scale and the number of farms ranked as average, good and very good on welfare scale was 49, 9, 31 and 11% respectively. About 58% of the farms in different areas of welfare needed improvements. Most of the large dairy farms (80%) were assessed as good to very good on welfare scale whereas most of the small (65%) and medium dairy farms (60%) were assessed as poor to average on welfare measurement scale.

Religion and animal welfare – an Islamic perspective

Abdul Rahman Sira
Commonwealth Veterinary Association, 123 7 B Main Road 4 Block West Jayanagar, Bangalore 560011, India; shireencva@gmail.com

There is considerable support for the importance of animal welfare in all religions whether henotheistic (Hinduism), non-theistic (Buddhism) or monotheistic (Sikhism, Judaism, Islam, Christianity).There is potential for poor welfare at every stage of animal production and this can include accidental or deliberate cruelty. There is considerable debate on the role of religion in animal welfare, with implications for the study of welfare (including applied ethology), for welfare assessment and for implementation of solutions to welfare problems. Here emphasis will be placed on welfare itself, with examples of what religious scriptures say and what happens in practice, taking Islam as an example. In Islam, the Qur'an is explicit with regard to using animals for human purposes and there is a rich tradition of the Prophet Mohammad's (pbuh) concern for animals to be found in the Hadith and Sunna, The Qur'an and tradition reveals teachings of kindness and concern for animals. Islam has also laid down rules for humane slaughter. Shackling and hoisting conscious animals violate the humane intent of Islamic slaughter law. In many countries and Islamic traditional slaughter, animals are killed without pre-stunning and some people believe this to be a significant welfare insult. And yet, humane slaughter of animals is strongly supported in the Islamic tradition, many people do not realise that to be classified as halal meat, the animal must be treated with compassion up to and including slaughter. Therefore, regardless of pre-stunning or not, such meat should not be treated as halal because the animals have been beaten or treated without compassion during production, handling, transport and slaughter. Many Muslims and Islamic religious leaders are not aware of the cruelty that is routinely inflicted on animals during transport, pre-slaughter and slaughter in many Islamic countries. There is an urgent need to sensitise all Muslims to the teachings on animal welfare in the Qur'an and the Hadiths. A campaign is needed to apprise religious leaders of the current cruelty which occurs during transport and slaughter, for example by slides and videos. This should be done by competent and knowledgeable individuals who are also aware of the Islamic principles of animal welfare, preferably by Muslims in order to give authenticity to their claims. This approach is bound to be more effective in influencing the majority of Muslims in the livestock trade, especially slaughter, in treating animals more humanely, thus making the meat halal for human consumption. This needs to be done by intervention at the highest level by religious bodies and organisations, which could be most effective in giving rulings (fatwas) on this issue.

Rehoming of research animals – analyzing animal welfare pitfalls and presenting possibilities

Christel Palmyre Henri Moons

Ghent University, Nutrition, Genetics and Ethology, Heidestraat 19, 9820 Merelbeke, Belgium; christel.moons@ugent.be

Belgium has ratified Directive 2010/63/EU in the Royal Decree of 29th May 2013, stipulating that an ethical committee can release a research animal for return to its natural habitat or to a husbandry system suitable for the species, or for rehoming. In order to make this decision, the health and welfare of the animal, as well as the safety of the public and the environment must be considered. In case of rehoming, an adoption programme must be in place that foresees in the 'socialization' of the animal. In this paper, I critically review the concept of rehoming animals used for research, by pointing out potential threats to animal welfare and discussing the research animal rehoming programme of the Faculty of Veterinary Medicine at Ghent University, which was developed and approved in 2014 and has been running since January 2015. From a welfare point of view, there are several issues to consider when rehoming animals formerly used for research. First, the risk of adjustment problems is greatest when significant discrepancy in the social and non-social environment exists between research setting and environment post rehoming, like when research animals are rehomed as companion animals. This risk involves not only the welfare of the research animal, but also that of humans and other animals in the new environment. Next, the expectations of the new owner towards the animal may be greater than its coping ability, resulting in additional stress for the animal and disappointment in the owner. Finally, when animals have a chronic ailment, due to old age or other, the adoption must be considered carefully in light of the required treatment to maintain sufficient quality of life for the animal. The rehoming programme of the Faculty of Veterinary Medicine at Ghent University consists of screening of research animals when available for adoption, screening of potential adopters by investigating the match with a particular animal in terms of housing facilities and adopter expectations, informing potential adopters, and the signing of a rehoming contract. Because dogs and cats live in close contact with humans (adults and children) and possibly other animals, the above-mentioned welfare risks are even greater. For these species, behavioural observation is an important additional step in the adoption procedure. The animal is observed in different contexts by an animal behaviour specialist, in an attempt to predict which behavioural strategies the animal is most likely to adopt post rehoming. The behaviour specialist then meets with the potential adopters to discuss the findings, to educating them about canine or feline stress signalling, and to motivate them to use this knowledge to facilitate the adjustment process.

'A man of science should attend only to the opinion of men of science' – animal research regulation 150 years after C. Bernard

I Anna S Olsson

IBMC, Institute for Molecular and Cell Biology, Laboratory Animal Science, Rua Alfredo Allen 208, 4200-135 Porto, Portugal; olsson@ibmc.up.pt

The cited words represent Bernard's reaction to antivivisectionist critics, but translated into our days sound like a call for self-regulation of research with animals. On the surface, there seem to be little room for this, as animal experimentation is probably the most regulated human activity using animals, where for each individual research project an external evaluation is required for authorization. Such heavy external regulation suggests public distrust in scientists' ability to self-regulate, a view going against the increasing focus on mechanisms other than legislation to regulate research through Responsible Research and Innovation (RRI). I propose a keynote lecture to review the different mechanisms for internal and external regulation of animal research in Europe, drawing on ongoing research in my group using systematic reviews and sociological methods. Since legislation cannot regulate research in detail, major determinants of how research is done are the licensing procedure – usually involving review by an animal ethics committee – together with scientists' own decisions when planning experiments. In Europe, the composition of ethics committees is increasingly technocratic (e.g. committee members are experts on topics such as animals, research, law or ethics) and the role of lay member participation is limited. Harm-benefit analysis is often pointed out as the review aspect where lay member contributions are particularly important. However, although legislation establishes this as a cornerstone of project evaluation, committees are given very little guidance on how to weigh harm against benefits, and preliminary data from interviews with ethics committee members suggest that many committees do not perform this analysis but rather focus on how to minimize animal suffering. On the other hand, when scientists in focus groups discussed their potential involvement in hypothetical research project with animals, scientific validity was a major factor determining whether or not to accept collaborating in a project. Through systematic reviews, we have demonstrated that the way animal research in several fields is described in international peer-reviewed publications is not in agreement with best practice in terms of 3Rs and experimental design. We are presently investigating how these issues are taken into account in editorial decisions. I will discuss these results from an RRI perspective. Overall, they indicate that scientists engage in self-regulation and critical analysis of their use of animals in research, but that there is room for improvement, and that the involvement of society in formal decision-making over animal experiments is quite limited.

Cold-blooded care: identifying novel methods of welfare assessment for reptiles

Sophie Moszuti, Anna Wilkinson and Oliver Burman
University of Lincoln, School of Life Sciences, Joseph Banks Laboratories, Beevor Street, Lincoln, LN6 7DL, United Kingdom; oburman@lincoln.ac.uk

Keeping reptiles as pets, both nationally and internationally, has become increasingly common, with estimates of UK captive reptile numbers (1.1-7 million) suggesting that populations are even greater than those of some typical pet species (e.g. hamsters and guinea-pigs). Because most species of reptile are considered unsuitable for private husbandry –- to the extent that it has been suggested that no reptiles are really suitable as pets – they represent a huge welfare concern. Yet, whilst a great deal of research has been focused on identifying ways to assess the welfare of captive mammals and birds, there is comparatively little knowledge on how reptilian species are affected by captivity, and the ways in which their welfare can be accurately assessed. Therefore, the present study investigated the behavioural response of two species of reptile to novelty – a commonly used approach to assess anxiety-like behaviour and negative affect in non-human animals – with the aim of determining whether the existing validated behavioural measures used to assess the welfare of mammalian and avian species could be translated for use in reptiles, and whether we could also identify novel reptile-specific and/or species-specific behaviours. Local ethical approval was obtained, and all animals formed part of the University research population. Red-footed tortoises (*Chelonoidis carbonaria*, n=8) and bearded dragons (*Pogona vitticeps*, n=17) were observed individually in both familiar and novel environments for 10 minute time periods, and their behaviour recorded. Tortoises were found to begin locomotion sooner in a famiiar, compared to a novel, environment (familiar: 1.8±0.7 s; novel: 58.5±26.8 s, Z=-1.960, d.f.=8, P=0.05), and extended their necks further in a familiar environment ($F_{1,7}$=39.241, P<0.05), with neck length consistently increasing over time in both environments ($F_{10,70}$=5.867, P<0.05). In contrast, bearded dragons showed no difference in latency to move when comparing the familiar and novel environments (Z=-0.450, d.f.=1, P=0.653), but exhibited more surface licking in the novel environment (familiar: 2.2±0.5; novel: 4.8±0.9, Z=-2.207, d.f.=16, P=0.027). This study has identified a combination of both existing and new species-specific anxiety-like behavioural responses to novelty that may have the potential to be used in future studies when assessing the welfare of reptiles in response to aspects of captive environments; the key first step when addressing concern for the welfare of captive reptiles.

Out like a light? The effects on sleep of being a nocturnal mouse in a diurnal lab

Amy Robinson-Junker[1], Bruce O'hara[2] and Brianna Gaskill[1]
[1]Purdue University, 725 Harrison St, West Lafayette, IN 47906, USA, [2]University of Kentucky, 675 Rose St, Lexington, KY 40546, USA; robin274@purdue.edu

Sleep disruption in humans, caused by shift work, can be detrimental to physical and behavioral health. Nocturnal lab mice may experience a similar disruption caused by human daytime activities, but it is unknown if mouse sleep is disrupted by these systems designed to meet diurnal human needs, or if this affects their welfare. We hypothesized that the timing of human disruptions would alter mouse sleep amounts and patterns. We predicted that mice disturbed during their normal rest period (light period) would either sleep less or spend more time sleeping during their typical active period (dark period) than those disturbed during the active period. We used non-invasive sleep monitoring cages that use a piezoelectric mat to detect mouse movement; these require solitary housing. The apparatus analyzes these movements to discern a waking mouse from a sleeping mouse with 90% accuracy, as determined in previous studies. 48 mice were used in a factorial design to test a sleep disruption treatment, where mice were disturbed with routine husbandry at either 10:00 or 22:00. All mice were exposed to each disruption treatment (after 4 days of acclimation to the cage and individual housing) for 1 week; the first treatment they experienced was balanced across groups to account for order effects. We also wanted to determine if there were any potential ameliorating effects of a refuge on sleep disruption; because amount of material correlates with nest quality, we tested 4 amounts of nesting material (3, 6, 9, or 12 g). We tested both sexes of 3 types of mice (CD-1, C57BL/6, and BALB/c); sex and type of mouse were balanced across all treatments. Sleep was continuously recorded over the 2 weeks. Data were analyzed as a repeated measures ANOVA using a GLM. C57BL/6 mice slept less overall compared to other mice (P<0.001). CD-1 and BALB/c male mice slept more during the active period than females when they were disturbed at 10:00 (P<0.001). Analysis of percentage of time spent sleeping over 24 hours, when divided into 2 hour intervals, showed multiple differences between types of mice throughout the day. Of particular interest were points after lights on (P=0.004) and immediately prior to and after AM and PM disturbance times (P<0.001). Amount of nesting material was only found to increase sleep bout length in CD-1 mice with 12 g compared to those with 3 g (P=0.02). These results suggest that disturbance timing affects sleep patterns but not the overall amount of sleep that mice get, and that the changes in sleep patterns vary between mouse type and sex. It is possible that our brief welfare checks may have been too predictable and minor to induce true sleep disruption.

The importance of burrowing, climbing and standing upright for laboratory rats

I. Joanna Makowska and Daniel M. Weary
University of British Columbia, Animal Welfare, 2357 Main Mall, V6T 1Z4, Canada;
makowska@interchange.ubc.ca

Rats (*Rattus norvegicus*) are one of the most commonly used animals in research. In the wild, rats construct burrows, climb to forage, and stand upright to stretch, explore and socialize with other rats. Standard laboratory cages prevent rats from performing these behaviours, but little is known about how this inability may affect their welfare. The first aim of this study was to record the propensity to burrow, climb and stand upright in 3-, 8- and 13-month old female Sprague-Dawley rats housed in semi-naturalistic environments. Because standard-housed rats are unable to stretch in the upright position, we hypothesized that they would perform more lateral stretches than rats housed in the semi-naturalistic environment; stretching is also a corrective response to stiffness and positional stress. Thus, the second aim of this study was to compare the frequency of lateral stretching in standard- vs semi-naturalistic-housed rats. Semi-naturalistic cages (n=6 cages each housing 5 rats) measured 91×64×125 cm, contained soil, and allowed climbing. Standard cages (n=6 cages each housing 2 rats) measured 45×24×20 cm and contained bedding and a PVC pipe. Rats were filmed continuously with infrared cameras; 24-hr footage from each cage was sampled at each age period and analysed using a mixed model (aim 1) or an independent samples t-test (aim 2) in SAS. Rats' propensity to burrow remained constant as they aged (approximately 30 bouts/day totalling 20-30 min; $F_{2,8}=0.41$, P>0.67), suggesting burrowing is important to rats. Climbing decreased from 76 to 7 bouts/day at 3 vs 13 months ($F_{2,8}=30.49$, P<0.001), likely because of declining physical ability. Upright standing decreased from 178 to 73 bouts/day ($F_{2,8}=20.52$, P<0.001), but continued to be frequently expressed even in older rats. Approximately 15% of bouts of upright standing included upright stretching. Standard-housed rats performed lateral stretches much more frequently than semi-naturalistic-housed rats (53 vs 6 bouts/day at 13 months; $t_{5.3}=-4.56$, P<0.01), perhaps in compensation for the inability to stretch upright and low mobility associated with standard housing. These findings suggest that standard laboratory cages interfere with important natural behaviours, likely compromising rat welfare. Providing burrowing substrate and increasing cage height are recommended.

Use of an outdoor run by dairy goats is reduced under adverse weather conditions

Nina Keil[1], Anette Lanter[2], Edna Hillmann[2], Lorenz Gygax[1], Beat Wechsler[1] and Joanna Stachowicz[1]
[1]*Center for Proper Housing: Ruminants and Pigs, Tänikon 1, 8356 Ettenhausen, Switzerland,* [2]*Ethology and Animal Welfare Unit, ETH, Universitätstrasse 2, 8092 Zürich, Switzerland; nina.keil@agroscope.admin.ch*

Access to an outdoor run is expected to be benefical for goats' welfare in respect to social behavior and claw wear in loose housing systems. To promote its use it is necessary to understand under which climatic conditions goats go outside. The aim of this study was to investigate the influence of different weather parameters on the use of an outdoor run by dairy goats under temperate weather conditions. Data was collected from February to April and in October 2014 on 14 commercial dairy goat farms in Switzerland (n=11) and Germany (n=3) for 10 days each. Weather parameters (temperature, humidity and solar radiation, which were combined in a measure for 'warmth' by PCA to reduce redundancy among higly intercorrelated paramters; wind speed; precipitation) were continuously measured using a weather station installed next to the outdoor run. The mean proportion of goats outside was recorded based on video data. The influence of warmth and wind speed was assessed in relation to the peak use hours of the outdoor run, that is, within the two hours after morning milking. The influence of precipitation was investigated by comparing two hour periods with rain and without rain at the same time of day on different days. Linear mixed effect models were used with a stepwise-backward selection of predictors (threshold $P<0.05$ based on parametric bootstrap). Within the measured temperature range (-0.8 to +16.5 °C) warmer conditions (more solar radiation, higher temperature, lower humidity) enhanced the use of an outdoor run ($P<0.001$). Even moderate wind speeds reduced the mean proportion of goats outside ($P<0.001$) with 25% (confidence interval: 21-29%) at 0 m/s to 0.01% (-0.09-0.1%) at 4 m/s (setting: 'warmth'=1). On days with rain (3%, 2-4%), but also in periods before the rain started and after it had stopped (7%, 5-11%), the outdoor run was used by a markedly smaller proportion of goats than during comparable days without rain (22%, 16-31%; $P<0.001$). We conclude that under temperate weather conditions goats react sensitively to wind, cold and rain. Provision of protection against rain and wind as well as the solar exposition of the outdoor run should be considered when constructing an outdoor run.

Effects of three different methods of restraint for shearing on behaviour and heart rate variability (HRV) in alpacas

Susanne Waiblinger[1], Franziska Hajek[1], Bianca Lambacher[2], Teresa Salaberger[3], Sebastian Becker[2] and Thomas Wittek[2]
[1]University of Veterinary Medicine, Vienna, Department for Farm Animals and Veterinary Public Health, Institute of Animal Husbandry and Animal Welfare, Veterinärplatz 1, 1210 Wien, Austria, [2]University of Veterinary Medicine, Vienna, Department for Farm Animals and Veterinary Public Health, Clinic for Ruminants, Veterinärplatz 1, 1210 Wien, Austria, [3]University of Veterinary Medicine, Vienna, Department for Biomedical Sciences, Unit of Physiology, Pathophysiology and Experimental Endocrinology, Veterinärplatz 1, 1210 Wien, Austria; susanne.waiblinger@vetmeduni.ac.at

Alpacas are increasingly kept in Europe for different purposes including fibre production. Yearly shearing is necessary to harvest fibre and for welfare reasons. Different methods of restraint are used during shearing, which may affect the welfare of the animals differently. The aim of the study was to compare three common restraint methods: STANDING, the animal is standing and held by hand; GROUND, the animal is laid down and its legs are stretched out by ropes; TABLE, the animal is restrained on a special table, legs are fixed by ropes. We recorded defence and stress-associated behaviour continuously during restraint and, afterwards, general activities at 10-min intervals from 30 to 120 min in alpacas sheared by always the same person during restraint (Exp1, 45 animals, 15 animals per treatment) or only restrained (Exp2, 15 animals, repeated measures). HRV was measured in Exp2. Large individual variation was found. During restraint, STANDING animals in Exp1 showed least flinching (Kruskal-Wallis $P=0.01$, Mann-Whitney U $P<0.05$), and tended to vocalise less (KW $P=0.09$), but most often showed flight attempts (KW $P<0.001$, MWU $P\leq0.001$). After shearing, GROUND animals were not observed lying or ruminating with the exception of one animal, so that STANDING animals lay more often (KW $P<0.04$, MWU $P_{S-G0}=0.02$) and tended to ruminate more often (KW $P<0.06$, MWU $P_{S-G0}=0.02$), while TABLE tended to lie more often ($P_{G-T}=0.10$) and ruminated more often ($P_{S-T}=0.03$). Results in exp2 were similar for STANDING but GROUND and TABLE changed order. Accordingly, HRV (SDNN) was higher in STANDING compared to TABLE. Shearing induced strong reactions (vocalizing continuously) in some of the animals in each of the treatments, indicating stress elicited by shearing itself. Nevertheless STANDING seems to be the least aversive. Effects of GROUND and TABLE might depend on the exact level of stretching legs and body position which merits further investigation.

Animals and us: history, exciting developments, cross-pollination and new directions for applied ethology research

Jennifer Brown[1], Yolande Seddon[2], Jean-Loup Rault[3], Rebecca Doyle[3], Per Jensen[4], Bas Rodenburg[5] and Jeremy Marchant-Forde[6]

[1]Prairie Swine Centre, Saskatoon, Saskatchewan, Canada, [2]Western College of Veterinary Medicine, University of Saskatchewan, Saskatoon, Canada, [3]Animal Welfare Science Centre, Faculty of Veterinary and Agricultural Sciences, University of Melbourne, Parkville, Australia, [4]IFM Biology, AVIAN Behaviour Genomics and Physiology Group, Linköping University, Linköping, Sweden, [5]Behavioural Ecology Group, Wageningen University, Wageninen, the Netherlands, [6]Livestock Behaviour Research Unit, USDA Agricultural Research Service, West Lafayette, Indiana, USA; jennifer.brown@usask.ca

Applied ethology is a relatively new field, which has grown rapidly in scope, geographical distribution and influence since the inception of the SVE/ISAE in 1966. There has been limited historical documentation of the field, and students new to the area have little information available to understand the history of this science beyond what is found in journal publications. To address this gap, several ISAE members and affiliated researchers have compiled a book, 'Animals and us: 50 years and more of applied ethology' which documents and celebrates the first 50 years of the ISAE. The book is written in four parts, including: (1) the history of the society and early pioneers; (2) research advances in behaviour; (3) global perspectives; and (4) future directions for applied ethology research. Contributing authors include many leading researchers in the field- too numerous to mention here. This presentation will explore highlights from the text related to achievements, trends in research, collaborative studies with other fields, and new directions for both basic and applied research, all in an attempt to explain why ethologists are so passionate about their work, and why this field remains more exciting now than ever before. Themes include human-animal interaction, personality, play behaviour, cognition, multi-level selection, polyvagal theory, and the relationship between applied ethology and animal welfare science. We conclude that other animals, with their amazing and various forms and habits, may be the perfect human enrichment.

Behaviour of working equids engaged in different types of work

N Bhardwaj, B Negi and D Mohite

Brooke Hospital for Animals, India, A Block, 223-226, Pacific Business Park, Sahibabad Industrial Area, Ghaziabad- 201010, India; nidhish@thebrookeindia.org

Different environments provide different behavioural opportunities; these signs enable us to recognise and assess an animal's mental state within that specific context. This study investigated frequency of different equid behaviours according to work type so as to understand potential effect of working environment on behaviour as an indicator of mental state. Working equids were selected in Muzaffarnagar, India in December, 2014. Through the entire duration of a welfare assessment exercise (approximately 10 minutes), previously standardised by NGO The Brooke, each equid's general attitude was qualitatively appraised by trained assessors using three Categories: (1) alert/relaxed (eyes open, ears moving, head level higher than withers, active interest in surroundings); (2) apathetic/dull (eyes partially closed, ears lowered, head level with or below withers, lack of interest in surroundings); (3) aggressive/avoidance (ears back, whites of eyes or teeth showing, avoiding observer). Equid behaviour for >50% of assessment indicated score awarded; this was not over-ruled by isolated instances of other behaviours. Work type equids were engaged in was recorded. Data were analysed in R. 489 horses, 34 mules and 4 donkeys (total 527) of age range 2-15 years were assessed. Overall, 468 (89%), 33 (6%) and 26 (5%) showed Category 1, 2 and 3 behaviour, respectively. Frequency for each work type of Category 1, 2 and 3, respectively, was 331 (92%), 18 (5%) and 9 (3%) from total 358 that transported bricks by cart; 52 (88%), 6 (10%) and 1 (2%) from total 59 that transport of goods or people by cart; 51 (72%), 9 (13%) and 11 (15%) from total 71 used for breeding and trading (BT); 9 (64%), 0 (0%) and 5 (36%) from total 14 ceremonial or riding (Cer/R) equids. Remaining 25 equids were non-working. Cer/R and BT equids were more likely to display Category 3 behaviour (χ^2=57.3; P<0.0001). Overall frequencies of behaviours indicating negative mental state (Categories 2 &3) were low compared to those indicating positive mental state (Category 1), suggesting positive animal-human interaction within this population. Cer/R and BT animals appear to have higher frequency Category 3 and lower frequency Category 1 behaviour than other types, although work type may act as a proxy indicator for other variables. Differences may be due to handling, training and other stimuli that equids encounter in their respective work environments. Understanding what is positive or negative from the animal's point of view enables improvements in handling and environment. Further work to identify handling and environmental factors associated with Category 2 & 3 behaviours may help to design appropriate interventions that improve mental state.

Opinion of applied ethologists about expectation bias and debiasing techniques

Frank Tuyttens, Lisanne Stadig, Jasper Heerkens, Eva Van Laer, Stephanie Buijs and Bart Ampe
Institute for Agricultural and Fisheries Research (ILVO), Animal Sciences Unit, Scheldeweg 68,
9090 Melle, Belgium; frank.tuyttens@ilvo.vlaanderen.be

Observers in applied ethological research are rarely blinded despite mounting evidence that expectation biases may affect, and even invalidate, research outcomes. We surveyed delegates of the ISAE-2014 congress shortly before (n=39 respondents) and after (n=51 respondents) a combined congress plenary lecture and workshop on expectation bias in applied ethology. During the lecture some examples of how expectancy effects influenced scores of animal behaviour and welfare were presented. The aims were to evaluate the effect of the plenary lecture and workshop on the opinion of the congress delegates on blinding observers and alternative debiasing techniques. Awareness about expectancy effects and debiasing techniques was lower before than immediately after the congress plenary lecture and workshop. Research situations considered as most susceptible to expectation bias – i.e. when the data-collector uses subjective methods and has strong expectations about the research outcome – were perceived to be more common in applied ethology than in other scientific disciplines (29.8 vs 16.9%, P<0.001). Non-blinded data collection in such research situations was viewed more disapprovingly after the plenary lecture and workshop as compared to before (lsmean of 6.8 vs 7.8 on a 10-point Likert scale, P<0.05). The main reasons why blinded observations are uncommon in applied ethology seem to relate to a limited awareness about expectancy effects and to logistic constraints of blinded observations rather than that the susceptibility of the research field is perceived to be low. In addition to the immediate effect of the plenary lecture and workshop, a more sustained and concerted effort throughout all stages of the research process seems warranted in order to avoid expectation bias affecting research findings and in order to safeguard the scientific credibility of the field of applied ethology.

Effects of zoo visitors on quokka (*Setonix brachyurus*) avoidance behaviour in a walk-through exhibit

Mark James Learmonth[1], Michael Magrath[2], Sally Sherwen[2] and Paul Hemsworth[1]
[1]*Animal Welfare Science Centre, The University of Melbourne, Parkville, 3010, Australia,*
[2]*Zoos Victoria, Wildlife Conservation and Science, Melbourne Zoo, Parkville, 3052, Australia;*
mlearmonth@student.unimelb.edu.au

Walk-through enclosures with free-ranging animals are becoming increasingly popular in zoos. Behaviour and welfare of zoo animals may be influenced by contact with zoo visitors. Visitor effects have been interpreted as stressful, enriching, or innocuous in many studies of many zoo species. Avoidance behaviours (avoidance of humans, avoidance of conspecifics) have been widely used as an indicator of fear and potential stress in animals. The aim of this research was to identify the effects of visitors on avoidance behaviour displayed by seven free-ranging quokkas (*Setonix brachyurus*) in a walk-through enclosure at Melbourne Zoo. We studied the quokkas in a controlled experiment, randomly imposing two treatments: enclosure open to visitors as normal (Open); and enclosure closed to visitors (Closed). Treatments were imposed for 2-day periods, with five replicates of each treatment (total of 10 2-day study periods). Cameras were used to monitor 4 previously identified areas frequently used by the quokkas and the adjacent visitor paths in the enclosure. Instantaneous point sampling at 1-min intervals from 09:00 to 16:00 h over each 2-day study period was used to record number of quokkas in each of the study areas and the proportion of these quokkas visible to visitors from the visitor paths. To examine treatment effects, we conducted a statistical analysis that allowed causal conclusions, using the experimental design paradigm, about the effect of treatment on the group of animals in the enclosure. Given that treatments were randomly allocated to blocks, this approach meant that the population considered was the group of animals in the enclosure and the experimental unit was the population over a 2-day period. All measurements were analysed using a two-treatment, one-way analysis of variance with 2-day summary values as the unit of analysis. There was no treatment effect on the number of quokkas in the study areas (Means: Open, 0.30 and Closed, 0.33 animals/sample point/2-day period, $P=0.690$). However, the proportion of quokkas visible from the visitor paths was lower in the Open treatment (0.35 vs 0.46, $P=0.037$). The results indicated that visitors are moderately fear-provoking for quokkas, but clearly further research is required to examine the effects of visitors on the behaviour and particularly stress physiology of quokkas in this enclosure.

Further development of a behavioural protocol for the assessment of stress in cats

Elin N Hirsch[1], Jenny Loberg[2,3] and Maria Andersson[2]
[1]*Swedish University of Agricultural Sciences, Department of Animal Environment and Health, P.O. Box 103, 230 53 Alnarp, Sweden,* [2]*Swedish University of Agricultural Sciences, Department of Animal Environment and Health, P.O. Box 234, 532 23 Skara, Sweden,* [3]*Foundation Nordens Ark, Åby säteri, 456 93 Hunnebostrand, Sweden; elin.hirsch@slu.se*

Behavioural stress in the domestic cat (Felis silvestris catus), housed in groups (GH) or singly (SH), has previously been assessed using the seven-level Cat-Stress-Score (CSS). The CSS consists of an extensive list of behavioural elements (BE) combined to score each cat on one of seven possible stress levels. Some BE are found on several score levels, hence, providing a score can be difficult. As a first step in the further development of a behavioural tool to assess stress, the aim of this study was to determine which behavioural elements are salient enough to be observed and if the same behaviours were recorded for GH and SH cats. A protocol containing the 65 BE from the CSS and 20 taken from the literature was used. Cats were observed using the original methodological setup with two 1 min observations, 15 min in between, during a morning and afternoon session, resulting in four scorings per cat and day. Instead of calculating a median score per cat, all BE were registered separately during the observations. Cats were observed at an animal shelter (Indiana, U.S.A.). For GH cats (n=64: 47 females, 18 males) 253 and for SH (n=13: 6 females, 7 males) 52 scorings were performed. Preliminary result show that of the 85 BE included in the protocol 22 (26%) were never recorded for GH and 36 (42%) were never recorded for SH cats. This difference between GH and SH cats was significant (Mann-Whitney test, P<0.05). Non-recorded BE could be found in all seven stress levels as well as the BE from the literature. Although, there is a difference in the number of recordings for GH and SH cats, our results indicate that housing can impact the observation of stress related behaviours and should therefore be considered.

Exhibit use, behaviour and activity budget of Yellow-throated martens (*Martes flavigula aterrima*) kept in zoos

Angelica Åsberg
Furuviksparken, P.O. Box 672, 80127 Gävle, Sweden; angelica.asberg@gmail.com

The Yellow-throated marten (*Martes flavigula aterrima*) population in European zoos is suffering from high neonatal mortality and is said to display high levels of stereotypic behaviours, which might indicate inappropriate husbandry routines. Little is known about its ecology and behaviour but it is thought to be more social and arboreal than other marten species. To contribute to efforts for increasing welfare for this species, a questionnaire was sent to all holding institutions in Europe and Russia which was followed up by behavioural observations in four institutions keeping marten pairs. A studbook analysis was carried out to evaluate the breeding problems in institutions and specific specimens. It showed that historically only a few animals have reproduced. However, in recent years an increasing number of institutions have been successful in breeding. The overall degree of inbreeding is high with a population inbreeding coefficient of 0.3986 and mean kinship of 0.3718 whilst genetic diversity is low, 0.6282. In conclusion, the reason for low reproductive success is mainly due to high neonatal mortality, where many institutions reported their females to eat their kits within the first days post-partum. Behavioural data on activity budget, exhibit use, social – and stereotypic behaviour was collected using instantaneous sampling every five minutes during a total time of 216 hours per pair. This was performed in two categories of exhibits: two open air >500 m^2 exhibits and two net-roofed <50 m^2 exhibits. Comparisons between the two categories were performed using Mann-Whitney U-test for un-paired samples. Selected pairs had similar activity budgets independent of exhibit category and were mainly diurnal. Martens in the <50 m^2 exhibits engaged significantly more in behaviours at the tree level than did martens in >500 m^2 exhibits ($P<0.05$). All pairs spent significantly more time together than on their own ($P<0.05$) always resting in the same nesting box. The amount of stereotypic behaviour did not differ significantly between exhibit categories ($P<0.05$). Based on these findings, recommendations can be given that the species should be kept in pairs instead of individually which is common today and exhibits should provide plenty of climbing opportunities.

Investigating daily changes in flamingo activity patterns

Paul Rose[1,2], Imogen Lloyd[2], Joe Linscott[3] and Darren Croft[2]
[1]WWT, Slimbridge Wetland Centre, Gloucestershire, GL2 7BT, United Kingdom, [2]University of Exeter, Centre for Research in Animal Behaviour, Washington Singer, Perry Road, Exeter, EX4 4QG, United Kingdom, [3]Sparsholt College Hampshire, HE Animal Management, Sparsholt, Winchester, Hampshire, SO21 2NF, United Kingdom; p.rose@exeter.ac.uk

Flamingos are one of the most numerously-kept zoo birds (around 9,000 individuals are thought to exist in European institutions alone) and investigations into their daily time budgets provide evidence for positive welfare. Information on the activity patterns of wild flocks do suggest some similarities with captive flocks- but there are notable differences too. Literature suggests that throughout the day, wild flamingos move between foraging and preening/resting flocks; this provides a useful guide for assessing behavioural normality in captive birds, and whether they are given the appropriate opportunities to undertake wild-type changes in behaviour. This study aimed to determine changes in flamingo activity over time to see how this may impact upon enclosure usage, and to provide a way of assessing captive flamingo welfare. Data from flocks of five flamingo species, from two UK zoos and one German zoo are presented here, being collection over spring and summer 2013 and 2014. Photographs were taken of the flocks at predetermined points during the day. Data were collected using an instantaneous scan sampling method, with the whole flock scanned for active and inactive behaviours at a morning, a midday, and two afternoon points (depending on weather and season). The photographs of the flocks were scored for the behaviour of the birds and their location within the enclosure (using a modified SPI approach). Behavioural categories of moving, foraging, and inactive (preening or standing/sitting) were used when assessing each photograph. Data were analysed using a mixture of ANOVA and Chi-Squared testing. Results show that flock behavioural repertoires and enclosure usage are more diverse in the late afternoon and evening, e.g. 47% inactive (am) to 29% inactive (evening) in some instances (F=29.8; df=3; P<0.001); and mean SPI 0.74 (±0.007) to mean SPI 0.63 (±0.01) (X^2=7.97; df=3; P=0.047). Therefore it is apparent that time of day has an effect on the types of state behaviour performed by captive flamingos. The natural follow-up to this project is to investigate the nocturnal habits of zoo-housed flamingos to determine their night-time activity and enclosure use. These differences in activity pattern can be used to guide future decisions on flamingo husbandry to allow birds to perform behaviours according to their normal circadian rhythms. By building-in opportunities for daily changes in behaviour pattern can positively impact on the welfare of many individual birds.

Validation of a questionnaire to evaluate the personality construct sensory processing sensitivity in dogs

Maya Braem Dube[1], Sibylle Furrer[1], Lucy Asher[2], Hanno Würbel[1] and Luca Melotti[1]
[1]University of Bern, Division of Animal Welfare, Länggassstrasse 120, 3012 Bern, Switzerland,
[2]Newcastle University, Centre for Behaviour and Evolution, Framlington Place, NE2 4AB Newcastle, United Kingdom; maya.braem@vetsuisse.unibe.ch

The personality construct 'sensory processing sensitivity' (SPS) in humans involves greater processing of sensory information, which can be associated with physiological and behavioral overarousal. Hence, this construct influences the manner in which people process information in their environment and, therefore, how the environment affects them. This study is a first attempt to explore this construct in dogs by means of a questionnaire, a commonly used tool for the assessment of personality in dogs. A 32-item questionnaire resulting in an 'SPS score' for dogs (SPS-d) was developed using the same inductive approach applied to create the validated questionnaire for humans. A large-scale, international online survey was then conducted, including the developed SPS questionnaire, questions on fearfulness and neuroticism and on 'demographic' (dog sex, age, weight; age at adoption; country of origin; previous owners; number of people in household) and 'human factors' (owner age, sex, profession; use of punishment and reinforcement; degree of environmental stimulation; active time), as well as the validated questionnaire for humans. Data were analyzed using linear mixed effect models with forward stepwise selection to test prediction of SPS by the above-mentioned factors with country of residence and dog breed treated as random effects. 3647 questionnaires were fully completed in the survey. The constructs SPS, fearfulness, and neuroticism of dogs and SPS of humans showed good internal consistency (Cronbach's alpha=0.90, 0.84, 0.78, and 0.93, respectively). SPS-d only moderately correlated, but did not fully overlap with fearfulness (r=0.37, P<0.001) and neuroticism (r=0.41, P<0.001) scores, as in previous findings in humans. Intra- (same owner, n=120) and inter-rater (other person familiar with same dog, n=446) reliability were good (r=0.83, P<0.001; r=0.65, P<0.001; respectively). Demographic (marginal pseudo R^2=0.04, conditional R^2=0.09) and human factors, including SPS of humans, (marginal pseudo R^2=0.03, conditional R^2=0.07) explained only a small amount of the variance of SPS-d. The SPS construct resulting from this questionnaire showed good internal consistency, partial independency from fearfulness and neuroticism, and good intra- and inter-rater reliability, suggesting good construct validity of the SPS questionnaire. Further, the fact that human and demographic factors only marginally affected the SPS-d suggests that the measured construct might be similar to the SPS personality construct in humans.

Animal welfare research in the last three decades in the Danube region – a bibliographic study

Miroslav Radeski[1], Tomislav Mikus[2], Manja Zupan[3], Katarina Nenadovic[4], Mario Ostovic[5], Viktor Jurkovich[6], Ludovic T. Cziszter[7], Radka Sarova[8], Ivan Dimitrov[9] and Marlene K. Kirchner[10]
[1]St. Cyril and Methodius University, Lazar Pop Trajkov 5/7, 1000 Skopje, Macedonia, [2]Croatian Veterinary Institute, Savska cesta 143, 10000 Zagreb, Croatia, [3]University of Ljubljana, Jamnikarjeva ul. 101, 1000 Ljubljana, Slovenia, [4]University of Belgrade, Bulevar oslobođenja 18, 11000 Belgrade, Serbia, [5]University of Zagreb, Heinzelova ul. 55, 10000 Zagreb, Croatia, [6]Szent István University, István utca 2, 1400 Budapest, Hungary, [7]USAMVB, Calea Aradului nr. 119, 300645, Timisoara, Romania, [8]Institute of Animal Science, Přátelství 815, 10400, Praha, Czech Republic, [9]Agricultural Institute, Stara Zagora, 6000 Stara Zagora, Bulgaria, [10]Institute of Animal Welfare and Disease Control, University of Copenhagen, Department of Large Animal Sciences, Grønnegårdsvej 8, 1870 Frederiksberg C, Denmark; mk@sund.ku.dk

The objective of this bibliographic study was to reveal and summarize published scientific literature on animal welfare (AW) over the last 30 years in the Danube region countries: Austria, Bulgaria, Czech Republic, Croatia, Germany, Hungary, Macedonia, Romania, Serbia, Slovak Republic and Slovenia. The literature taken into consideration comprised AW assessment protocols, their indicators and local AW topics in domestic animals. Prior the searching and the collecting process, 14 keywords were identified, including animal behaviour, human-animal relationship and animal's emotional state. Defining and translating of the keywords in different languages resulted in a multilingual matrix. Total 497 publications with first or last authors from respective countries were found for the Danube region. The predominant were scientific papers published in local/national scientific journals (40%), followed by considerably lower amount in international scientific journals (19%) and the least among Master, Graduation theses and Monographs (<3%). According to the animal species observed, most publications were for cattle, pigs and chickens (43, 17 and 11%, respectively), particularly in Serbia, Romania, Croatia, Czech Republic, Austria and Hungary. Less than 3% of the total findings referred to horses (Romania and Serbia), sheep (Romania, Bulgaria and Macedonia) and goats (Austria). Exponential growth of the publishing timeline in 30 years was indicated by only 9% within 1985-1999; 37% in the first decade of 21st century, while in the last 6 years, 54% of the total literature was published. This trend indicates future even greater involvement of the Danube region on topics like animal behaviour and welfare. A review and database on AW assessment in the Danube region will follow, enabling visibility and solid ground for future applied ethology and AW research in this region.

Behavioural measures in the welfare assessment of urban working horses

Tamara Tadich and Rodrigo Lanas
Universidad de Chile, Depto de Fomento de la Producción Animal, Santa Rosa 11735, La Pintana,
Santiago, 8820000, Chile; tamaratadich@u.uchile.cl

Welfare assessment of working horses commonly includes animal-based and resource-based measures. The behavioural assessment is an important component of animal-based measures. In the present study animal based and resource-based measures were applied in order to determine risk factors for two common welfare issues found in working horses (lesions and poor body condition scores (BCS)). A welfare assessment protocol including 6 resource-based indicators, speed or draught type conformation, the response (friendly, aggressive, no-response) to 4 behavioural tests (walk down beside horse, contact with head, approximation, and lifting hoof tests), and 7 qualitative behavioural descriptions of horses (nervous, self-confident, friendly, calm, aggressive, shy and) was applied to 154 working horses. Behavioural tests were performed by one observer and the owner to determine if horses respond in the same manner towards a known and unknown subject, and avoid erroneous negative interpretation of responses in relation to welfare. The observer also assessed the presence of lesions and BCS (1 to 5 scale) of horses. The Chi Square test was applied to determine association between behavioural responses of horses towards their owners and the observer. A multivariate logistic regression model was used to determine risk factors for to the presentation of lesions and poor BCS. Working horses had a mean age of 8.5±5 years of age (1.5 to 22 years). From the 154 horses 29.8% presented lesions and 20.6% were in poor BCS (scores 2 and 1). A significant association between the behavioural responses showed by horses towards owners and towards the observer was found for two of the four tests (walk down beside horse, P=0.01; and contact with head, P=0.001). Most owners defined their horses as self confident (92.7%), friendly (87.5%) and calm (82.3%); only 12.5% defined them as aggressive. None of qualitative descriptors resulted to be a risk factor for poor BCS or presence of lesions. Draught type conformation was a protection factor for BCS (P=0.02, OR=0.24) and speed conformation a risk factor for the presence of lesions (P=0.04, OR=2.39). A low frequency of work was a protection factor for the presence of lesions (P=0.02, OR=0.32), and the use of blinkers a risk factor for the presence of lesions (P=0.05, OR=2.26). Although behavioural parameters can be useful when assessing welfare of working horses, differences in the responses in relation to who performs the test should be considered, since horses respond differently to unknown subjects. The qualitative behavioural description of horses did not result in risk factors for common welfare issues, but could determine the owner-horse bond. Funding Source: FONDECYT No. 11121467.

Impact of weaning on grazing and drinking behavior in pastured lambs

Allison N. Pullin, Braden J. Campbell, Magnus R. Campler and Monique D. Pairis-Garcia
The Ohio State University, Department of Animal Sciences, 2029 Fyffe Road, Columbus, OH 43210, USA; pullin.4@osu.edu

Early weaning of lambs in the Eastern United States takes place commonly around 60 days of age. Weaning lambs prior to the natural weakening of the ewe-lamb bond can be stressful for the lamb due to deviations in nutritional sources, social relationships, and environment. As a result of the stress associated with this process, lamb productivity may also be negatively affected as demonstrated by decreased grazing behavior. The objective of this study was to assess the impact of early weaning on grazing and drinking behavior in pastured lambs during the first two days post-weaning (15 h/day). A total of 36, 60 day old lambs (19.1±2.8 kg) were blocked by body weight and randomly allocated to one of two treatments. Treatment one: lambs weaned at 60 days of age and placed with similar aged lambs (n=18 lambs; weaned treatment, WE); Treatment two: lambs kept with their ewes and weaned at 116 days of age (n=18 lambs, 18 ewes; ewe-lamb treatment, E). Grazing and drinking behavior were collected continuously using one color wireless outdoor IP video camera per pasture for 15 hours/day (0600-2100) for two consecutive days. Data were analyzed as total duration and frequency using PROC MIXED in SAS. The treatment and day interaction had no effect on grazing frequency (Day 1: WE: 130.9±14.6, E: 123.3±14.6 bouts/day; Day 2: WE: 126.7±14.6, E: 120.8±14.6 bouts/day, $P>0.05$) or duration (Day 1: WE: 3.6±0.5, E: 2.5±0.5 hours/day; Day 2: WE: 3.9±0.5, E: 2.8±0.5 h/day, $P>0.05$). The treatment and day interaction also had no effect on frequency (Day 1: WE: 7.2±1.4, E: 3.6±1.4 bouts/day; Day 2: WE: 4.8±1.4, E: 3.4±1.4 bouts/day, $P>0.05$) or duration (Day 1: WE: 1.8±0.4, E: 1.2±0.4 min/day; Day 2: WE: 1.1±0.4, E: 0.9±0.4 min/day, $P>0.05$) of drinking behavior for lambs. The results from this study suggest that the absence of the lactating ewe did not impact duration and frequency of lamb grazing and drinking behavior within the first two days post weaning. However, additional factors including lamb gender, pasture size and quality, and lamb visibility were not taken into account for this study. These factors may have had a significant effect on grazing and drinking behaviors and thus require further evaluation.

The relation between stereotypies and recurrent perseveration in laboratory mice depends on the form of stereotypy

Janja Novak, Jeremy D. Bailoo, Luca Melotti and Hanno Würbel
VPH Institute, University of Bern, Division of Animal Welfare, Länggassstrasse 120, 3012 Bern, Switzerland; j.novak2@gmail.com

Stereotypies are highly prevalent in laboratory mice and are associated with impaired inhibitory control of behaviour termed 'recurrent perseveration' (repetition or a non-random pattern of behavioural responses). Overall, animals with high levels of stereotypy show higher levels of recurrent perseveration; however, in mice this relation is inconclusive and may at least in part be explained by different forms of stereotypy, indicating different underlying mechanisms. We tested CD-1 (n=44) and C57BL/6 (n=40) female mice with different forms and expression levels of stereotypic behaviour on a two-choice guessing task to assess recurrent perseveration. The test consisted of 80 trials in which mice had to choose either left (L) or right (R) adjacent compartment to receive a reward. The sequences of choices made were analysed using 3rd order Markov chains, which measures higher order patterns of sequential dependence (perseveration score) and the number of pure repetitions made (LLLL or RRRR). Stereotypy level was not correlated with perseveration score in either strain (CD-1: $F_{1,43}$=1.42, P>0.05; C57BL/6: $F_{1,39}$=0.22, P>0.05); however, C57BL/6 mice with higher levels of route-tracing ($F_{1,39}$=9.13, P<0.05), but not bar-mouthing ($F_{1,39}$=0.91, P>0.05), made more repetitive responses in the guessing task. Our findings confirm previous research indicating that the implications of stereotypies for animal welfare may strongly depend on the strain of animal as well as on the form and expression level of the stereotypy, highlighting the need to differentiate between different forms of stereotypies in future studies. Furthermore, they indicate that variation in stereotypic behaviour may represent an important source of variation in many animal experiments.

Rumination in pre-weaned goat kids and its relation to lying time

Gosia A Zobel[1], Hannah Freeman[2] and Mhairi A Sutherland[1]
[1]AgResearch Ltd., Ruakura Research Centre, 10 Bisley Road, Private Bag 3123, Hamilton 3214, New Zealand, [2]Waikato University, Department of Science, Gate 8 Hillcrest Road, Private Bag 3105, Hamilton 3240, New Zealand; gosia.zobel@agresearch.co.nz

To ensure a smooth transition off milk, it is important that goat kids are ruminating prior to weaning. Adult dairy goats ruminate approximately 7 h/d, typically while lying down. We hypothesized that goat kids near weaning would have similar time budgets of rumination. The goals of this observational work were to: (1) describe rumination behaviour in young goat kids prior to weaning; and (2) determine the relationship between rumination and lying behaviour. Fifteen Saanen X goat kids were artificially reared from within 24 h of birth on milk replacer. The kids were offered *ad libitum* milk in bar feeders (2 teats allotted per kid) for two 30 min periods each day and had free access to water, hay, straw and meal. When kids were 13 weeks old, they were moved into three pens of five kids each and were bedded on shavings; this was done one week prior to observation for habituation. The behaviour of four randomly selected kids per pen (n=12) was observed for two continuous days via video before weaning at 14 weeks of age. Data for lying time, rumination time, rumination bouts, boli chewing time, and total number of rumen boli were calculated and summarized (SAS) for each kid on a daily, and on a two-hour period, basis. Results are descriptive and presented as mean±SD. Goat kids spent 3.0±0.5 h/d ruminating, of which 1.8±0.7 h/d were spent chewing rumen boli. The mean number of boli regurgitated was 205±25 boli/d over 19±3 rumination bouts/d. Mean rumination time per two-hour period peaked between 04:00-05:59 (43±6 min/period), and a smaller peak was observed between 12:00-13:59 (30±6 min/period). Lying time was 14.7±0.5 h/d and the kids were lying for 96±4% of the time during the rumination events. These results indicate that at 14 weeks of age, the goat kids had not reached previously reported adult dairy goat rumination times. Identifying lying time, and focusing on these time periods instead of the entire day, appears to be an efficient way of capturing rumination behaviour in kids.

Behavioural and metabolome characterization in socially defeated mice for understanding psychiatric disorders

Atsushi Toyoda[1,2,3], Tatsuhiko Goto[1,2], Hikari Otabi[2], Shozo Tomonaga[4], Daisuke Kohari[1,2,3] and Tsuyoshi Okayama[1,2,3]

[1]Ibaraki Univ, Cooperation between Agriculture and Medical Science, 3-21-1, Chuo, Ami, Inashiki, 300-0393 Ibaraki, Japan, [2]Ibaraki University, Agriculture, 3-21-1, Chuo, Ami, Inashiki, 300-0393 Ibaraki, Japan, [3]Tokyo University of Agriculture and Technology, United Graduate School of Agricultural Science, 3-8-1 Harumi-cho, Fuchu, 183-8538 Tokyo, Japan, [4]Kyoto University, Graduate School of Agriculture, Kitashirakawa Oiwake-cho, Sakyo-ku, 606-8502 Kyoto, Japan; atsushi.toyoda.0516@vc.ibaraki.ac.jp

We have established a depression model using mice subjected to subchronic mild social defeat stress (sCSDS). sCSDS mice (C57BL/6J male) showed increased body weight gain, polydipsia, and social avoidance. Also, nest building in sCSDS mice was retarded compared to control mice. Interestingly, the social avoidance in sCSDS mice were affected by diet conditions such as purified- and non-purified-diet feedings, namely sCSDS mice fed non-purified diet (MF, Oriental Yeast, Japan) showed more resilience compared to sCSDS mice fed purified diet (AIN-93G, Oriental Yeast, Japan). To elucidate these phenomena, metabolic profiles in the tissues (plasma, liver, and cecum digesta) of sCSDS mice were analyzed by metabolome techniques using a gas chromatography-mass spectrometry system under two different feeding conditions, purified pellet diet (AIN-93G) and non-purified pellet diet (MF). Four test groups were set as 'sCSDS + AIN-93G diet' (n=7), 'sCSDS + MF diet' (n=6), 'control (without sCSDS) + AIN-93G diet' (n=8), and 'control + MF diet' (n=8). Production of sCSDS mice and tissue sampling were carried out as previously. Two-way ANOVA were used to compare the factors 'stress', 'food', and 'stress × food'. To control the p-value for multiple comparisons, the false discovery rate was determined using the methods of Benjamini and Hochberg. The significance threshold was set at $Q<0.1$. Metabolome analysis revealed that the diet effect on metabolites is larger than the stress effect. Namely, 22, 27, and 31 of metabolites in plasma, liver, and cecum, respectively, were significantly changed by diet, while 8 and 5 of metabolites were significantly changed by social defeat stress in plasma and liver, respectively. Also, in the interaction of stress and diet, 5 and 1 of metabolites were significantly changed in plasma and cecum, respectively. These metabolites may have crucial roles in resiliency and vulnerability to psychosocial stress, which will need to be elucidated in the future study.

Sheep welfare in stud and meat farms in South Brazil using the Animal Welfare Indicators protocol

Fabiana de Orte Stamm[1], Luana Oliveira Leite[2] and Carla Forte Maiolino Molento[1]
[1]Animal Welfare Laboratory (LABEA), Federal University of Paraná (UFPR), Agrarian Sciences Sector, Rua dos Funcionários, 1540, 80035-050, Juvevê, Curitiba, Paraná, Brazil, [2]Department of Veterinary Medicine, Federal University of Paraná (UFPR), Agrarian Sciences Sector, Rua dos Funcionários, 1540, 80035-050, Juvevê, Curitiba, Paraná, Brazil; carlamolento@ufpr.br

Considering the lack of information about sheep welfare in Brazil, our objective was to assess ewe welfare in stud (S) and meat (M) farms. Seven S and 10 M farms were assessed during spring 2015, in Southern Brazil, using the Animal Welfare Indicators (AWIN) protocol for sheep, which has four welfare principles: good feeding, good housing, good health and appropriate behaviour. Total number of ewes was 164 in S and 267 in M farms. Results were compared with Mann-Whitney test at 0.05, and are presented as median (min-max) values, in the order S followed by M; AWIN indicator names are marked in italic. No differences were observed for indicators within good feeding and appropriate behaviour principles, including social isolation, stereotypies, excessive itching, familiar human approach test and Qualitative Behaviour Assessment. Overall, 6.9 (0.0-21.9)% of evaluated ewes were emaciated and the automatic drinker was not functioning in one M farm. There were differences within good housing principle, in the category fleece clean and dry with 74.2 (9.4-96.9)% vs 4.7 (0.0-73.9)%; and within good health principle, categories some fleece loss with 4.2 (0.0-45.8)% vs 17.3 (3.1-78.9)%; soiling and dags with 0.0 (0.0-0.0)% vs 17.4 (0.0-44.8)% and minor lesions 8.4 (3.1-27.6)% vs 0.0 (0.0-13.0)%; these differences probably relate to different shearing schedules (twice vs once a year) and to differences in breed (28.6 vs 90% of predominantly wool-blood). Head lesions were observed in 3.4 (0.0-8.3) vs 10.4 (0.0-26.3)%, probably due to tears caused by ear tagging, less common in S (50%) than in M (80%) farms. Indicators that seemed to restrict welfare but did not differ between groups were hoof overgrowth in 44.8 (11.5-75.0)% and tail docking 95.5 (3.4-100)%. We were able to identify main welfare restrictions in both stud and meat farms; overall welfare seems higher in stud farms due to fleece characteristics and shearing practices.

Goats use an outdoor run more with more resources provided

Joanna Stachowicz[1], Lorenz Gygax[1], Beat Wechsler[1], Edna Hillmann[2] and Nina Keil[1]
[1]Center for Proper Housing: Ruminants and Pigs, Tänikon 1, 8356 Ettenhausen, Switzerland,
[2]Ethology and Animal Welfare Unit, ETH, Universitätsstr. 2, 8092 Zurich, Switzerland;
joanna.stachowicz@agroscope.admin.ch

Access to an outdoor run has potential benefits for the activity behaviour of dairy goats in loose housing systems. To be attractive for goats an outdoor run should provide resources necessary for its use, that is good accessibility (number, size of exits), sufficient space, protection against adverse weather conditions (roofing, exposure of the run) as well as items to occupy the goats (climbing possibility, hayrack, brush, trough). If indoor housing offers enough space, adequate structures to retreat and climb, food/water and brushes, this could counteract the use of the outdoor run. On 13 dairy farms, the use of the outdoor run, the activity and lying behaviour of goats was examined during the winter-feeding period in relation to the quality of outdoor run and indoor housing of the farms. As outdoor runs and indoor housing varied in respect to the resources provided between farms, a quality index for each of these two aspects was defined. For each available resource in the outdoor run and in indoor housing, respectively, a point score was summed up. A high score indicated a farm having several resources of outdoor run and of indoor housing. 3-7 days per farm with comparable weather conditions (3.5-16.5 °C, 39-93% humidity, no rain) were selected. On these days the mean daily proportion of goats outside between the two milkings was assessed by video data. The activity and lying behaviour of 15 focal goats per farm was measured with 3-D acceleration loggers. Age was used as an indicator of rank. Direct observation (on 2 days, 2 h in the morning and 2 h in the afternoon on each farm) allowed assessing the activities performed in the outdoor run. Linear mixed effect models were used with a stepwise-backward selection approach (threshold P<0.05 based on a parametric bootstrap). Almost all focal goats visited the outdoor run for lying, interacting with herd members and used the items if provided. With increasing outdoor quality index (range: 1-8) the mean proportion of goats outside increased from 17% (confidence interval: 13-22) to 38% (25-52; P=0.023). An increasing outdoor quality index led to a decrease in activity in lower ranking goats and an increase in activity in higher ranking goats (P=0.028). An effect of the indoor quality index on outdoor run use could not be detected. A higher indoor housing index (range: 2-5) had no influence on lying duration but lead to a lower number of lying bouts (P=0.023). Lower ranking goats had more lying bouts than higher ranking ones (P<0.001). Our results show that goats use an outdoor run more the more resources are provided.

Effect of topical healing agents on scrotal lesions after surgical castration in 4-5 month old beef calves

Sonia Marti[1,2], Daniela Melendez[1,2], Eugene Janzen[1], Désirée Soares[1,2], Diego Moya[1,2], Ed Pajor[1] and Karen Schwartzkopf-Genswein[2]
[1]University of Calgary, Production Animal Health, 2500 University Dr. NW, T2N 1N4, Canada, [2]Agriculture and Agri-Food Canada, Beef Welfare, 5403 1st Ave S, T1J 4B1 Lethbridge, Canada; sonia.marti@agr.gc.ca

Beef cattle are castrated in North America to reduce aggressive and sexual behaviour, and to improve carcass quality. However, surgical castration results in lesions that require several weeks to heal. The objective of the present study was to determine the efficacy of commercially available topical healing agents (HA) to improve wound healing and reduce inflammation and secondary infection. Forty-eight Angus bulls (187±4.9 kg BW and 4-5 mo of age) were randomly assigned to control (CT, castrated without the application of a post-operative HA), or surgical castration followed by either the application of a topical germicide (GR), aluminium powder spray (AL), or liquid bandage (LB). At the time of castration, all calves were administered an anesthetic (xylazine at a dose of 0.07 mg/kg via epidural). Wound healing was assessed in all calves over a 77 d period post-castration. Indicators of wound healing assessed included scrotal area temperature (ST; maximum; °C) using infrared thermography, scrotal circumference (SC; cm), visual evaluation of swelling (SW, 5 point-scale) and healing rate (HR, 5-point scale) collected on d -1 and immediately before castration, d1, d2, and d7 post-castration, and weekly thereafter until the end of the study. Pain sensitivity of the wound and surrounding skin to manual pressure was evaluated using a digital algometer (DA) on d -1, 0, 1 and 2 and a Von Frey anesthesiometer (VA; kg) weekly for the first 42 d post-castration. Animal BW and rectal temperature (RT) were also recorded weekly. Data were analyzed using a mixed-effect model and categorical data were analyzed using a Chi-square test using repeated measures. No treatment differences were observed for maximum ST, SW or SC, and for BW and RT over the course of the study. Castrated calves tended (P=0.06) to have increased pain sensitivity compared to CT for the first 2 d after castration as witnessed by lower DA readings and for up to 42 d following castration with lower (P=0.09) VA readings. Both GR and LB calves were more sensitive (P<0.05; required less pressure to respond to the DA) than CT and AL calves on d 2 and 5 post-castration. Although no statistical differences were observed in HR between the treatments, calves treated with HA had numerically greater HR scores on d 35 and 42 than CT calves. The HA used in this study did not improve indictors of healing such as swelling and healing rate scores or indicators of inflammation including scrotal temperature and circumference of surgical castration lesions.

Assessment of time of administration of subcutaneous Meloxicam on indicators of pain after knife castration of weaned calves

Daniela M. Melendez[1,2], Sonia Marti[1,2], Eugene D. Janzen[2], Diego Moya[1,2], Desiree R. Soares[1,2], Edmond Pajor[2] and Karen S. Schwartzkopf-Genswein[1]
[1]*Agriculture and Agri-Food Canada, 5403 1 Ave S, T1J4P4, Lethbridge, AB, Canada,* [2]*Univeristy of Calgary, Production Animal Health, 2500 University Dr. NW, T2N1N4, Calgary, AB, Canada;* dani88ms@gmail.com

The newly revised Canadian Codes of Practice for the management of beef cattle requires that as of 2018, calves older than 6 months of age must be castrated using pain control. To date, an optimal strategy to mitigate castration pain has not been identified. The aim of this study was to identify the optimal time of administration for the analgesic meloxicam (Metacam®) prior to castration, treatments were selected based on peak meloxiam serum concentrations after oral administration. Thirty four Angus bull calves (282.23±4.80 kg BW, 7-8 mo old) were randomly assigned to one of three treatments receiving a single s.c. injection of meloxicam (0.5 mg/kg BW): 6 h (6H; n=11); 3 h (3H; n=12); or immediately (0H; n=11) before knife castration. Measurements included visual analog scale (VAS), head movement (HM), hind stride length (SL), lying and standing behaviour, salivary cortisol (CL), haptoglobin (HP), serum amyloid A (SAA), substance P (SP), and scrotal area temperature (SAT). Samples were collected on d -7, -5, -2, -1, and immediately before castration; and 30, 60, 120, and 240 min, 1, 2, 5, 7, 14, 21 and 28 d after castration, except for VAS and HM obtained during castration. Data were analyzed using mixed-effects model including administration time as main effect and animals as a random effect. Mean SL was shorter (P=0.05) in 3H (47.6±0.9 cm) compared to 0H (49.1±0.9 cm) but no differences were seen between these two groups and 6H (47.9±0.9 cm) calves up to 240 min after castration. Percentage of standing was greater (P=0.03) in 0H (40.5±1.3%) and 6H (42.6±1.3%) compared to 3H (38.0±1.3%) calves, while mean standing duration was greater (P=0.03) in 6H (63.0±5.0 min) compared to 0H (45.8±5.0 min) and 3H (43.8±5.0 min) calves 8-13 d after castration. No treatment differences (P>0.20) were seen in VAS or HM at the time of castration. A time × treatment effect (P=0.01) was observed for SP, where 0H (106.4±6.6 pg/ml) had lower concentrations than 3H (119.2±6.6 pg/ml) and 6H (122.8±6.3 pg/ml) calves one day after castration. Mean SAT were greater (P=0.006) in 6H compared to 0H and 3H calves 0-240 min and 1-28 d after castration. No treatment differences (P>0.50) were observed for mean CL, SAA and HP. Based on these results, the optimal time to administer s.c. meloxicam in 7-8 mo old knife castrated calves, is immediately before castration (0H) as evidenced by fewer indicators of pain and inflammation compared to 3H and 6H.

Lameness, productivity, and cow behaviour in dairy herds with automated milking systems

Meagan T. M. King[1], Ed A. Pajor[2], Stephen J. Leblanc[3] and Trevor J. Devries[1]
[1]University of Guelph, Department of Animal Biosciences, 50 Stone Road E, N1G 2W1, Canada, [2]University of Calgary, Faculty of Veterinary Medicine, 2500 University Dr NW, T2N 1N4, Canada, [3]University of Guelph, Department of Population Medicine, 50 Stone Road E, N1G 2W1, Canada; tdevries@uoguelph.ca

The objective of this study was to evaluate how herd management, barn design, and lameness relate to productivity and cow behaviour in automated milking systems (AMS). We visited 41 AMS farms in Canada (Ontario: n=26; Alberta: n=15), averaging 105±56 (mean±SE) lactating cows and 2.2±1.3 AMS units. Details of barn design, stocking density, and herd management were collected. At each farm, 40 cows were gait scored (or 30% for herds >130 cows) using a 5-point numerical rating system (NRS; 1=sound to 5=lame). Cows were defined as clinically lame with NRS≥3 (26.2±13.0%) and severely lame with NRS≥4 (2.2±3.1%). For 6-d periods we collected milking data from AMS units and lying data from electronic data loggers, which were later analyzed in multivariable mixed-effect regression models. At the herd level, an increase of 1 percentage point (p.p.) in the prevalence of severe lameness was associated with production losses of 0.6 kg of milk produced/cow/d (P=0.05) and 32 kg milk harvested/AMS/d (P=0.03), while a 10 p.p. increase in clinical lameness prevalence was associated with 0.1 fewer milkings/cow/d (P=0.05). One additional cow/AMS unit was associated with 32 kg more milk harvested/AMS/d (P<0.001) but also reduced milking frequency (-0.2 milkings/cow/d with 10 additional cows/AMS; P<0.001). Daily lying time was positively associated with the frequency of feed push-ups (+3 min/cow/d/push-up; P=0.05) and negatively associated with the placement of neck rails from the rear curb of lying stalls, such that cows lied down less as neck rails were placed farther forwards (-23 min/cow/d/10 cm; P=0.03). Lying bouts were 12 min longer in deep-bedded stalls vs mattresses (P=0.003), and 5 min longer with each 10 p.p. increase in the prevalence of clinical lameness (P=0.001). In a cow-level comparison (30 cows/ farm) of lame (NRS≥3: n=353) and sound cows (NRS<3: n=865), lame cows were fetched more often (P=0.002), produced 1.6 kg/d less milk (P=0.002) in 0.3 fewer milkings/d (P<0.001), and spent more time lying down (+38 min/d; P<0.001) in longer bouts (+3.5 min/bout, P=0.03). In conclusion, lameness is especially problematic for AMS herds, reducing productivity at the cow and herd level. Although few cows in our study were severely lame, producers need to identify and reduce clinical lameness. Widening lying stalls, providing deep-bedded stalls, and scraping alleys more frequently were factors associated with reduced lameness prevalence and are potential ways to optimize productivity in AMS herds.

Cow-level risk factors for lameness in transition dairy cows – a study on grazing dairy herds in southern Brazil

Rolnei R Daros[1], Jose A Bran[2], Alexander J Thompson[1], Stephen Leblanc[3], Maria J Hötzel[2] and Marina A G Von Keyserlingk[1]
[1]*University of British Columbia, Faculty of land and food systems, Animal welfare program, 2357 Main Mall 248, Vancouver, V6T 1Z4, Canada,* [2]*Universidade Federal de Santa Catarina, Depto de zootecnia e desenvolvimento rural, laboratório de etologia aplicada e bem-estar animal, Rodovia Admar Gonzaga, 1346, Florianópolis, 88034-000, Brazil,* [3]*University of Guelph, Population Medicine, Ontario Veterinary College, 50 Stone Road E., Guelph, N1G 2W1, Canada; rrdaros@gmail.com*

Lameness is considered to be one of the greatest animal welfare challenges facing the dairy industry. Previous work has focused on confinement dairy systems with less information available on the risk factors for lameness on grazing dairies. Common hoof ailments often develop in the period around calving and may contribute to lameness. The aim of the current study was to evaluate the prevalence, and risk factors for lameness in grazing dairy cows during the first 3 wk after calving. We visited 40 farms located in Santa Catarina State in Brazil between February and September 2015. All transition cows (n=170) were live locomotion scored using a 1 to 5 scale as they exited the parlour (score ≥3 as lame). All cows were also body condition scored and measured for hoof overgrowth, metritis, and ketosis. Cows were diagnosed with ketosis when the blood BHBA was ≥1.2 mmol/l. Metritis was assessed using a vaginal discharge score. Parity and presence of retained placenta were reported by producers. Univariate logistic regression models were used to select variables associated with lameness (P<0.2). The multivariable logistic regression model of lameness, specifying farm as random effect was built using manual backwards elimination. The final model included variables with P<0.05. The overall prevalence of lameness was 30% for these farms. Parity (≥ 3rd lactation) and hoof overgrowth were associated with lameness (OR: 6.1; 95% CI: 2.7-15.8 and OR: 3.5; 95% CI: 1.7-7.2, respectively). The prevalence of lameness was highest in the ≥ 3rd lactation cows (45%) compared to 11% in the primiparous and second lactation cows. Cows in ≥ 3 lactation had a higher prevalence of hoof overgrowth (55%) compared to the other two groups (19%). Although no association was found between lameness and evaluated transition period diseases, the association between hoof overgrowth and parity for lameness in transition cows on grazing dairies in this region suggests that older cows may benefit from regular hoof trimming, which could reduce the prevalence of lameness.

Passive immunity of suckling calves and associations with colostrum management routines on organic dairy farms

Julie Føske Johnsen[1], H. Viljugrein[1], K.E. Bøe[2], S.M. Gulliksen[3], A. Beaver[4], A.M. Grøndahl[1], T. Sivertsen[5] and C.M. Mejdell[1]
[1]Norwegian Veterinary Institute, Pb 750 Sentrum, 0106 Oslo, Norway, [2]University of Life Sciences (NMBU), Basic sciences and Aquatic medicine, box 5003, 1432 Ås, Norway, [3]Norwegian Agriculture Agency, P.O. Box 8140 Dep, 0033 Oslo, Norway, [4]Cornell University, College of Agriculture and Life Sciences, 149 Morrison Hall, Ithaca, NY 14853, USA, [5]NMBU, Production Animal Clinical Sciences, P.O. Box 8146 Dep, 0033 Oslo, Norway; julie.johnsen@vetinst.no

Suckling dairy calves are known to be at risk of ill health because the acquisition of passive immunity from colostrum may be compromised. The aim of the study was to explore the prevalence of failure of passive transfer (FPT; serum immunoglobulin (IgG)<10 g/l 24-28 h post natum) in suckling organic dairy calves and associations with management routines to ensure colostrum intake. From a total of 20 herds, 156 calf blood samples (mean±SD; 7.8±1.24 per herd) and 141 colostrum samples from the dams were analysed. All calves suckled the dam. Factors known to affect serum and colostrum IgG were evaluated, including the method the farmer used to ensure calf colostrum intake and whether colostrum was ensured by routine or non-routine management practice. Separate statistical regression models were used to predict FPT, calf serum IgG and colostrum IgG. Herd was included as a random term. Prevalence of FPT was 31%. Mean serum and colostrum IgG (± SD) was 16.0±10.03 g/l and 39.4±26.44 g/l, respectively. Only colostrum IgG had a statistically significant influence on FPT prevalence (P<0.001) and serum IgG variation was also explained mainly by colostrum IgG (P=0.005). Of calves receiving colostrum according to farm routine, calves given additional colostrum with a bottle had lower serum IgG levels than did calves not given additional colostrum (13.8±1.04 vs 17.9±1.56 g/l, P=0.028). However, no within herd effect was found. With a high between-herd variation, colostrum IgG ranged from 2 to 135 g/l on cow level, and only 23% of the samples had a IgG content >50 g/l. Colostrum IgG was higher in the spring vs winter (50.7±4.6 vs 29.6±1.93 g/l IgG, P=0.003), and higher in Norwegian Red vs Swedish breeds (35.6±2.52 vs 23.5±2.46 g/l IgG, P=0.03). The results indicate that for calves capable of finding the udder and suckling independently, there is no direct benefit in routinely hand feeding colostrum although herd level factors may play an important role. FPT prevalence in this study was high, but comparable to that of non-suckling calves at conventional farms in Norway. Still, the high proportion of calves with FPT and inferior colostrum quality indicates a need for improved awareness among dairy farmers practicing cow-calf suckling.

Pain in primates: using behaviour and facial expressions as a welfare tool

K Descovich[1,2], H Buchanan-Smith[2], S Richmond[3], M Leach[4], P Flecknell[5], D Farningham[6] and SJ Vick[2]
[1]Unitec Institute of Technology, Auckland, 1142, New Zealand, [2]University of Stirling, Psychology Division, Stirling, FK9 4LA, United Kingdom, [3]SRUC, West Mains Rd, EH9 3JG, United Kingdom, [4]Newcastle University, School of Agriculture, Food and Rural Development, Newcastle upon Tyne, NE1 7RU, United Kingdom, [5]Newcastle University, Comparative Biology Centre, Newcastle upon Tyne, NE2 4HH, United Kingdom, [6]Medical Research Council, Centre for Macaques, Salisbury, SP4 0JQ, United Kingdom; kdescovich@unitec.ac.nz

Animals used in bioscience undergo potentially painful procedures. Alleviation of pain is critical for good welfare and the implementation of the 3Rs. Primates are common biomedical models but pain identification in these species is challenging. Opportunistic video was collected from 33 rhesus macaques (Macaca mulatta) undergoing 1 of 5 procedures as part of unrelated neuroscience research. Analgesia (meloxicam, dexamethasone, buprenorphine or methadone) was given during the procedure and the following morning. Animals were filmed in four 20-minute periods: pre-procedure (PreO), post-procedure once normal ambulation occurred (PostO), and the following day before (PreA) and 1 h after additional pain relief (PostA). Behaviour was recorded and the Macaque Facial Action Coding System (MaqFACS) was used to measure facial movement. The observer used was MaqFACS accredited and checked for reliability in behavioural coding against an independent coder. The observer was partially blind to experimental condition as periods PostO, PreA and PostA were visually identical, but had unavoidable differences from PreO. As several non-pain variables differed between PreO and other periods (e.g. skin shaving, jackets), differentiation from PreO was not considered diagnostic of pain. Pain behaviours were defined as those with a significant omnibus test (Friedmans Tests), and post-hoc tests (Wilcox signed rank tests) showing highest prevalence in the PreA period, lowest in PreO, and moderate in PostO and PostA, when pain should be controlled. Friedmans tests suggest that several behaviours [aggression ($P=0.042$), hair pulling ($P=0.038$) and shivering ($P<0.001$)] may change with pain states in this species. Shivering, in particular, mapped closely with expected pain, being more prevalent in the PreA period (89 ± 24 shivers/hr) than in the PostA (61 ± 19 shivers/hr, Wilcox Signed Rank: $P=0.02$), PostO (22 ± 13 shivers/hr, $P=0.01$), or PreO period (11 ± 6 shivers/hr, $P<0.001$). Discriminant function analysis determined if facial movements predicted experimental condition. One function, explaining 60% of the variance, mapped closely with expected pain state. Cross validation predicted condition 60% of the time. We will discuss practical implementation of pain identification for primates in the biosciences.

Effects of transport on the clinical condition of sows destined for slaughter

Katrine K. Fogsgaard, Karen Thodberg and Mette S. Herskin
Aarhus University, Department of Animal Science, Blichers Allé 20, 8830 Tjele, Denmark; katrine.
kopfogsgaard@anis.au.dk

Each year over 400,000 Danish sows are transported by road to abattoirs. Only very limited knowledge about these animals is available, as most research on pig welfare on the day of slaughter has focused on finishing pigs. This study is part of a larger study examining sow fitness for transport. Here, we describe the clinical condition of a sample of 395 sows before and after transport to slaughter and examine effects of transport duration and transport breaks on changes in the clinical condition of the sows. Clinical condition was assessed as number of wounds (lesions involving dermis of >1 cm) and scratches (elongated lesions restricted to epidermis >5 cm) – counted 1-15, and if more noted as 15+; and lameness on a score from 0-3. All procedures were approved by the Danish Animal Experiments Inspectorate. The study included sows from 11 private farms sampled randomly, and stratified according to distance to a large slaughterhouse (4 categories, expected duration: 0-2, 2-4, 4-6 or 6-8 h). Selection of animals to be slaughtered was done by the farmers taking into account the EU Council Regulation. Clinical registrations were made on-farm before transport and immediately upon arrival at the abattoir. Transport duration ranged from 72-469 min with a mean of 250 ± 101 min and 77% involved at least one break. Differences between clinical conditions of the sows on farm and at the abattoir were analysed by Signed Rank Test in SAS. Effects of breaks and duration of transport were analysed by PROC Glimmix in SAS. Transport to slaughter led to an increased number of wounds (from 1.0 ± 0.1 to 2.5 ± 0.2; S=11,183, P<0.001) and scratches (from 2.2 ± 0.3 to 9.3 ± 0.5, S=19952, P<0.001). In addition, lameness score increased from 0.1 ± 0.02 to 0.2 ± 0.03 (S=684, P<0.001). Transport duration influenced the odds of having more scratches upon arrival at the abattoir ($F_{3, 218}$=5.97, P<0.001; 48, 91, 89 and 76% for sows transported 0-2, 2-4, 4-6 and 6-8 h, respectively). Transport duration also influenced the odds of getting wounds ($F_{3,205}$=3.7, P=0.01), but no clear connection between wound risk and transport duration was found. When breaks were included, the risk of getting wounds was increased (67% with breaks vs 45% without breaks, $F_{1, 195}$=5.02, P=0.02). Lameness score at the abattoir was not affected by transport duration or breaks. These preliminary results show that the clinical conditions of sows are worsened by transport to abattoir, and that the duration of the transport (less than 2 hours compared to more than 2 hours) influenced these changes. Future analyses will clarify whether these differences are affected by factors such as climate, type of housing and an eventual stay in on-farm pick-up facilities.

Systematic review of on-farm factors affecting leg health in broilers

Ida Just Pedersen and Björn Forkman
University of Copenhagen, Department of Large Animal Science, Grønnegårdsvej 8, 1870
Frederiksberg, Denmark; idajp@sund.ku.dk

Factors affecting leg health in broilers have been addressed in narrative reviews. However, systematic reviews use a standardized approach to answer a well-defined and explicit study question, and are often more focused and transparent in their aim than narrative reviews. No systematic reviews are available on the subject. Several factors have been shown to affect leg health in broilers, these can roughly be divided into two main categories; factors that are related to the chicken such as growth rate and nutrition, and factors that are related to the external production environment. The focus of our systematic review is factors that are related to production environment, under the assumption that these factors are directly linked to the behavior of the broilers. The review question for our systematic review holds the structure of a PECO question, which seeks to investigate the effect of an exposure and contains four key elements; the population (P), the exposure (E), the comparator (C) and the outcome (O). The population (P) in question is broiler chickens. The exposure (E) is a list of potential risk factors e.g. types of environmental enrichment, lighting regimes and stocking density. The comparator (C) is the situation that the exposure (E) is being held against, for example a control group. In this review the outcome (O) is leg problems defined both as pathological conditions such a tibial dyschondroplasia, foot pad dermatitis and hock burns, as well as behavioral conditions such as changes in gait score, latency to lie and incidence of lameness. For each risk factor a systematic literature search was performed using the databases Pubmed, CAB Abstracts and Web of Science. Only articles which were in compliance with the selected key elements were included in the review. A total of 71 articles were included in the review, with the main focus on lighting regime and stocking density (54 articles). Fewer studies were available on the effect of environmental enrichment (17). Low intensity of light and continuous lighting programs have been reported as risk factors for leg problems in several studies. Some effect of stocking density has also been reported, with high stocking density being a risk factor. Environmental enrichment in the form of perches does not seem to be a risk factor with only 1 of 10 studies reporting a significant effect on leg problems. Limited studies are available on environmental enrichment in the form of barriers between food and water or straw bales (7). However, the majority of these report a significant effect on leg problems. More studies are needed to investigate the potential benefits of enrichment in the broiler production. This research was funded by the EU FP7 Prohealth project (no. 613574).

Assessing the physical well-being of dogs in commercial breeding facilities

Judi Stella[1], Moriah Hurt[2], Paulo Gomes[3], Amy Bauer[2] and Candace Croney[2]
[1]USDA-APHIS, Purdue University, 625 Harrison St, W Lafayette, IN, 47907-2042, USA, [2]Purdue University, Comparative Pathobiology, 625 Harrison St, W Lafayette, IN, 47907-2042, USA, [3]Purdue University, Veterinary Clinical Sciences, 625 Harrison St, W Lafayette, IN, 47907-2042, USA; judith.l.stella@aphis.usda.gov

In the U.S., dogs are raised in commercial breeding (CB) facilities for sale as pets. This study aimed to assess the physical well-being of dogs housed in CB kennels. A random sample of dogs (n=111, 86 female, 25 male, mean age 35.7 months) representing 14 breeds were assessed at five facilities that volunteered for the study. Assessment included body condition score (BCS) using a 1-5 scale, paw health (e.g. toenail length, paw pad abnormalities), elbow/hock health (e.g. alopecia, calluses, inflammation), ear health (e.g. debris, excessive hair, infection), and periodontal disease (PD) using a visual scale of severity (0-4). Published scales were used to assess BCS and PD, while the others were developed for this study based on the literature and veterinary input. Descriptive statistics (e.g. means, standard deviations) were used within facilities since variation in breed and management prohibited direct comparisons across facilities. Relative risk (RR) was calculated for PD by weight and age. Mean BCS was 3.2 (3 being ideal). Mean toenail length was 1.3 (1 being ideal). Matted hair was the most common paw problem, more common in long-haired breeds (n=69). Sores (n=1), inflammation (n=7), and interdigital cysts (n=1) were rare. Elbow/hock exams found no inflammation or calluses while the presence of alopecia was more common in dogs housed on concrete flooring (concrete n=5, other n=2, RR=15.7, P<0.0001). Ear exams revealed overgrown hair (n=21), debris (n=121), and erythema (n=106) most commonly in long-haired breeds. Periodontal disease was more common in dogs <20 kg (n=12) than in larger dogs (n=0) but neither size (RR=0.0, P=0.1) nor age (RR=0.65, P=0.5) significantly increased the RR of PD. Significant physical welfare problems were rare at these facilities. Ongoing studies, including assessment of the dogs' behaviour, should provide a more comprehensive understanding of their well-being in CB operations.

Effect of prepartum physical activity on behaviour and immune competence of non-lactating dairy cows

Randi A. Black, Gina M. Pighetti and Peter D. Krawczel
University of Tennessee, Department of Animal Science, 2506 River Drive, Knoxville, TN 37923,
USA; rblack12@vols.utk.edu

The objective was to determine the effect of prepartum exercise, pasture turnout, or total confinement on activity and immune competence of dairy cows. Sixty pregnant, non-lactating dairy cows were assigned to control (Holstein=19; Jersey × Holstein=1), exercise (Holstein=19; Jersey × Holstein=1), or pasture (Holstein=20) treatments using rolling enrollment from Jan to Nov 2015 at cessation of lactation. Cows were balanced by parity (1.8±0.9), projected metabolizable energy fat-corrected milk yield (13,831±2,028 kg per lactation) and projected due date. Cows were housed in a naturally ventilated, 4-row deep-bedded sand freestall barn at the University of Tennessee's Research Unit (Walland, TN, USA). Cows were moved to a maternity pen with a rubber mattress to calve. Fitted 3 d before cessation of lactation, accelerometers determined daily lying time (h/d), daily lying bouts (n/d), lying bout duration (min/bout), and daily steps (n/d) at 1-min intervals. Data were averaged by four periods relative to actual calving date: -58 to -15 d (FO), -14 to -1 d (CU), d 0 (CA), and 1 to 14 d (PP). Exercise was done on five consecutive days per wk for 1.4±0.1 h/d (targeted 1.5 h/d), at a pace of 1.88±0.58 km/h. Pasture turnout occurred on a grassy paddock five consecutive days per wk for 1.8±0.3 h/d (targeted 1.5 h/d). Control cows remained in the home pen throughout the non-lactating period. Blood was sampled on d -3 and 42, relative to cessation of lactation and projected calving date to assess immune competence via reactive oxygen species generation using phorbol myristate acetate. A mixed model determined the effects of treatment, period, and treatment × period on daily lying behaviour and steps and the effect of treatment, day, phorbol myristate acetate level, and their interactions on reactive oxygen species generation. Cow within treatment was the random subject. Exercise cows lay down less frequently at CA (11.6±1.0 bouts/d) compared to control cows (14.6±0.9 bouts/d; P=0.03). However, lying bout duration and daily lying time did not differ among treatments at CA (P>0.31). Exercise cows were more active at FO, CU, and CA (2,895.4±107.6, 2,614±125.2, and 2,824.6±224.4 steps/d, respectively) than control (1,788.8±103.9, 1,840.8±120.7, and 1,969.3±216.2 steps/d), and pasture (2,132.0±103.6, 1,951.6±120.9, and 2, 234.9±216.3 steps/d; P<0.01). Reactive oxygen species production was not affected (P=0.63). Exercised cows took more steps, but had fewer lying bouts during calving, suggesting more comfort during calving. Furthermore, physical activity did not alter immune competence. Prepartum exercise may be a viable management strategy to improve calving performance.

Can gastric health issues in veal calves be linked to poor welfare?

Laura Webb[1], Harma Berends[2], Kees Van Reenen[1], Walter Gerrits[2] and Eddie Bokkers[1]
[1]Wageningen University, Animal Production Systems group, P.O. Box 338, 6700 AH Wageningen, the Netherlands, [2]Wageningen University, Animal Nutrition group, P.O. Box 338, 6700 AH Wageningen, the Netherlands; laura.webb@wur.nl

Abomasal damage (AD) and poor rumen development (RD) are 2 common gastric impairments in veal calves. The question is: do they negatively impact welfare? We searched for links between gastric health and behavioral indicators of welfare in 2 experiments (Exp). Exp1: 48 calves (8 to 26 wk, 3 calves/pen) were fed milk replacer with 1 of 4 solid feed (SF) amounts (0, 40, 80, 120 kg DM/calf provided over 4 months, twice daily and increased on a weekly basis) with 50% concentrate, 25% maize and 25% straw on DM basis. Exp2: 160 calves (12 to 29 wk, 5 calves/pen) were fed milk replacer with 1 of 4 SF amounts (20, 100, 180, 260 kg DM/calf fed over 4 months, once daily and increased on a weekly basis) and 1 of 2 SF compositions (80% concentrate, 10% maize, 10% straw or the one used in Exp1). Behavior (ruminate, tongue play, manipulate object, lie, groom, play, idle) was noted using instantaneous scan sampling (5 min intervals for 30 min, 7 scans/d) at 24 wk. At slaughter, AD (number and size of lesions) and RD (score from 1 to 4, with 4 being 'well developed') were noted. For each Exp, a principal component analysis with behavioral and gastric health variables was done. Loadings >0.40 were considered of interest. The different diets ensured diversity in gastric health and behavioral variables but were not included in the analysis. RD and AD were compared across Exp using genralized linear models. RD was similar in both Exp (1.9 ± 0.05; $P=0.381$), whereas AD was worse in Exp2 (4.2 ± 0.41 vs 2.2 ± 0.37; $P=0.011$). The first 2 principal components (PC) had eigenvalues >1 and, together, explained 35.8% and 33.8% of the variance for Exp 1 and 2 respectively. Exp1: PC1 did not include gastric health variables. PC2 loaded high on RD and AD, and low on tongue play. This was expected because RD occurs with high amounts of SF, which increase rumination and decrease tongue playing. Exp2: PC1 loaded high on RD and ruminate, and loaded low on manipulate object and idle: 2 potential indicators of poor welfare. PC2 loaded high on AD, lie and idle, and loaded low on play. RD was linked with indicators of good welfare: i.e. high rumination and low abnormal behaviors, i.e. tongue play and manipulate object. AD was linked with indicators of good welfare (i.e. low tongue play) and poor welfare (i.e. low play). Tongue playing may protect against AD by reducing stress. The association between AD, lie and idle may indicate pain, because animals in pain are less likely to perform active behaviors. Finally, this study does not support the idea that a developed rumen protects against AD.

Shoulder lesions in sows: a behavioural review and needs assessment for future research

Fiona C. Rioja-Lang[1], Yolande M. Seddon[2] and Jennifer A. Brown[1,3]
[1]Prairie Swine Centre, Ethology, 8 St E, Saskatoon, SK, S7H 0T8, Canada, [2]University of Saskatchewan, Western College of Veterinary Medicine, 52 Campus Drive, Saskatoon, SK, S7N 5B4, Canada, [3]University of Saskatchewan, Dept. of Animal and Poultry Science, 51 Campus Driive, Saskatoon, SK , S7N 5A8, Canada; fionarlang@hotmail.com

Shoulder lesions/ulcers in sows are most commonly observed in the intitial weeks post-farrowing, and are comparable to human pressure sores (bed sores). A change in sow behaviour during farrowing and early lactation, along with prolonged lying bouts in lateral recumbancy, decreased activity, and decreased body condition are all risk factors. Continued pressure on the shoulder leads to restricted blood flow, and necrosis of the skin and underlying tissue. The severity of lesions can vary greatly, from superficial sores to deep, subcutaneous ulcers, and are believed to cause pain. Shoulder lesions represent a significant welfare concernm and an economic cost to producers. The purpose of this review was to assimilate the available literature regarding how the behaiour of the sow affects the occurrence and severity of shoulder lesions, and to provide a needs assessment for future research. It is likely that while decreased activity is one of the major causes of sow' developing shoulder ulcers, it is also suggested that sow's display increased activity in response to these injuries (as a pain related response). In one study investigating the associations between sow lying behaviour ad sking lesions in farrowing crates, it was determind that thin sows were more likely to get shoulder lesions, and that moderate to severely lame sows also had increased risk. Lameness increases time in recumbancy and can impede the sows' ability to stand up and relieve the pressure on their shoulders. However, a more recent study reported that sows with ulcers showed an increased level of activity, spent more time in active postures, and performed a high frequency of postural changes. This could indicate that the ulcerated sows were showing signs of restlessness: a sign of motivational conflict and an indicator of pain. Despite the existence of a large body of literature, much of the information is epidemiological and thre is still important information lacking on the condition. Future studies should include how sow lying behaviour and the frequency of nursing and postural changes influence ulcer development. Research in this area, inclding the interaction of sow behaviour, body condition, and flooring on ulcer formation would benefit both sow welfare, and producers.

Pain related to piglet tail docking and/or castration

Valérie Courboulay[1], Morgane Gillardeau[1], Marie-Christine Meunier-Salaün[2] and Armelle Prunier[2]
[1]*IFIP, Institut du Porc, BP 35104, 35651 Le Rheu cedex, France,* [2]*INRA 1079, UMR SENAH, Domaine de la Prise, 35590 Saint-Gilles, France; valerie.courboulay@ifip.asso.fr*

In French commercial piggeries, piglets are often submitted to tail docking shortly after birth and to surgical castration a few days later with meloxicam used to alleviate pain due to surgical castration. A series of experiment was performed to evaluate different scenarios of analgesic treatment and age at docking and castration. In the present part of the study, we have compared tail docking alone, castration alone and tail docking + castration. The protocol was approved by the French ethical committee n°07 and authorized by the French ministry of Education and Research. Three piglets per litter were either tail docked with hot cautery iron (T) or castrated (C) or submitted to both procedures (TC) at 2 days of age after i.m. injection of meloxicam. Movements and intensity of vocalizations were registered during the interventions in 35 piglets/group. Behaviour and tail movements were registered at 2-min interval during 60 min at H0 (the first hour following the intervention), H4 and H24 in 24 piglets/group. A jugular blood sampling was drawn at 30 min after the intervention in 20 additional piglets/group in order to measure plasma cortisol. Piglets were weighed at birth, just before the intervention and at weaning. Treatments were compared by use of linear model with treatment as fixed effect and litter as random effect. Behavioural data were transformed and analysed the same way, taking into account the period of observation as a repeated effect. Movements during the procedures did not differ among treatments. Mean intensity of vocalisations was higher during castration than tail docking (104.7 ± 0.4 vs 94.2 ± 1.0 dB, respectively) and was significantly higher at castration when piglets were previously tail docked (106.0 ± 0.8 vs 103.5 ± 1.0 dB, for TC and C respectively, $P<0.01$). Cortisol was higher in TC than in T pigs (185.1 ± 17.6 vs 96.2 ± 6.8 ng/ml respectively, $P<0.001$) with C pigs being intermediate (148.5 ± 12.3 ng/ml, $P<0.1$). After the procedure, TC piglets spent more time sitting than T piglets ($P<0.05$) with C pigs being intermediate; this was particularly the case during the first hour post-intervention. T and C piglets spent more time standing than TC piglets ($P<0.01$) and exploring was more frequent in T piglets than in C piglets ($P<0.05$). C piglets presented more ample tail movements and less trembling tails than T piglets ($P<0.001$). Our results suggest that acute pain was the highest in TC pigs, slightly lower in C pigs and more clearly lower in T pigs.

Associations between a behavioral attitude score and health events in group-housed, preweaned dairy calves

M. Caitlin Cramer[1] and Theresa L. Ollivett[2]

[1]University of Wisconsin-Madison, Dairy Sci, 1675 Observatory Drive, Madison, WI 53706, USA, [2]University of Wisconsin-Madison, School of Vet Med, 2015 Linden Drive, Madison, WI 53706, USA; mccramer@wisc.edu

Recognizing behavioral changes during illness may improve the ability of producers to identify sick calves. While research has been conducted regarding the effect of respiratory disease on behavior, disease was typically defined only using clinical signs. Thoracic ultrasonography has been validated for the identification of pneumonia in dairy calves and allows for the classification of subclinical pneumonia, which may have important behavioral implications. The objective of this study was to determine associations between a behavioral Attitude Score (AS; reflects animals' demeanor) on the Wisconsin Calf Health Scoring App and four different disease categories: diarrhea, clinical pneumonia (CP), subclinical pneumonia (SCP), and upper respiratory tract infection (URT) in preweaned group-housed dairy calves. Preweaned dairy calves (n=110) on a commercial dairy farm in Ohio, USA were enrolled upon entry to an automated milk feeder barn, in which calves were housed in groups. Calves were 29±19 (median±stdev) days of age at the time of study enrollment and were followed for 4 weeks. During biweekly health exams, the AS was recorded, and consisted of normal (0), dull but responds to stimulation (1), depressed/slow to stand/ reluctant to lie down (2), and unresponsive to stimulation (3). Health exams included use of the Wisconsin Calf Scoring System, which yields a clinical respiratory score (CRS), a thoracic ultrasound score (US), and fecal score (range 0-1; 0 normal/ formed feces; 3 watery feces). For CRS, the nose, eyes, ears, cough, and rectal temperature were assigned a score ranging from 0 (normal) to 3 (abnormal); scores for all categories were summed to obtain an overall CRS. A calf was considered CRS+ when the CRS was 4 or greater. The US ranged from 0 (normal) to 5 (abnormal); calves with US of 3 or greater were considered to have pneumonia, and thus were US+. The CRS and US were used to categorize respiratory disease into four types: CP (US+,CRS+), SCP (US+,CRS-), URT (US-,CRS+), and none. A generalized linear model (PROC GENMOD in SAS) was used to determine if AS (outcome) was associated with respiratory disease (four categories), fecal score, and age. Calf was included as a repeated measure. Calves with CP were more likely to have a greater AS compared to calves with no respiratory disease (P<0.001). Calves with diarrhea were more likely to have a greater AS compared to calves without diarrhea (P<0.01). Age was not significant (P>0.05). Calves with CP and diarrhea have behavioral changes during illness. More research is needed to determine the impact of different categories of respiratory disease on behavior.

Animal personality: what does it mean for our understanding of animal welfare?

Lisa Collins
University of Lincoln, School of Life Sciences, Joseph Banks Laboratories, LN7 6DL, Lincoln, United Kingdom; lcollins@lincoln.ac.uk

Animal welfare is at heart a science of the individual. The propensity of individuals to suffer is central to our understanding and quantification of welfare issues. However, the scales of the systems studied in animal welfare science can be wildly different – from individual dogs presenting with behavioural problems at a clinic, to welfare hazards impacting an entire species, breed, or line that can add up to billions of animals worldwide. We are only just starting to uncover the myriad ways in which animal personality alters our understanding of how animals perceive the world around them. The individual differences highlighted by results so far suggests a level of internal complexity previously unconsidered. Measuring welfare at this level is a clear reminder that individuals are indeed individual, and assumptions of uniformity are unfounded. In this presentation, I will highlight the common pitfalls in personality assessment, and I will consider how developments in the field of animal personality have led us to uncover personality-related differences across a wide range of species in response to welfare hazards, in the extent to which individuals express, or potentially even experience pain, in how they interact with other individuals within a social group, leading to changes in group dynamics, and in how personality interacts with mood to influence information processing biases. There are clear implications of animal personality for group- or population-level measures, and we must consider how we can avoid losing the individual when considering welfare at scale. Experimental design and approaches to statistical analysis can impact hugely on the extent to which we are able look in detail across multiple scales, as well as having serious implications for the reliability and accuracy of our conclusions. The extent of individual variation has consequences for experimental design in all areas of animal science, particularly relating to the calculation of sample sizes, but also more broadly throughout the experimental process. Finally, the issues and benefits of a multi-scale perspective in animal welfare, and the ways in which we could approach an individualised approach to population-level measures of welfare will be considered.

Fearfulness and access to the outdoor range in commercial free-range laying hens

Hannah Larsen[1], Greg Cronin[2], Paul Hemsworth[1], Carolynn Smith[3], Jean-Loup Rault[1] and Sabine Gebhardt-Henrich[4]
[1]*Animal Welfare Science Centre, University of Melbourne, Monash Rd, 3010, Parkville, Australia,* [2]*University of Sydney, 425 Werombi Rd, Camden, NSW, 2570, Australia,* [3]*Macquarie University, Balaclava Rd, North Ryde, NSW, 2109, Australia,* [4]*ZTZH Division of Animal Welfare, VPH Institute, University of Bern, Burgerweg 22, 3052, Zolikofen, Switzerland; hlarsen@student.unimelb.edu.au*

Recent studies on free-range laying hen behaviour indicate that upwards of 80-85% of the flock will access the range at least once, and many will do so on a daily basis. Few studies have investigated the relationship between range access and fearfulness, a factor suggested to be negatively correlated with range access, and none have investigated this on commercial free-range farms. This study investigated the relationship between range access, as indicated by time spent outside, and fearfulness based on a tonic immobility test, in free-range laying hens on a commercial farm. We hypothesised that fearfulness would be negatively associated with range access, and hens that utilised the outdoor range most would be less fearful than hens accessing the range less or not at all. An 18,000 hen commercial flock was subdivided to isolate a group of approximately 2,000 40-wk old HyLine-Brown hens which had access to the outdoor range from 21 wks. Radio frequency identification (RFID) antennas were placed at all pop holes from the shed to the wintergarden and from the wintergarden to the outdoor range. Hens were randomly chosen (n=129) and tested for tonic immobility (TI) duration prior to tagging with RFID chips. TI tests were ended at 5 min if the bird did not right itself, with a maximum of five attempts to induce. Ranging data were collected over subsequent 13 days and hens were grouped into three categories; no rangers (NR) low rangers (LR) and high rangers (HR). TI durations were analysed using Kaplan-Meier survival analysis in SPSS with ranging duration (NR, LR and HR) as the fixed factor. No rangers (n=22) never accessed the outdoor range, LR (n=34) spent up to 41 h outside over the course of the study and HR (n=73) spent from 41 to 82 h outside. TI durations were higher for NR (254.3 ± 14.0 s) than LR (159.4 ± 18.0 s; $X^2=8.55$, P=0.003) and HR hens (185.6 ± 13.0 s; $X^2=5.13$, P=0.024). However there was no significant difference in TI durations between LR and HR hens ($X^2=1.66$, P=0.198). These findings indicate that hens that never entered the outdoor range may be more fearful, as indicated by response to TI test. However, hens that varied in their duration of range access did not significantly differ in terms of fearfulness. Fearfulness, as based on TI duration, may be an independent or mediating variable in relation to accessing the range, but not the duration of range access.

The relationship between heat stress tolerance and temperament of dairy cows

Maria Jorquera Chavez and Ellen Jongman
University of Melbourne, Animal Welfare Science Centre, Alice Hoy Building, Parkville 3010,
Australia; ejongman@unimelb.edu.au

The response to heat stress, resulting in a reduction in milk production, varies between dairy cows. While both milk production and stress responses affect metabolic load it is not known if temperament of dairy cows affect their tolerance to heat stress. Cows were selected from a herd of 125 cows on pasture and were classified as heat-tolerant or heat-sensitive based on their reduction in milk production after 3 distinct heat events over a 6-week period in summer. The responses of 12 heat-tolerant and 12 heat-sensitive cows were observed in two common temperament tests, the 'Exit velocity test' and 'Novel object test'. In addition cows were fitted with heart rate monitors and their eye temperature was measured with an infrared camera while confined to a crush. The tests were performed in winter when the cows were not lactating. Data were analysed using a one-way ANOVA and linear regressions after appropriate transformations. Heat-sensitive cows reduced milk production by an average of 2 l after a heat event, while heat-tolerant did not lose milk production (P<0.05). Exit velocity did not differ between groups (4.2 vs 6.4 sec; P>0.05). Latency to interact with the novel object and number of interactions was not significantly different for sensitive and tolerant cows (62.1 vs 32.7 sec, P>0.05 and 2.9 vs 1.6, P>0.05). However, there was a slight negative correlation between latency to approach the novel object and heat sensitivity (r=-0.1, P=0.05). Sensitive cows tended to have a higher heart rate than the tolerant cows during the whole testing procedure (97.6 vs 73.6 bpm, P=0.064). In addition sensitive cows had higher eye temperatures than tolerant cows during confinement in the crush (38.3 vs 39 °C; P<0.01). This pilot study, with a limited number of animals, indicates that individual physiological responses to a general stressor may relate to tolerance to heat stress and the underlying physiological mechanism warrants further research.

Stereotypic behaviour in standard cages is an alternative to a depression-like response, being awake but motionless, in mice

Carole Fureix[1,2], Michael Walker[2], Laura Harper[2], Kathryn Reynolds[2], Amanda Saldivia-Woo[2] and Georgia Mason[2]
[1]University of Bristol (present address), Bristol, BS405DU, United Kingdom, [2]University of Guelph, Ontario, N1G2W1, Canada; carole.fureix@bristol.ac.uk

Depressive-like forms of waking inactivity occur in primates, horses and mice. We tested the hypotheses that for laboratory mice, being awake but motionless within the home-cage (during the dark phase, when mice are normally active), is a depression-like symptom; and that in impoverished housing, it represents an alternative response to stereotypic behaviour. We raised C57BL/6 ('C57') and DBA/2 ('DBA') females to adulthood in non-enriched (n=62 mice) or enriched (n=60 mice) cages. In-cage behavioural data, i.e. stereotypic behaviour and being 'still but awake' (immobile, but with open eyes) were collected by trained observers, using a mix of live scan- and focal-sampling in two daily 4 h blocks over repeated days. We also took a well-accepted measure of helplessness (one symptom of depression): immobility in Forced Swim Tests (FST). Mice are placed individually once in an inescapable cylinder filled with water and videotaped for 6 min. Every tape was observed by one of us and an assistant (blind to treatment and hypothesis) to score each mouse's total duration of immobility: floating with at least 3 legs motionless (a variable known to be amplified by stressors, alleviated by antidepressants and correlated with other depression-like signs, e.g. anhedonia). Data were analysed using general linear mixed models. We predicted that being still but awake would: reduce with environmental enrichment; be more pronounced in C57s, the strain most prone to learned helplessness; predict longer immobility times in FSTs; and negatively covary with stereotypic behaviour. As predicted, non-enriched subjects spent more time than enriched mice being still but awake ($F_{1, 34}$=20.369; P<0001), especially if displaying relatively little stereotypic behaviour ($F_{1, 54}$=20.532; P<0.0001). C57 mice also spent more time still but awake than DBAs ($F_{1, 56}$=6.993; P=0.011). Furthermore, even after statistically controlling for housing type and strain, this behaviour very strongly tended to predict increased immobility in FSTs ($F_{1, 100}$=3.849, P=0.052), while high levels of stereotypic behaviours in contrast predicted low immobility in FSTs ($F_{1, 101}$=7.378, P=0.0078). For non-enriched mice, being motionless during the active phase of the day, despite being awake, is thus an alternative to stereotypic behaviour, and seems to be an alternative reaction style. It also predicts helplessness in Forced Swim Tests, and as such could reflect depression-like states. These findings have implications for mouse welfare, and potentially also for the 'coping hypothesis' of stereotypic behaviour.

Is leaner meaner? The effect of leanness on pig-pig aggression and pig-human interactions in finisher gilts

Lorri Jensen, Carly O'malley, Sarah Ison, Kaitlin Wurtz, Juan Steibel, Ronald Bates, Cathy Ernst and Janice Siegford
Michigan State University, Animal Science, 474 S. Shaw Ln, East Lansing, MI 48824, USA; jensenl9@msu.edu

Pigs in production have typically been selected for traits, such as growth rate and leanness, without consideration for possible negative behavioral side-effects including social aggression and fearfulness of humans. The purpose of this study was to determine if there was a relationship between leanness and aggression in finisher pigs, and also to determine if pigs that were more aggressive to each other interacted with humans more frequently or in different ways than less aggressive pigs. We predicted that pigs that were more aggressive to each other would be leaner and more likely to interact with humans. A total of 270 pure Yorkshire gilts were mixed into new groups of ~14 pigs at the finisher stage in 4.83×2.44 m pens. Lesions were defined as fresh, red marks on the skin and were counted approximately 24 h after pigs were mixed into groups upon entry into the finishing unit. Total numbers of lesions on the skin of each pig were used to measure pig-pig aggression. To measure leanness, backfat measurements were taken from individual pigs via ultrasound. Pigs' interactions with humans were assessed by having an observer enter the pigs' home pen for 9 min and record which pigs were biting, levering, or nosing the observer at 30-second intervals. A generalized linear mixed model was used to investigate the relationship between lesions and backfat and also between lesions and total interactions with humans. The fixed effects in the model were observer, repetition, and either backfat or number of human interactions. Pen was included as a random effect and weight was a covariate. The response variable was the number of lesions on the pig 24 h after being mixed into a new group at the finisher stage. Results suggested there was no relationship between lesions and backfat (P=0.584) or between backfat and interactions with humans (P=0.607). There was also no relationship between lesions and total interactions with humans (P=0.509) or with total number of bites (P=0.196). Leaner pigs were not necessarily more aggressive than pigs with more fat, and pig-pig aggression was not related to the how pigs interacted with humans. This suggests that pig-pig aggression is not related to exploratory behavior in pigs or fear of humans. These results are relevant to the field of animal breeding and welfare since they indicate that producers can select for less aggressive animals without impacting leanness or fearfulness of humans. An additional 270 pigs will analyzed to add more power to these results. Future research investigating strains of pigs with backfat variation could confirm these results.

Personality, plasticity, and predictability of shelter dogs during interactions with unfamiliar people

Conor Goold and Ruth C. Newberry
Norwegian University of Life Sciences, Animal and Aquacultural Sciences, Ellingsrudlia 15, 1400, Ski, Norway; conor.goold@nmbu.no

Dog personality assessments are often used to make predictions about future behaviour based on behaviour at one time point. The predictive accuracy of such assessments can be low, suggesting a need for improved knowledge of behavioural consistency and variability. The behavioural reaction norms framework divides individual differences in behaviour into personality (individual differences in average behaviour), plasticity (individual differences in behavioural change) and predictability (individual differences in residual variability). We explored these measures in shelter dogs based on behaviour during interactions with unfamiliar people and examined the influence of dog age and sex. Data on 474 dogs (mean±SD age = 3.51±2.91 years; 254 males; 220 females) in the first 10 days post-admission were collected from a shelter's longitudinal assessment protocol. Each day, shelter staff recorded how dogs responded towards strangers (8.82±0.98 observations/dog) during spontaneous interactions using a 13-point graded ethogram from the most (1=Friendly) to least (13=Aggressive) sociable. Observations on each dog were not all completed by the same staff member, although agreement between staff on the use of the ethogram in dedicated video coding sessions was high (consensus=0.76; $P<0.001$). Data were analysed with an ordinal Bayesian mixed model using JAGS in R. Results were considered significant when 95% Highest Density Intervals (HDI) of regression parameters did not include zero. Standard deviations of personality (HDI: 0.71, 0.88), plasticity (HDI: 0.05, 0.09) and predictability (HDI: 0.33, 0.49) parameters were all non-zero. Behaviour became more sociable with days post-admission (HDI: -1.6, -0.6). Increasing age was associated with more sociable personalities (HDI: -0.32, -0.13), a greater increase in sociability over time (increased plasticity; HDI: -0.07, -0.04), and larger residual variances (reduced predictability) in sociability (HDI: 0.002, 0.14), although the latter HDI approached zero. No sex differences in these measures were detected. Our results demonstrate that, in addition to showing personality differences, dogs also differ in plasticity and predictability, and that these measures vary with age. Conducting brief assessments over multiple time points allows integration of these measures, which should facilitate accurate predictions of future behaviour.

Individual-based variation in grazing behavior of rangeland-raised beef cows

Laura E. Goodman[1], Andres F. Cibils[2] and Lyndi Owensby[2]
[1]Oklahoma State University, 303F Agriculture Hall, Stillwater, OK 74078, USA, [2]New Mexico State University, P.O. Box 30003, MSC 3I, Las Cruces, NM 88003, USA; acibils@nmsu.edu

Individual variation in grazing behavior of rangeland-raised cows was characterized using the behavioral syndromes framework. Behavioral syndromes are suites of correlated behaviors that are consistently different among individuals across situations (e.g. foraging in confinement vs pasture), context (e.g. foraging, mating, avoiding predators) and time. The objective of this study was to test cross-situation consistency of behavioral differences among individuals belonging to two breeds of beef cattle. We first measured feed supplement consumption rate in individual stalls by recording the time it took individual cows to consume 1 kg of cotton seed cake while confined. Multiple SCR tests were conducted on each cow. We then selected: (1) 30 Angus × Hereford (AH) mature lactating cows with highest, intermediate, and lowest SCR (n=10/group) from a herd of 138 cows; and (2) 12 Brangus (BR) mature lactating cows with highest and lowest SCR from a herd of 99 cows (n=6/group). All 42 animals were fitted with GPS collars set to log animal positions at 5 min intervals, and were tracked for either 2.5 months (AH at a grassland site in 2012) or 1.5 months (BR at a desert site in 2013). Collared animals grazed together with the rest of the herd in pastures that ranged in size from 1,800 to 2,500 ha. Spatial behavior variables measured included: distance traveled (km/day); area explored (ha/day); spatial search pattern (m/ha); mean and maximum distance traveled from water (km/day); time spent at water (min/day); and time spent in each of three vegetation types (min/day). We recorded calf weaning weights for all cows and lifetime calf production (# calves) for BR cows. We calculated Pearson correlation coefficients between these variables and SCR for each individual. In AH cows, SCR was positively correlated with daily area of the pasture explored (r=0.41; P<0.10), distance traveled from water (r=0.52; P<0.05), avoidance of bottomlands (r=-0.63; P<0.05), preference for side slopes (r=0.43; P<0.10) and ridgetops (r=0.47; P<0.05). SCR of AH cows was also positively correlated with calf weaning weights (r=0.45; P<0.05). In BR cows, SCR was positively correlated with daily distance traveled (r=0.90; P<0.05), and with the total number of calves weaned over the cow's lifetime (r=0.53; P<0.10). Differences in foraging behavior among cows persisted across situations (confinement vs rangeland) and were apparently of consequence to the cow's reproductive performance regardless of the breed considered.

Sociability in grazing dairy cows is related to individual social network attributes and behavioural synchrony

Kees Van Reenen[1], Joop Van Der Werf[1], Lysanne Snijders[2], Bas Engel[3], Inma Estevez[4,5] and Ane Rodriguez-Aurrekoetxea[5]
[1]Wageningen University, Livestock Research, P.O. Box 338, 6700 AH Wageningen, the Netherlands, [2]Wageningen University, Behavioural Ecology Group, P.O. Box 338, 6700 AH Wageningen, the Netherlands, [3]Wageningen University, Biometris, P.O. Box 16, 6700 AA Wageningen, the Netherlands, [4]IKERBASQUE, Basque Foundation for Science, Maria Diaz de Haro 3, 48013 Bilbao, Spain, [5]Neiker-Tecnalia, Department of Animal Production, P.O. Box 46, 01080 Vitoria-Gasteiz, Spain; kees.vanreenen@wur.nl

Sociability is a personality trait that refers to the motivation of individuals to remain close to conspecifics. Previous research in housed dairy cows suggests that sociability may affect behaviour at group level. In the present study the relationship between sociability, individual social network attributes and behavioural synchrony in grazing dairy cows was examined, involving three groups of 20 cows each, kept in three grazing systems. All cows were individually subjected to a social runway test (SRT) twice, with 8 weeks between tests. Measures of sociability recorded in the SRT were latencies to reach 5 m (Lat5m) and 2 m (Lat2m) from a group of herd mates. Repeatabilities of Latm5 and Latm2 (log transformed) were 0.48 and 0.30 respectively ($P<0.01$, likelihood ratio test, REML mixed model analysis). Locations (XY coordinates) of individual cows on pasture were obtained during 105 visual scans over a 14-day period with the use of Chikitizer software. Location data transformed into proximity data were used for social network analysis by UCINET and SOCPROG software, providing individual connectivity metrics for each cow, including 'Strength' and 'Eigenvector centrality' (EVC). All cows were equipped with a sensor attached to the leg (IceRobotics, UK), recording standing/lying behaviour. Leg sensor data were used to calculate the level of synchrony between the behaviour of each cow and the behaviour of the rest of the herd while on pasture per day, and daily values were averaged over the 6-month grazing season. Average log transformed values of Lat5m and Lat2m across SRTs were introduced as dependent variables in analysis of covariance with fixed effects for grazing system, lactational stage and parity, and either level of behavioural synchrony or individual network properties as a covariate. Lat5m and Lat2m were negatively associated with 'Strength' ($P<0.05$) and EVC ($P<0.01$). Latm2 was negatively associated with level of behavioural synchrony ($P<0.05$). Thus, short SRT latencies, putatively reflecting high sociability, corresponded to close proximity, high connectedness to herd mates, and high behavioural synchrony. Our findings suggest that sociability is a stable personality trait in grazing cows that influences behavioural dynamics at group level.

Fear and its relationship to social aggression in group-housed swine

Carly O'malley[1], Kaitlin Wurtz[1], Juan Steibel[1,2], Catherine Ernst[1], Ronald Bates[1] and Janice Siegford[1]
[1]*Michigan State University, Animal Science, 474 S Shaw Ln, E Lansing, MI 48824, USA,* [2]*Michigan State University, Fisheries & Wildlife, 480 Wilson Rd, E Lansing, MI 48824, USA;* omalle50@msu.edu

As swine producers move towards group housing of sows during gestation, it is increasingly important to identify traits influencing levels of aggression in social groups. However, as we select for improved sociability, we do not want to unintentionally cause negative outcomes, such as increased fear. Pigs' reaction to a novel object was compared to post-mixing lesion scores in 257 Yorkshire barrows (~4 mo of age at testing) to determine if fear responses were related to social aggression. Individual pigs were moved to a novel arena (4.3×3.4 m, lines on the floor 0.5 m and 1 m away from the novel object). Pigs were allowed a 1-min acclimation period in the arena, followed by 5-min exposure to the novel object (basketball). Reaction to the novel object was quantified using the following measures: latency to approach within 1 m, latency to approach within 0.5 m, latency to touch novel object, number of 1 m line crosses, number of 0.5 m line crosses, number of times object touched and number of vocalizations. These responses were compared to number of new skin lesions (defined as fresh, red marks on the skin) obtained 24 h after being mixed into a new social group at ~10 wk of age. To explore the relationship between social aggression and fear response to novelty, two linear models were initially tested for fit to the data. For both, the response variable was total number of lesions 24 h after mixing, weight at mixing was a covariate, and home pen was a random effect. The first model simply included fixed effects of pre-and post-mix lesion scorers and repetition (R^2=0.34); the second also incorporated the variables quantifying reaction to the novel object described above (R^2=0.35). The increase in R^2 value due to including novel object test variables in the model was not significant (ANOVA; P=0.52); therefore the model including novel object variables was used in subsequent analyses. Novel object test variables were found to be unrelated to skin lesion scores: latency to 1 m (P=0.19), latency to 0.5 m (P=0.94), latency to touch (P=0.97), number of 1 m crosses (P=0.93), number of 0.5 m crosses (P=0.94), number of touches (P=0.97), number of vocalizations (P=0.88). The covariate of weight was also found to be unrelated (P=0.21). The results suggest that there is no relationship between fear response to a novel object and social aggression, and therefore fear and inter-pig social aggression may be regulated by different biological processes. Therefore, swine breeding programs could select for pigs, including gestating sows, that are less socially aggressive in group housing systems without unintentionally affecting their fear response.

Breed, gender and age variation in behavioural traits in domestic cats (*Felis silvestris catus*)

Bjarne O. Braastad[1], Silja C. Brastad Eriksen[1] and James A. Serpell[2]
[1]*Norwegian University of Life Sciences, Dept. of Animal and Aquacultural Sciences, P.O. Box 5003, 1432 Ås, Norway,* [2]*University of Pennsylvania, School of Veterinary Medicine, Center for the Interaction of Animals and Society, 3900 Delancey Street, Philadelphia, PA 19104, USA; bjarne.braastad@nmbu.no*

New cat owners need guidance on breed-typical behaviour in the domestic cat. Therefore, we characterized breed variation in behavioural traits in Norwegian cats, while controlling for gender and age. Owner-reported assessments of behaviour and temperament (n=1,204) were obtained from an online survey with 99 items scored on a 1 (never) to 5 (always) scale using a Norwegian translation of the Fe-BARQ, Feline Behavioural Assessment and Research Questionnaire. Iterated factor analyses on the item scores revealed 22 factors or behavioural traits, accounting for 66.3% of the overall variance. Effects on these traits by breed (random effect) and gender and age (fixed effects) were analysed by Generalized linear mixed models (GLMM: n=1,137; 15 breeds with N ranging 12-80 plus domestic shorthair (DSH, n=410) and longhair (DLH, n=287). Significant gender differences were found for 11 traits, the largest difference was found for 'Sociability with non-household cats' (males: 2.7 ± 1.2 (M$\pm$SD), females: 2.2 ± 1.2; F=29.2, P<0.0001). Seven traits showed significant age effects, which was most pronounced for 'Activity/playfulness', decreasing with age (F=31.2, P<0.0001). Significant breed effects were found for 21 of the 22 behavioural traits. The following five traits showed the most pronounced breed differences: 'Fear of unfamiliar dogs or cats' (mean 2.60 ± 1.24) was highest in DSH and DLH, and lowest in Burmese, Egyptian Mau and Oriental cats (Kruskal-Wallis test: χ^2=103.4, P<0.0001). 'Sociability with people' (mean 3.40 ± 1.13) was highest in Burmese, Cornish Rex and Siamese cats, but lowest in DSH and DLH (χ^2=96.0, P<0.0001). 'Touch sensitivity/owner-directed aggression' (mean 1.38 ± 0.56) was most pronounced in DSH and DLH, Maine Coon and Siberian cats, and particularly infrequent in Abyssinian, Egyptian Mau and Oriental cats (χ^2=93.5, P<0.0001). 'Directed vocalization' (mean 4.05 ± 0.97) was most frequent among Burmese, DSH, DLH, Maine Coon, Bengal, Norwegian Forest Cat, Ragdoll and Siberian cats, and least in Devon Rex, Orientals, Sacred Birman and Persian (χ^2=81.5, P<0.0001). 'Activity/playfulness' (mean 3.37 ± 0.71) was most frequent among Abyssinian, Bengal, Burmese, Devon Rex, Oriental, and Siamese cats, and least frequent among British Blue, DLH and, in particular, Persian cats (χ^2=70.2, P<0.0001). These results may aid prospective cat owners in selecting a cat breed most suited to their expectations, and they may serve as a base when breeders wish to modify the breed-typical behaviour.

Patterns of range access of individual broiler chickens in commercial free-range flocks

Peta Taylor[1], Peter Groves[2], Paul Hamilton Hemsworth[1] and Jean-Loup Rault[1]
[1]The University of Melbourne, Animal Welfare Science Centre, Parkville, 3010 Victoria, Australia,
[2]The University of Sydney, Poultry Research Foundations, Camden, NSW 2570, Australia;
petat@student.unimelb.edu.au

Accessing an outdoor range may affect broiler welfare. The degree of these effects is likely related to individual ranging behaviour. However little is known about the ranging behaviour of individual free-range broiler chickens or the effects of seasonal variation. This study investigated ranging patterns of ROSS 308 broiler chickens on a commercial free-range farm. Individual range use was monitored via radio frequency identification (RFID) on two winter (W) and two summer (S) flocks of 6,000-12,000 mixed sex chickens; 300 tagged individuals per flock. Frequency and duration of range visits were recorded from day 1 of range access (21 days of age) until first slaughter pick up (32 days of age, W flocks) or final pick up (42 days of age, S flocks). Weight and sex were recorded at the same time points. Ranging data were analysed using chi square analysis and Spearman correlations, and weight data using GLM in SPSS. The total number of tagged birds that accessed the range was 32% and 41% in W flocks and 76% and 87% in S flocks. In all flocks a consistent population of the tagged birds accessed the range only once (6-14%). Less than 5% of tagged birds accessed the range every day when the doors were opened, but 25% and 47% of tagged birds accessed the range every other day, in W and S flocks respectively. For S flocks, 93% of tagged birds that ranged did so prior to first pick up despite an additional 8-9 days of range access, suggesting that ranging patterns may be established by 8-10 days of range access. There was no effect of sex or weight on range use (sex: $X^2=3.63$, $P>0.05$; weight: $X^2=0.02$, $P>0.05$). However birds that used the range more than once had lower body weights than birds that only accessed the range once and birds that never accessed the range at first pick up (accessed range more than once: 1.58±0.02 kg; accessed range once: 1.72±0.05 kg; never ranged: 1.76±0.02 kg). Frequency and duration of individual range access varied; between 1 and 34 visits/day (average 4.02±0.06 visits) and between 10 s and 11 h spent in the range/day (average 82.79±1.41 min). Our results show that there are large inter-individual differences in range access. Although there is a clear seasonal effect, there is still unexplained variance in individual ranging behaviour within seasons. A better understanding of the relationship between individual ranging behaviour, bird characteristics and temperament is required so that optimal range use in commercial broiler flocks can be achieved.

Links between owner adult attachment style and the behaviour of the dog during challenging situations

Therese Rehn and Linda J. Keeling
Swedish University of Agricultural Sciences, Department of Animal Environment and Health, P.O. Box 7068, 750 07 Uppsala, Sweden; therese.rehn@slu.se

The aim of this study was to investigate if an owner's adult attachment style (AAS) influences how their dog interacts and obtains support from them during challenging events. A person's AAS describes how they perceive their relationship to other people, but it may also reflect their caregiving behaviour. In this study, we measured the AAS of 51 Golden retriever owners, using the Adult Attachment Style Questionnaire (ASQ). Also, we observed the dog-owner dyads in response to four different challenging events (selected from the Swedish Dog Mentality Assessment and the Dog Behaviour and Personality Assessment tests) as well as in a separation and reunion test. The challenging events included two different sudden surprises (visual (V) and auditory (A)) and two tests where a strange-looking person slowly approached the dyad (a person dressed as a 'ghost' (G) and a person dressed in a coat, hat and sunglasses (C)). The dog was then left alone in a novel environment for three min and its behaviour observed. The interactions between the dog and owner were observed both before and after separation. Spearman rank correlation tests were performed using the data on the owner's AAS and the behaviour of their dog during the challenging events as well as the behaviour of dog and owner before and after separation. The more confident the owner (according to their score in the ASQ), the longer the dog was oriented to the two sudden stressors (V: $P=0.01$, $R^2=0.35$; A: 0.03, $R^2=0.30$) and the shorter its latency to approach the visual stressor ($P=0.001$, $R^2=-0.28$). Also, these dogs were more active upon reunion with the owner ($P=0.04$, $R^2=0.28$). The more ambivalent the owner, the longer the dog oriented to the owner during challenging events (V: $P=0.005$, $R^2=0.38$; C: $P=0.02$, $R^2=0.33$), and the dog showed less tail wagging upon reunion ($P=0.04$, $R^2=-0.29$). The more avoidant the owner, the longer the latency for the dog to approach the auditory stressor ($P=0.02$, $R^2=0.34$). These dogs also barked more during separation ($P=0.004$, $R^2=0.40$), and such owners initiated more physical contact with their dogs upon reunion ($P=0.04$, $R^2=0.28$). To conclude, links between owner AAS and dog behaviour in the tests were found. This implies that dogs may develop different strategies to handle challenging situations together with the owner, based on the type of support they get from their owner. Moreover, it indicates that a person's caregiving behaviour towards their dog is linked to how they relate to and interact with other people.

Changes in autonomic balance of pigs in different behavioral contexts with special focus on coping type

Annika Krause, Birger Puppe and Jan Langbein
Leibniz Institute For Farm Animal Biology, Behavioural Physiology, Wilhelm-Stahl-Allee 2, 18196
Dummerstorf, Germany; krause@fbn-dummerstorf.de

Analysis of heart rate (HR), blood pressure (BP) and their respective variability (HRV, BPV) may provide a sensitive measure of autonomic activity. This allows the analysis of changes in sympathovagal balance related to diseases, psychological stressors or individual characteristics such as coping strategies. Characterization of piglets based on the degree of resistance they display in a so-called Backtest is, to a certain extent, predictive of their coping responses to several challenges in later life. It is unknown, however, whether these individual coping characteristics of pigs are also reflected in their autonomic balance. To get deeper insight into the relationship between the coping type and physiological features, the autonomic state of 14 pigs with either reactive (RE; n=7) or proactive (PR; n=7) coping type was studied in different behavioural contexts. The pigs were equipped with a telemetric device and ECG, BP and behaviour were recorded for 1 hour daily (10 days). Three different behavioural contexts (resting, feeding and exploration), each 5 minutes in length, were chosen for the calculation of HR, HRV (long term variation (SDNN) and short-term variation (RMSSD) of HR; indicative of vagal activity), BP and BPV (standard deviation of BP; indicative of sympathetic activity). Data were statistically analysed using the GLIMMIX procedure in SAS applying a model with main effects experimental day, behavioural context, coping type and two way interactions. Repeated measurements on the same animal were taken into account. The results reveal an impact of the interaction between coping type and behavioural context on HR ($F_{2,297}$=12.3, P<0.001), RMSSD ($F_{2,288}$=6.2, P<0.01) and BPV ($F_{2,245}$=2.9, P<0.05). PR-pigs showed higher HR (P<0.05, 108.5±8.9 bpm) and lower RMSSD (P<0.01, 39.5±19.1 ms) during resting compared to RE-pigs (HR: 92.5±6.1 bpm; RMSSD: 75.1±24.5 ms). During feeding, PR-pigs displayed elevated BPV (P<0.05, 12.8±2.3 mmHg) compared to RE-pigs (10.1±2.1 mmHg). The data suggest that the vagal tone determines the differences between the coping types during resting, whereas sympathetic tone determines the differences during feeding. This is the first investigation of changes in the autonomic balance, concerning coping types in pigs using telemetry. PR- and RE-pigs differed in their autonomic balance, underlining the importance of taking individual differences into account. These findings may serve as a basis for future studies concerning autonomic control during various behavioural situations in the context of stress and animal welfare.

Responses in a social isolation test are related to spontaneous home pen activity in dairy cattle

Jill R D Mackay[1], Kees Van Reenen[2] and Marie J Haskell[1]
[1]SRUC, West Mains Road, EH9 3JG, United Kingdom, [2]Wageningen University, P.O. Box 65, 8200 AB Lelystad, the Netherlands; jill.mackay@sruc.ac.uk

In order to be meaningful, tests of animal personality must not only be repeatable but also reflective of an underlying trait that is persistent outside of the test environment. In this study we investigated whether a well-used test of social motivation (SOC) in dairy cattle reflected their spontaneous home pen behaviours, e.g. did more sociable cows behave differently in the home environment? Ninety nine dairy cattle housed at the Dairy Research Centre, Leeuwarden (NL) were tested. Each cow was separated from the herd, with the herd confined at the opposite end of a 12 m runway. The subject was released and latency to rejoin the herd and agitation behaviours shown (such as vocalisations, frequency crossing a line 5 m from the herd, etc.) in a 5 min period were recorded. Home pen activity using tri-axial accelerometer based activity monitors and milk production traits were recorded for 40 days prior to the start of SOC testing (n=74 cows). Multiple regressions were used to test associations between behaviours relating to social motivation and long-term spontaneous home pen activity prior to testing. Most behaviours in the SOC test were affected by the cow's lactation stage at testing and, after accounting for lactation stage at testing, were not significantly associated with spontaneous home pen behaviours. However there was a positive association for cows who frequently crossed the threshold line 5 m with shorter standing bouts, longer overall standing duration in a day, producing less milk per day and were less variable in their daily number of lying bouts (Model: $R^2_{adj}=0.19$, $F_{4,68}=5.08$, P=0.001; average standing bout duration $t^{(64)}=-3.08$, P<0.003; average daily standing duration $t^{(64)}=2.63$, P=0.011; average milk yield per day $t^{(64)}=-3.53$, P<0.001; standard deviation of number of lying bouts per day $t^{(64)}=-3.01$, P=0.004). This model explained 19% of the variation in the number of times the cow crossed the 5 m line. These results indicate a relationship between some SOC test responses and spontaneous home pen behaviour but may reflect an underlying activity motivation rather than sociability. The results suggest that measurement of reproductive stage and lactational status need to be taken into account when assessing some aspects of personality in cows and question the validity of the runway test in lactating dairy cattle.

Why and how should we assess pet rabbit personality?

Clare Frances Ellis[1,2], Wanda Mccormick[2] and Ambrose Tinarwo[3]
[1]University of Northampton, Science and Technology, Boughton Green Rd, Northampton, NN2 7AL England, United Kingdom, [2]Moulton College, Animal Welfare and Equine, West Street, Moutlon, Northamptonshire, NN3 7RR, United Kingdom, [3]Hadlow College, Animal Management, Tonbridge, Kent, TN11 0AL England, United Kingdom; clare.ellis@moulton.ac.uk

Domestic rabbits (*Oryctolagus cuniculus*) have received relatively little attention for personality and temperament research to date, despite being a popular pet in the UK. The field of animal personality research is still in its infancy and there is much discussion around appropriate methods of assessing personality, due in part to the many reasons for exploring this phenomenon, for example, for cross species comparisons or assessing an animal's suitability for a particular role or job. Domestic rabbits make an interesting candidate for personality studies due to their domestication being predominantly influenced by a desire for morphological or physical traits, rather than for behavioural characteristics, as with domestic dogs. Additionally, rabbits appear to be relinquished by their owners in high numbers and a recent survey suggested that one reason for this was difficulty in bonding a newly acquired rabbit to a current pet. Personality studies may help to raise owner awareness of a rabbit's individual characteristics, which may support the formation of a human-animal bond and in turn reduce relinquishment. A deeper understanding of a rabbit's profile may also support owners to select a suitable rabbit when acquiring a companion for one currently owned. In addition to making a case for personality assessments being needed at the point of acquiring pet rabbits, the presentation will review companion animal personality studies and describe methods currently being developed for the assessment of personality and temperament in pet rabbits.

Individual variation in concentrate consumption rate of ewes

Stine Grønmo Vik and Knut Egil Bøe
Norwegian University of Life Sciences, Department of Animal and Aquacultural Sciences, P.O. Box 5003, 1432 Ås, Norway; stine.vik@nmbu.no

Ewes housed during the winter season are usually provided concentrate in a feeding trough where all animals eat simultaneously. Here, we assume that each individual ewe eats the targeted amount of concentrate and do not consider the variation of individual feed intake. To examine the variation within and between individuals in concentrate consumption rate and the effect of moderate social competition, two experiments were conducted. Eight ewes (four primiparous and four multiparous), were fed 250 grams of concentrate twice daily for 60 seconds. The ewes ate their ration of concentrate individually for three days to examine individual variation in consumption rate (experiment 1). In experiment 2, the same ewes were fed in pairs (7 tests per ewe during 5 days) to examine the effect of competition. Feed consumption rate was significantly higher when ewes were tested in pairs (mean 183.0±8.9 g/min) than when tested individually (mean 172.0±10.5 g/min). In fact, all the ewes increased their consumption rate when tested in pairs. The variation in consumption rate between individuals was higher when ewes were fed in pairs (CV=9.4%) than when fed individually (CV=5.6%). Within individuals, the variation in feed consumption rate decreased when the ewes were fed in pairs (CV=4.9%) compared to individually fed (CV=6.0%). Parity and body weight had no significant effect on feed consumption rate when ewes were fed individually or in pairs, but seems to rather be an individual characteristic. When tested in pairs, no instances of vocalizations and displacements were recorded.

Backtest behavior in pigs – more than noise but less than a fixed strategy

Manuela Zebunke[1], Gerd Nürnberg[1] and Birger Puppe[2]
[1]*Leibniz Institute for Farm Animal Biology, Institute of Genetics and Biometry, Wilhelm-Stahl-Allee 2, 18196 Dummerstorf, Germany,* [2]*Leibniz Institute for Farm Animal Biology, Institute of Behavioural Physiology, Wilhelm-Stahl-Allee 2, 18196 Dummerstorf, Germany; zebunke@fbn-dummerstorf.de*

Since several years applied ethologists gained more interest in the concept of animal personality and its practical implementations in animal breeding, housing and welfare. Referring to the concept of coping, Hessing et al. introduced the backtest in young piglets and hypothesized that it can detect coping strategies. Following studies revealed ambivalent results and the backtest was severely criticized. One possible explanation for ambivalent results might be the way of classification that is used in many studies. Hence, we investigated in a large number of piglets (n=3,555) several classification methods, that use different parameters (latency, duration, frequency) and different numbers of test repetitions (1, 2, 3, 4). The effect of these methods on the detection of differences between the categories was tested in a subset of 120 piglets where we also studied the inter-situational consistency of coping behavior (backtest; mixing/dominance; group testing: human approach test (HA), novel object test (NO), open door test (OD); individual testing: open field test (OF)). Our results showed that the correlation between backtest behavior and dominance was low (r_S<0.18) and not significant (P>0.07). The correlation between backtest behavior and group test behavior (HA, NO, OD) was moderate till low (r_S<0.38) with some correlations reaching significance, especially with the HA. The correlation between backtest behavior and individual test behavior (OF) was also moderate till low (r_S<0.39) with several correlations reaching significance, e.g. with locomotion, exploration, object contact. Finally, comparing the classification methods show a great similarity, i.e. 81% of the pigs were assigned to a certain category (HR=high reactive, LR=low reactive, I=intermediate) by at least 30 out of 45 classification methods. Nevertheless, the application of different methods has an impact on the results, i.e. depending on the method we found differences between the different backtest categories or not. All in all, the backtest behavior of piglets seems to be more than just noise or random behavior but less than a fixed strategy as originally supposed. It seems to show rather a coping disposition that is environmentally modulated. Finally, the results implicate that care has to be taken when classifying animals into categories. Nevertheless, the backtest could be a useful method to investigate individual adaptations to the different challenges of a farm environment.

Individual differences and the social environment affect risk of endometritis in dairy cows

Kathryn Proudfoot[1,2], Becca Franks[2] and Marina Von Keyserlingk[2]
[1]*The Ohio State University, Veterinary Preventive Medicine, 1920 Coffey Road, Columbus, OH 43214, USA,* [2]*University of British Columbia, Animal Welfare Program, 180-2357 Main Mall, Vancouver, BC, V6T1Z6, Canada; proudfoot.18@osu.edu*

Dairy cows face a number of social stressors and are at high risk of disease after giving birth. Little is known about how individual differences in behaviour interact with the social environment to influence disease risk. To determine whether individual differences in behaviour affect endometritis risk under different social environments, cows were randomly assigned to 'Control' (8 groups, 4 cows/group) or social 'Stress' (8 groups, 4 cows/group) groups. Cows were cared for according to a protocol approved by the University of British Columbia's Animal Care Committee (Canadian Council on Animal Care, 2009). Four weeks before calving, each group was moved into a free stall pen with access to Insentec electronic feed bins; cows could either be assigned to eat from one or multiple feed bin. The 4 cows in each Control group were given free access to 6 feed bins and 12 lying stalls, whereas the 4 cows in each Stress group were housed with 4 non-experimental cows, and the group was given access to one of 4 feed bins (2 cows/bin) and 8 lying stalls. Stress groups also experienced either an unpredictable feeding time or a new feed bin assignment daily, and were moved into a new pen with new non-experimental cows after 2 wk. Dry matter intake (DMI) and feeding behaviours were automatically collected from electronic feed bins. Endometritis (>5% neutrophils) was diagnosed via uterine cytology smear 3-5 wk post-calving. Multilevel models revealed stable individual differences across a number of aspects of feeding behaviour. Cows varied consistently in how much they ate: total DMI (Intraclass-correlation coefficient [95% Confidence Interval]=0.71 [0.62, 0.79]) and in how they ate: average feeding rate (0.56 [0.44, 0.66]) and proportion of non-nutritive visits (0.50 [0.39, 0.62]). In Control cows, consistently low DMI was a risk factor for endometritis (logistic regression: P<0.001), but low DMI was a protective factor in Stress cows (P<0.05). Similarly, low feeding rate was a risk factor in Control cows (P<0.05), but was unrelated to disease in Stress cows (P>0.8). Across both conditions, a high proportion of non-nutritive visits to the feed bins were associated with a high likelihood of endometritis: 58% of cows with higher (>25%) non-nutritive visits became ill, whereas, only 26% of cows with fewer non-nutritive visits became ill (P<0.05). These results provide the first evidence that individual differences in feeding behaviour influence endometritis risk in dairy cows, but the direction and magnitude of these effects is dependent on the social environment.

Test-retest reliability of buffalo response to humans

Giuseppe De Rosa[1], Fernando Grasso[1], Ada Braghieri[2], Andrea Bragaglio[2], Corrado Pacelli[2] and Fabio Napolitano[2]
[1]Università di Napoli Federico II, Dipartimento di Agraria, Via Università 133, 80055 Portici (NA), Italy, [2]Università della Basilicata, Scuola di Scienze Agrarie, Forestali, Alimentari ed Ambientali, Via dell'Ateneo Lucano 10, 85100 Potenza, Italy; giderosa@unina.it

This study aims to assess the test-retest reliability of buffalo response to humans and the relationship between the behaviour of stockpeople and buffalo. The research was carried out in 27 buffalo farms located in southern Italy. Milk production was recorded only in the 14 farms enrolled in the milk recording scheme. Before assessment informed consent was obtained by stockpeople. Human-animal relationship was assessed through two different tests: behavioural observations of stockperson and animals during milking and avoidance distance at the manger. These tests were repeated within one month to assess test-retest reliability. Stockpeople attitude was evaluated using a questionnaire divided into four sections (general beliefs about buffaloes, general beliefs about working with buffaloes, behavioural intentions with respect to interacting with buffaloes and job satisfaction) including 21 statements. A high degree of test-retest reliability was observed for all the variables concerning the behaviour of stockpeople and animals. The values of Spearman rank correlation coefficient (r_s) ranged from 0.578 (P<0.01) for the number of kicks performed by the animals during milking to 0.937 (P<0.001) for the percentage of animals that move when approached by ≤0.5 m. Overall the coefficients of the variables measured on the stockpeople were lower than those obtained for the animal-based variables. A negative correlation between job satisfaction of stockpeople and number of steps performed by the animals during milking was found (r_s=-0.372, P<0.05). During milking negative interactions performed by stockpeople both in absolute number and in percentage terms positively correlated with the number of kicks performed by the animals (r_s=0.421 and r_s=0.430, P<0.05; respectively). A positive correlation was also found between the number of negative interactions and the percentage of buffaloes treated with oxytocin (r_s=0.424, P<0.05). Milk yield expressed as kg of milk/buffalo/year was positively correlated with the number of positive interactions (r_s=0.588, P<0.01) and negatively correlated with the number of steps performed by the animals during milking (r_s=-0.820, P<0.001). Finally, the higher was the mean avoidance distance at manger the larger was the percentage of animals treated with oxytocin (r_s=0.387, P<0.05). The present study showed that test-retest reliability of the variables used to assess human-animal relationship in buffalo was high and that stockpeople interactions may affect animal behaviour and production.

Can paw preference be an indicator of personality traits in dogs? A pilot study
Shanis Barnard, Deborah L. Wells, Peter G. Hepper and Adam D. S. Milligan
Queen's University Belfast, School of Psychology, Malone road, Belfast, BT7 1NN, United Kingdom;
s.barnard@qub.ac.uk

This pilot study aims to investigate if lateral behaviour is linked to personality traits in pet dogs. Behavioural laterality reflects the divergent processing of each brain hemisphere. Previous studies have shown that directionality and strength of laterality are associated with emotional stress, problem solving and personality in a range of vertebrate species. A sample of 24 privately owned pet dogs were recruited to participate in this study. Individual laterality scores were recorded by assessing the dog's preferred paw used to hold a toy filled with food. Binomial tests were used to classify dogs as lateralised (left- or right-pawed) or ambilateral and to assess a handedness index and strength of laterality. Further, a previously validated standardised test was used to assess six personality traits (Playfulness, Curiosity/Fearlessness, Chase-proneness, Sociability, Aggressiveness, Distant-playfulness) and a broader Shy-Boldness dimension. Mann-Whitney U-tests revealed significant differences in the behavioural response of lateralized dogs compared to ambilateral. The lateralised group (independent of direction) scored lower for the traits of Playfulness (U=24, P=0.005), Distant-playfulness (U=28, P=0.01), Curiosity/Fearlessness (U=38, P=0.05) and the Shy-Boldness dimension (U=32, P=0.02) than ambilateral dogs. No differences emerged between left and right biased dogs. A larger sample size is needed to confirm these results, nevertheless, the strength of laterality appears to be a potential novel indicator of dog's personality.

The visiting pattern of cows to the milking unit in an automatic milking system with free cow traffic

Shigeru Morita, Narumi Tomita, Takahisa Nodagashira, Mana Kato, Shinji Hoshiba, Michio Komiya and Keiji Takahashi
Rakuno Gakuen University, Agriculture, Food and Environment Sciences, Ebetsu, Hokkaido, 069-8501, Japan; smorita@rakuno.ac.jp

The automatic milking system (AMS) is based on cows voluntary visiting the automatic milking unit (AM unit) in free cow traffic. The objective of this study was to examine the visiting pattern of the cows to the AM unit. The data were collected for seven days from a commercial dairy farm that kept 52 cows in an automatic milking system with free cow traffic. One automatic milking unit was installed. Basal mixed ration was offered once a day at around 1300. The average of parity was 2.2 (from 1 to 4), and the average days in milk was 206 (from 28 to 450) days. The visits to the AM units were with milking (M) and without milking (R), and the sequence of M and R, and the length of the interval preceding M and R were calculated individually. The average of the daily visits to the AM unit was 8.8 visits. The frequency of visits to the AM unit was ranged from 2.0 to 27.4 visits/day. There was no difference of parity, daily milk yield, and the amount of daily concentrate offering in AM units with the number of visits. The average number of the milking was 3.7 times/day, and ranged from 1.9 to 6.6 milking/day, individually. The interval length of the milking (MM interval length) was widely varied in less visiting cows. The coefficient of variation (CV) of the length of the MM interval was decreased with the number of the visits to the AM unit. The MM interval length was shortened with the number of visit to AM unit. In the first half of MM interval, there was about twenty percent of visits without milking (R). Eighty percent of R was in the second half of MM intervals. The frequency of occurrence of R (trial of milking) was increased with the time after the last milking. It was concluded that the frequent visiting was need for the regular interval of milking in the automatic milking system. And cows visited the AM unit frequently around the time of the finishing the milking interval. It was suggested that the cow might feel the timing of the permission of the next milking, roughly.

Evidence of contrafreeloading in cattle: dairy heifers work for food even when it is freely available

Thomas Ede, Joao H. C. Costa, Heather W. Neave and Marina A. G. Von Keyserlingk
Animal Welfare Program, University of British Columbia, Faculty of Land and Food Systems,
2357 Main Mall, Vancouver, BC, V6T 1Z4, Canada; jhcardosocosta@gmail.com

Animals are known to work to gain access to food when the same food is freely accessible, a phenomenon known as contrafreeloading. This has been demonstrated in many species, but as yet no study has documented contrafreeloading in cattle. In this study, we presented twelve Holstein heifers with two *ad libitum* food sources: freely accessible or through closed electronic feed troughs (Insentec, Marknesse, the Netherlands). After a 12 d habituation phase to the environment, heifers were presented with three types of TMR (Regular TMR, TMR+5% straw, and TMR+20% grain) in a preference test over a period of 4 d in the open electronic feed troughs. Heifers showed a clear preference for the TMR+Grain followed by RegularTMR and lastly by TMRStraw (intake of 6.8±1.3, 3.75±0.4 and 0.85±0.9 kg/d, respectively). After this habituation period, we introduced a freely available source of the preferred food, TMR+Grain, placed in a 300 l feed trough at the back of the pen, offering 130% of the preferred free food consumed during the first test. The following day we also closed the gate of the electronic feed trough to a height of 92 cm. Heifers were only able to access the feed by reaching over the closed gate in order to consume the same 3 types of TMR. Throughout the duration of the study the animals always had access to the preferred food (TMR+Grain) at the back of the pen. Heifers showed clear contrafreeloading behaviour: they ate 1.0±0.3 kg/d of TMR+Grain, and interestingly 0.3±0.2 kg/d of RegularTMR and 0.1±0.1 kg/d of TMR+Straw from the closed feed troughs. Heifers varied in their intake from 0-8.1 kg/d for the TMR+Grain, 0-3.5 kg/d for the RegularTMR and 0-0.2 kg/d for the TMR+Straw. In the second phase of the contrafreeloading test, we tested the influence of the gate height, and consequently the work required to access the food. We provided RegularTMR in both the free and work-dependent food sources, with the gate set at 92 cm. The quantity of feed consumed through contrafreeloading was considered as 100%, and averaged 3.3±0.6 kg/d per heifer on the first test day. The height of the gate was then adjusted to the medium (100 cm) and highest settings (108 cm). Contrafreeloading averaged 1.6±0.5 kg/d when set at 100 cm and 0.6±0.4 kg/d when set at 108 cm, a reduction of 50.3 and 81.3%, respectively. This paper constitutes the first evidence that dairy heifers engage in contrafreeloading and this behavior is related to the animals' feed preferences and level of work required.

Stability of dominance rank deferentially affects exploratory behaviour in male and female mice

Justin A. Varholick, Jeremy D. Bailoo and Hanno Würbel
University of Bern, Division of Animal Welfare, Veterinary Public Health Institute, Vetsuisse Faculty, Längassstrasse 120, 3007, Bern, Switzerland; justin.varholick@vetsuisse.unibe.ch

The mouse, Mus musculus, is by far the the most widely used animal in biomedical research. Despite its prevalence, the behavioural biology of the mouse is often neglected. For example, in order to control for cage effects, one mouse per cage may be randomly selected. However, mice may form a dominance hierarchy where phenotypic differences may emerge as a consequence of dominance rank. Thus, randomly selecting an individual per cage may inadvertently exacerbate variation in experimental results if individuals of different ranks are tested between cages. The current study investigated whether (1) dominance hierarchies are present and stable across three time points and, (2) whether mice of different dominance ranks exhibit different behavioural phenotypes. Fifty male and 45 female C57BL/6byJ mice were separately assigned to groups of 5 (males n=10, females n=9). Using the competitive exclusion test, the dyadic dominance relationship for each pairing of mice per group was measured once per week over 3 consecutive weeks. Five days after the second week of dominance tests, mice were individually tested for distance travelled, time spent in open arms, and frequency of open arm entry, in an elevated zero maze. Clear dominance hierarchies (Landau's h >0.8) were only present in 6 cages of males and 6 cages of females. Assessment of the stability of individual ranks across the three time points, using Spearman correlations as an index (r_s>0.7), indicated that males had stable individual ranks from week 1 to week 2 (r_s=0.712), and week 2 to week 3 (r_s=0.833), while females had stable individual ranks from week 2 to week 3 (r_s=0.782). However, frequent changes in rank were observed except for alpha (α, 1st position) and delta (Δ, 4th position) ranks in both males (α, n=6; Δ, n=7) and females (α, n=7; Δ, n=6). Preliminary analyses indicate that stable α males showed significantly more exploratory behaviour in the elevated zero maze compared to stable α females (distance travelled: $F_{1,11}$=7.501, P=0.019). Furthermore, females with a stable rank spent significantly less time in the open arms compared to females with frequent changes in rank ($F_{1,43=}$5.279, P=0.027). Consistent with the literature, approximately 60% of groups showed linear hierarchies across time. Overall, we conclude that the relation between rank and phenotype is difficult to analyse because of unstable hierarchies. Nevertheless, we did find evidence that rank and rank stability may affect behaviour, and that this effect may depend on sex.

Genetic parameters for arena behaviour of lambs from commercial and indigenous South African sheep breeds

Ansie Scholtz, Jasper Cloete and Schalk Cloete
Western Cape Department of Agriculture, Private Bag X1, Elsenburg, 7607, South Africa;
AnsieS@elsenburg.com

This paper reports the response to humans from two different studies of lambs in an arena test. Firstly, data of 2798 8-month-old Merino lambs from Elsenburg were used. Data of 1,017 4-month-old Dorper, Namaqua Afrikaner, South African Mutton Merino and crosses between these breeds maintained at Nortier were also used. The test involved the placing of an individual in an arena marked out in evenly sized squares. At one end of and outside the arena was a pen containing six to seven contemporaries of the tested animals. A human operator sat on a chair directly in front of this pen. A second operator introduced the test sheep at the furthest point from the human seated inside the arena. The test sheep remained in the arena for three minutes and was observed by two recorders. The presence of the animal in a specific square was recorded every 15 seconds. The following traits described the behaviour of the sheep: the mean distance from the person, the total number of boundaries crossed between squares (as indication of the total distance travelled), the number of bleats, high- and low pitched, as well as the number of times an animal urinated or defecated. Genetic parameters and their standard errors were estimated with ASREML software, a value of twice the corresponding standard error being indicative of significance. The distance lambs maintained from the human operator was not heritable at Nortier (0.04±0.04), and lowly heritable at Elsenburg (0.06±0.02). The number of crosses and bleats were highly heritable at both locations (Nortier respectively 0.23±0.08 and 0.40±0.09; Elsenburg respectively 0.21±0.03 and 0.34±0.04). Urinating events (Nortier 0.13±0.06; Elsenburg 0.13±0.03) were moderately heritable, while defecating events were lowly heritable (Nortier 0.11±0.06; Elsenburg 0.07±0.03). The heritability estimate for defecating events was only significant at Elsenburg. The highly positive genetic correlation of 0.65±0.12 between the mean distance from the human operator and number of squares crossed suggested aversive animals were likely to travel larger distances. Such animals were also more likely to defecate (0.49±0.23). Animals likely to urinate also tended to be more likely to defecate at both locations (Nortier 0.54±0.31; Elsenburg 0.37±0.20). Selection on heritable behavioural traits indicating lower levels of stress during unfamiliar procedures (such as a willingness to approach the seated operator, fewer squares crossed, fewer bleats as well as fewer urinating and defecating events) may reduce adverse reactions to human handlers. Animal-stockperson relations are set to benefit from such an intervention.

Testing the reliability and validity of a 12-item welfare questionnaire to study chimpanzees

Lauren Robinson[1], Drew Altschul[1], Emma Wallace[2], Yulan Ubeda[3], Miguel Llorente[3], Zarin Machanda[4], Katie Slocombe[2], Matthew Leach[5], Natalie Waran[1] and Alex Weiss[1]
[1]*University of Edinburgh, 7 George Square, Edinburgh, Edinburgh, EH89JZ, United Kingdom,* [2]*University of York, Psychology department, University of York, York, YO10 5DD, United Kingdom,* [3]*Fundació Mona, Carr C-25 (Cassa-aeropuerto), S/N, 17457 Campllong, Girona, Spain,* [4]*Harvard University, 1 Oxford St, Cambridge, MA 02138, USA,* [5]*Newcastle University, King's Road, School of Agriculture, Food & Rural Development, Agriculture Building, Newcastle upon Tyne, NE1 7RU, United Kingdom; l.robinson@ed.ac.uk*

Behavioural observations and physiological measures are commonly used for welfare assessment but may be time-consuming and/or invasive. An additional tool, which may be both quick and non-invasive, are welfare questionnaires as they make use of staff knowledge without disturbing animals. In this study we aimed to test a 12-item welfare questionnaire as a tool for chimpanzee (Pan troglodytes) welfare assessment. The questionnaire includes items relating to stress, psychological stimulation, and indicators of negative and positive welfare states. We tested the reliability, association between items, and validity of welfare ratings as well as their correlation with personality, measured with the Hominoid Personality Questionnaire (HPQ). We analysed welfare ratings of the chimpanzees at the Edinburgh Zoo (n=18) and Fundació Mona (n=13) alongside previously collected behavioural observations and personality ratings of the Edinburgh Zoo chimpanzees. All the questionnaires were completed by staff, researchers, and volunteers familiar with the chimpanzees and raters were asked not to discuss their answers with each other. We found 11 welfare items to demonstrate interrater reliability (n=31, average=0.75, 9.35 ratings per chimpanzee); the unreliable item was removed from further analyses. A principal components analysis revealed that 10 items loaded onto a single component. For the Edinburgh chimpanzees (n=18), welfare ratings were negatively associated with the rate of observed 'negative behaviours' (e.g. regurgitation, cacophagy), using LASSO penalized linear modelling and covariance tests, indicating the welfare ratings were valid. The HPQ ratings (n=18, three ratings per chimpanzee) were used to create component scores for six personality dimensions based the previously published chimpanzee personality structure. In Spearman-rank correlation (n=18) we found higher welfare ratings were significantly correlated (P<0.05) with the Dominance (r=0.52), Neuroticism (r=-0.60), and Openness (r=0.52) personality traits. Our results suggest that chimpanzee welfare can be reliably assessed with this questionnaire and staff ratings relate to observed welfare states. Additionally, these ratings may be associated with personality traits.

Heifers' preference for feed variety and its relationship to fear and exploratory behaviour as calves

Rebecca K. Meagher, Daniel M. Weary and Marina A.G. Von Keyserlingk
University of British Columbia, Animal Welfare Program, 2357 Main Mall, Vancouver, BC V6T 1Z4, Canada; rkmeagher@gmail.com

Intensively housed dairy cattle are typically fed monotonous diets consisting of a mixture of grain and forages, which limits opportunities for exploratory foraging. The welfare effects of this limitation may depend on individuals' motivation to explore vs their fear of novelty, including novel feed. We tested whether Holstein heifers (287-343 days of age; n=10) preferred to consume a food that varied day to day and was available only during the daily 10 minute tests over their regular total mixed ration (TMR), available throughout the day. In the first test period, the varying feed was one of four forages (straw and three hays; Forage tests). A second test period was then conducted in which the varying feed was TMR with one of four powdered flavours added (Flavour tests). Varying feed and constant, unflavoured TMR were offered simultaneously in bins placed in consistent locations within each test period. Preference was assessed as the proportion of time eating spent at the bins containing the varying feed. In addition, the number of switches between feed bins was recorded as a measure of exploratory behaviour, and latency to eat during the first two trials was used as a measure of neophobia. Large individual differences in preference were observed: they ranged from an average of 0.1% to 46% of time eating at variable bins in the Forage tests, and from 0 to 93% in the Flavour tests. Some consistency in preference was observed between the two phases (Spearman r 0.47). The number of switches between bins was not strongly correlated with preference for varying feed. Preference and switching behaviour were then correlated with behaviour in novel object and food neophobia tests that had been conducted when the heifers were 35 and 45 days of age, respectively, to determine whether there were consistent individual differences in fearfulness or exploratory behaviour. Preference for varying forages was strongly positively correlated with intake of novel feed as calves (r_s 0.72, n=8), and preference for varying flavours negatively correlated with latency to contact a novel object (r_s -0.65, n=9). The number of switches between bins was also positively correlated with intake of novel feed as calves (r_s 0.71). These results suggest that feeding behaviour and preferences may reflect consistent individual traits that are already evident during the milk-feeding period, although the relative roles of exploratory motivation and fear of novelty are not yet clear. Offering a choice of a rotating feed may be an effective form of enrichment for individuals with a high tendency to explore novel stimuli.

Understanding variation in rat responses to CO_2

Lucia Améndola, I. Joanna Makowska and Daniel M. Weary
University of British Columbia, Animal Welfare Program, 2357 Main Mall, Vancouver, BC, V6T 1Z4, Canada; luciaame@mail.ubc.ca

The use of carbon dioxide (CO_2) for euthanasia remains controversial. Variability in how animals respond to gas exposure may account for the lack of consensus. Aversion and approach-avoidance tests have been used to evaluate rats' strength of aversion to CO_2; although all rats find this gas aversive, the strength of aversion is variable. The aim of this study was to evaluate individual differences in the strength of aversion to CO_2 within and between avoidance tests. Sprague Dawley female rats (n=12) were exposed to CO_2 twice in each of two avoidance tests. First, rats could choose between exposure to CO_2 gradual fill (18.3% of the test cage volume per minute) in a dark compartment vs escape to an aversive bright compartment (aversion-avoidance test). Then, the same rats could avoid CO_2 gradual fill (18.4%) by escaping to a safe cage, but at the cost of losing access to a valuable food reward (approach-avoidance test). The strength of aversion within each test, measured as CO_2% avoided, was analyzed using Linear Mixed Models with rat identity as a random intercept. The percentage of CO_2 avoided was compared within and between tests using Pearson correlation, with individual rat as the experimental unit. Again, rats varied in their responses (rat identity explained 86%, and 64% of the variation in aversion- and approach-avoidance tests, respectively), but rats avoided CO_2 regardless of test used. Rats were reasonably consistent in their responses within tests (aversion-avoidance: r=0.71, P<0.01; approach-avoidance: r=0.55, P<0.01), but not between tests (r=0.07, P=0.83). Consistency in individual differences within tests, and the lack of between-test agreement suggests that the two tests assess different personality traits; understanding this variability may allow for stronger inferences regarding how different animals respond to CO_2.

Assessment of franches-montagnes FM horse personality by standardized tests: a first step to temperament genes identification

Alice Ruet[1], Sabrina Briefer-Freymond[1], Léa Lansade[2], Marianne Vidament[2] and Iris Bachmann[1]
[1]Agroscope, Haras National Suisse, HNS, Les Longs-Prés, 1580 Avenches, Switzerland, [2]INRA, Centre de recherche Val de Loire, 37380 Nouzilly, France; alice.ruet@agroscope.admin.ch

Personality is defined as reactions towards environmental stimuli that are expressed by behavioural patterns and remain relatively stable across time and situations. Genetic background is at the basis of personality, also called temperament. A better knowledge of the link between genes and personality could allow better selection of horses. An international project aims to create a database of personality of different breeds in order to identify temperament genes. In France, behavioural tests called 'Tests de Tempérament Simplifiés (TTS)' have been developed to assess quickly, reliably and with little equipment, two dimensions of personality (tactile sensitivity and fearfulness) in various conditions. These tests meet the required validity criteria (responses are stable over time and across situations and measured dimensions are independent). We carried them out in accordance with the current laws of Switzerland. These two traits can be linked to the horse's rideability and performance. In this study, 184 FM horses, from 3 to 23 years old and of various genetic origins, were tested with TTS in 18 horse stables in Switzerland. For tactile sensitivity test, 4 Von Frey filaments were applied on the horse withers skin, and we recorded the muscle reaction (trembling/no trembling). Fearfulness test consist in opening a black umbrella in front of the horse and observing the intensity of the reaction (mark from 1 to 6 and flight behaviour). Linear mixed effect models were used to analyse the data. FM horse seemed to have a medium tactile sensitivity and a low level of fearfulness. We investigated which factors (individual characteristics, location, date and weather) could influence these two personality dimensions. It seemed that fearfulness increased with an increasing level of admixture with warmblood horses (χ^2, P=0.003) and decreased with age (χ^2, P=0.005). Sex seemed to have an influence on tactile sensitivity (χ^2, P=0.04). We also compared results of FM horses to those of a French sport horse breed and a French draft horse breed which were tested with the same tests in 2014. FM horse's tactile sensitivity was situated between the draft horse breed (P<0.001) and the sport horse breed (P=0.003). Fearfulness was significantly higher in FM horse than in the French draft breed (P=0.005) but did not differ from the French sport breed (P=0.34). Regarding these results, it seems that both studied dimensions are promising for the research of temperament genes, because they allow to highlight phenotypic differences between individuals as well as between breeds.

Flight speed as a predictor of sexual precocity in Nellore heifers and calf growth performance

Desiree Soares[1], Karen Schwartzkopf-Genswein[2], Joslaine Cyrilo[3] and Mateus Paranhos Da Costa[4]

[1]*University of Calgary, 2500 University Dr. NW, T2N1N4 Calgary, AB, Canada,* [2]*Agriculture and Agri-Food Canada, 5403 1 Ave S, T1J4P4 Lethbridge, AB, Canada,* [3]*Instituto de Zootecnia, SP 333 km 94, 14160900 Sertaozinho, SP, Brazil,* [4]*UNESP, Prof Paulo Donato Castellane S/N, 14884900 Jaboticabal, SP, Brazil; soares.desiree@hotmail.com*

Although flight speed (FS) is widely used as a measure of cattle temperament and indicator of performance in Bos indicus cattle, results can vary depending on when the FS measure is obtained and how it is summarized. The objective of this study was to determine how timing of FS collection and variants of the FS measure effect growth and reproductive performance of Nellore heifers and their offspring. A total of 229 pregnant heifers (525 ± 32.1 d of age and 425 ± 36.4 kg BW) were assessed over two years: 2012 (n=105) and 2013 (n=124). Heifers were initially housed in feedlot pens (40 m^2/animal) for 83 (Yr 1) and 88 d (Yr 2), and fed a diet consisting of 12% CP and 68% TDN on DM basis and provided water *ad libitum*. While housed in the feedlot, FS (m/s) was measured four times (28 d apart), and two measurements were calculated including the arithmetic mean of FS (FSmean) and the difference between the first and the last FS measurement (FSdiff). The FSdiff was used to define three FS categories: Acclimated: FSdiff > 1SD; Consistent: FSdiff ± 1SD; Sensitized FSdiff < -1SD. The first FS measurement (FS1) was used as a predictor of herd replacement based on growth and reproductive traits. At the end of the feedlot period heifers were moved to an adjacent pasture (Brachiaria spp.) for calving. At the time of calf weaning (~210 d of age) the following parameters were assessed: calf weight (CW, kg/animal); heifer weight (HW, kg/animal); age at first calving (AFC, d); days to calving (DC, d) and calving interval (CI, d). Data were analyzed using mixed-effects model, including year as fixed effect, feedlot pen within year as random effect, and initial BW and FS1 (FSmean, or FSdiff) as a covariate. A post-hoc (Tukey) test was used to compare the adjusted means. FSmean had a positive (P=0.04) linear effect on DC and tended (P=0.07) to show a positive linear effect on AFC, while FSdiff tended (P=0.07) to show a positive linear effect on CW, and FS1 showed a positive quadratic effect (P=0.04) on HW. Calves of Acclimated heifers had greater weaning weights (221 ± 7.7 kg/animal; P<0.05) than Consistent or Sensitized heifers (202 ± 3.0 and 198 ± 4.5 kg/animal, respectively). The results suggest that FSmean has potential value in predicting sexual precocity, while FSdiff has potential in predicting calf growth performance at weaning. FS1 was not a good predictor of either growth or reproductive traits.

Self-injurious behaviour and locomotor stereotypy are distinct behavioural presentations in laboratory-housed Rhesus macaques

Caroline Krall[1] and Eric Hutchinson[2]
[1]Royal (Dick) School of Veterinary Studies, Easter Bush Campus, EH25 9RG Edinburgh, United Kingdom, [2]National Institute of Health, Division of Veterinary Resources, 9000 Rockville Pike, 20892 Bethesda, MD, USA; s1146158@sms.ed.ac.uk

A challenge of managing laboratory-housed Rhesus macaques (*Macaca mulatta*) is understanding and alleviating abnormal behaviours, e.g. self-injurious behaviour (SIB) and locomotor stereotypy (LST). We explored their relationship to each other while applying a novel pharmacological therapy: guanfacine (a selective α2-adrenergic receptor agonist), which has previously proven efficacious in reducing SIB and LST in Rhesus macaques and Olive baboons (*Papius anubis*). Thirteen 7-year-old male Rhesus macaques were studied. The subjects were included based on their use in an unrelated longitudinal study, which also required single housing from a young age (average 1.1-years-old). In addition, they all possessed a history of SIB (including wounding) and LST prior to the current study. Subjects were randomly allocated to treatment (5 mg guanfacine b.i.d.; n=9) or vehicle-only (Skittle b.i.d.; n=4) groups. For each subject, a total of 900 minutes of video footage was obtained by continuously recording from 09:00 – 10:00 each day over the course of 12 weeks. This was further sub-divided into three 4-week phases (Baseline, Treatment, Post-treatment). Behaviour analysis was performed by a blind observer using continuous sampling for SIB frequency (i.e. number of bites per minute) and one-zero sampling at 5 minute intervals for LST. The effect of guanfacine was tested by comparing the frequency of SIB and LST between phases, and baseline SIB and LST levels were likewise correlated. We found that SIB frequency (Baseline M=30.8, SD=30.9) was significantly reduced following guanfacine treatment (M=8.6, SD=12.8, W=43.0, P=0.008). Contrarily, guanfacine engendered a significant increase in LST (Baseline [M=107.4, SD=75.2], Post-treatment [M=148.3, SD=61.9], W=45.0, P=0.004). There was no correlation between baseline levels of SIB (M=28.9, SD=27.9) and LST (M=105.1, SD=71.9) on the individual level (rs=-0.043, P=0.887). These results suggest that, within our cohort, SIB and LST may represent distinct behaviours rather than variants on a spectrum of a single abnormality. Given the anxiolytic effect of guanfacine, its efficacy upon SIB indicates that the motivation to self-bite likely derives from anxiety or stress. Conversely, the paradoxical response of LST demonstrates the existence of an unrelated origin, e.g. habit or self-induced enrichment. Future treatment efforts should therefore adopt a personally tailored approach, which considers both the individual animal and specific behavioural repertoire that requires modification.

Responses in a food neophobia test reflects individual differences in dairy calves

Heather W. Neave, Joao H.C. Cardoso Costa, Daniel M. Weary and Marina A.G. Von Keyserlingk
Animal Welfare Program, University of British Columbia, 2357 Main Mall, Vancouver, BC, V6T
1Z4, Canada; hwneave@gmail.com

Commonly used tests for assessing individual differences, including 'open field', 'novel object' and 'human approach' tests, have been criticized for lack of biological relevance to the animal. These tests may induce fear in animals, but can also provide opportunity for interacting or exploring that are less commonly measured. The objective of this study was to assess food neophobia responses as a more biologically relevant method of assessing individual differences in responses to novelty. We correlated within-individual responses across a food neophobia test and with each of three behavioural tests: novel environment (NE), human approach (HA) and novel object (NO) tests. Thirty-six group housed, milk-fed dairy calves (100±12 d of age), with ad libidum access to calf-starter concentrate and hay, were observed individually in a NE test (30 min in a novel arena), in a HA test (10 min with an unknown stationary human), and in a NO test (15 min with a black 110 l bucket). In the 30 min food neophobia test, calves were exposed to two identical white 20 l buckets, one empty and the other filled with a novel food [a total mixed ration (TMR)]. Agreement within and across tests was described using Spearman Rank correlations. Intake of the novel feed (mean±SD; 96.4±96.7 g) was correlated with time in contact with the food bucket (73±112 s; r=0.45), number of bucks (2.1±1.8; r=0.42) and time exploring the floor (256±449 s; r=0.36) during NE test, number of bucks during the HA test (1.4±0.6; r=0.87), and number of events interacting with the human (1.8±1.2; r=0.41) and object (3.7±3.3; r=0.36). Time spent eating the novel food (125±97 s) was similarly correlated with time in contact with the food bucket (r=0.54), number of bucks during the HA test (r=0.86), and playing events in the NE test (3.4±2.4; r=0.32). Time in contact with the food bucket was correlated with number of bucks in HA (r=0.58) and NO (1.8±1.5; r=0.68) tests, and time spent touching the object (125±100 s; r=0.54). However, latency to eat the novel food (361±582 s) showed no correlation with latency to touch the human (391±682 s; r=-0.01) or object (56±109 s; r=-0.09). Our results suggest that time spent eating and intake of a novel food, and interactions with the food bucket are related to exploratory behaviour in the NE, HA and NO tests, but latency measures are poorly related across tests. We suggest that these latency measures may not be reliable as indicators of an individual's response to novelty.

A comparative science of emotion

Elizabeth Paul and Michael Mendl
University of Bristol, School of Veterinary Sciences, Langford House, Langford, Bristol, BS40 5DU,
United Kingdom; e.paul@bristol.ac.uk

The past decade has witnessed a tremendous expansion of interest in the study of animal emotion, most especially within the fields of behavioural and pharmacological neuroscience. Terms such as 'feelings', 'emotions' and 'moods' may simply be too person-centred to be used comfortably in the context of non-human animals, but numerous recent studies of reward valuation, learning and decision-making in both animals and humans mean that a comparative review of emotional and emotion-related processes is now both timely and possible. In this presentation we start by considering the comparative study of emotion in historical context – from the publication of Darwin's Expression of Emotions in Animals and Man (1872), through behaviourism to the cognitive revolution of the 1970s and 80s, to present day interest in the roles played by dopamine, serotonin and opioids in emotion and emotional disorders. We set out a framework for investigating valenced affective states in animals and consider the extent to which current evidence points to the existence of both analogous and homologous processes in humans and non-human animals. In particular, we will draw from our own and others' studies of species such as dogs, rats and mice to argue that there is evidence of conservation of many attributes of affective processing across the mammalian phylum. Data from birds, reptiles, fish and invertebrates is much more diverse but even within these groups it is possible to find systems of reward and punishment processing that show both continuities and discontinuities with the properties seen in humans and other mammals. We finish by considering the most difficult and controversial question of all – whether and in what way consciously experienced, subjective affective states might ever be detectable in animals – and look forward to future developments in this field.

In your face: indications of emotional facial expressions in pigs

Inonge Reimert, Lotte Stokvis, Monique Ooms, Fleur Bartels and J. Elizabeth Bolhuis
Wageningen University, Adaptation Physiology Group, De Elst 1, 6708 WD, the Netherlands;
inonge.reimert@wur.nl

Knowledge on how farm animals are feeling has been recognized as crucial to assess and improve their welfare. However, methods to objectively measure diverging emotional states in these animals on a large scale are lacking. As in humans facial expressions are used as indicators of emotional state, we studied the presence of emotional facial expressions in pigs in diverging valenced (i.e. positive vs negative) situations. Ten 17-week-old pigs were all subjected to four different treatments. One of the treatments consisted of 5 sec of anticipation of 4-min access to a room (28 m^2) with peat (350 l), wood-shavings (275 l), straw (6 kg) and six different toys and was expected to evoke a positive affective state. The three other treatments, i.e. delayed (40 sec) access to the room; being startled by receiving ice cubes (500 g) on their backs; and 1-min confinement in a weighing crate (1.4×0.5×1 m (l×w×h)) were expected to evoke a negative affective state in the pigs. The faces of the pigs were filmed during the treatments. Thereafter, each video was studied frame by frame and frames with clear and complete faces were used further. In each of these frames, facial features of the eyes, nose, ears, cheeks and mouth were scored using specific scales. For instance, ears were scored on a scale from 0 to 2, with 0 indicating that ears were directed towards the front and 2 ears directed actively towards the back of the pig. Scoring was performed blind with respect to the treatments, except for weighing. Subsequently, differences in eye, nose, ear, cheeks and mouth features between the supposed positive treatment and each negative treatment were analysed using mixed linear models. Compared to the positive treatment, we found that pigs in the ice treatment had their eyes more widely open (0.11±0.03 vs -0.05±0.05, P<0.05), tended to have their ears more backwards (0.36±0.10 vs 0.80±0.31, P<0.1) and tended to have the sides of their nose more wrinkled (0.83±0.30 vs 0.50±0.12, P<0.1). Moreover compared to the positive treatment, delayed access to the room resulted in mouth corners being drawn more upwards (1.95±0.10 vs 2.39±0.15, P<0.05) and tended to result in more nose wrinkles (0.50±0.12 vs 1.01±0.28, P<0.1) and the nose being in a more upward positon (i.e. actively drawn up) (1.94±0.03 vs 2.36±0.24, P<0.1). Weighing also caused the nose of the pigs to be drawn more upwards as compared to the positive treatment (2.31±0.12 vs 1.94±0.03, P<0.01). These results indicate that pigs may express their emotional state with their face. Further research is needed to support these results and to study differences in facial expressions between different emotional states diverging in valence and intensity.

Face-based perception of emotions in dairy goats using 2-D images

Lucille G A Bellegarde[1,2,3], Marie J Haskell[2], Christine Duvaux-Ponter[1], Alexander Weiss[3], Alain Boissy[4] and Hans W Erhard[1]
[1]INRA, AgroParisTech, Université Paris-Saclay, UMR Modélisation Systémique Appliquée aux Ruminants, Paris, 75005, France, [2]SRUC, Edinburgh, EH9 3JG, United Kingdom, [3]The University of Edinburgh, School of Philosophy, Psychology and Language Sciences, Edinburgh, EH8 9JZ, United Kingdom, [4]INRA, UMR Herbivores, Saint-Genès-Champanelle, 63122, France; lucille.bellegarde@sruc.ac.uk

Faces of conspecifics are a major source of information in an individual's social life. Identity is displayed through the face, but facial expressions also convey gaze, and attentional or emotional state. In this study we investigated whether dairy goats (n=32) would show different responses to 2-D images of faces of familiar conspecifics displaying different levels of positive or negative emotional states. The faces of four of those goats were photographed in a positive and a negative situation. Three types of images of intermediate facial expressions were then created using morphing software (75% positive, 50% positive, and 25% positive). In a test-pen, each goat was exposed to each type of image, obtained from the same goat and displayed on a computer screen for 3 seconds. All goats were shown real faces first. The goats were then shown the three types of morphed faces, balanced for order. Finally, the first real face was shown again. Spontaneous behavioural reactions including ear postures and interactions with the image (time spent looking or touching) were recorded during the 3 seconds. Results were analysed using REML with repeated measurements for subject. When a negative face was displayed, goats spent longer in forward ear postures (negative: 2.8±0.2 s; positive: 1.9±0.6 s; $F_{55.8}$=6.0, P=0.02) and tended to interact more with the screen (negative: 2.9±0.1 s; positive: 2.6±0.2 s; $F_{48.4}$=3.0, P=0.09), indicating greater interest in faces displaying a negative emotional state. Overall, goats did not react differently to images of real and morphed faces. Identity of the photographed goat influenced the total time spent with the ears forward (($F_{57.1}$=7.4, P=0.001). Goats also spent more time interacting with the screen during the first session, when the stimulus was entirely novel, than during any other session (3.0±0.1 vs 2.5±0.1 s; $F_{9.3}$=4.9, P=0.02). We conclude that goats can discriminate finely between images of faces displaying different emotional states and that they perceive the emotional valence expressed. Their reactions to faces are however influenced by parameters such as the novelty of the stimulus and the identity of the individual photographed.

Facial indicators of positive emotions in rats

Kathryn Finlayson, Jessica Lampe, Sara Hintze, Hanno Würbel and Luca Melotti
University of Bern, Division of Animal Welfare, Länggassstr. 120, 3012 Bern, Switzerland;
finlayson.kathryn@gmail.com

The identification of facial indicators of positive emotions in rats could expand on existing methods to assess positive animal welfare. Previous studies found that specific rodent facial expressions occurred in situations inducing negatively valenced emotional states (e.g. pain, aggression and fear). This study aimed to investigate if facial expressions of positive emotions are exhibited in rats during heterospecific play. Fifteen adolescent male Lister Hooded rats were individually subjected to a 2 min Positive Treatment (PT) and a 2 min mildly aversive Contrast Treatment (CT) over two consecutive days. PT consisted of playful manual tickling administered by the experimenter, while CT consisted of exposure to a novel test room with intermittent bursts of white noise. PT always preceded CT to avoid carry-over of the mildly aversive experience in the test arena during CT, and the number of positive (frequency-modulated 50 kHz) ultrasonic vocalisations were recorded to ensure that rats had divergent emotional states in PT and CT. High-speed photos of the rats' faces in a profile or three-quarters view were taken during both treatments, in between tickling bouts or white noise bursts, respectively. Novel qualitative and quantitative measures, and also the established Rat Grimace Scale, were used to detect fine changes in facial expression. Photos were scored blind to treatment, and Bonferroni correction was used for multiple comparisons. The number of positive vocalisations was greater in PT (M=115.87, SEM=10.29) compared to CT (M=1.07, SEM=0.63; t(14)=- 10.96, P<0.001), indicating the experience of differentially valenced states in the two treatments. The main findings were that Ear Colour (0-2 scale) became significantly pinker in PT (M=1.68, SEM=0.10) than in CT (M=0.52, SEM=0.11; z=-3.41, P=0.001), and Ear Angle was wider (ears more relaxed) in PT (M=127.12°, SEM=2.23) compared to CT (M=112.10°, SEM=2.53; t(14)=4.24, P=0.001). All other quantitative and qualitative facial measures, which included Eyeball height/width Ratio, Eyebrow height/width Ratio, Eyebrow Angle, visibility of the Nictitating Membrane, and the Rat Grimace Scale, did not show significant differences between treatments. This study contributes to the exploration of positive emotions, and thus good welfare, in rats as it identified potential indicators of positive facial expression resulting from a positive heterospecific play treatment. Pinker Ear Colour and wider Ear Angle, both accompanied by the emission of positive vocalisations, may reflect internal physiological changes associated with positively valenced emotional arousal and muscle relaxation, respectively.

Emotion recognition in dogs, orangutans, and human infants

Min Hooi Yong[1,2,3], Ted Ruffman[1] and Neil Mennie[3]
[1]University of Otago, Psychology, Box 56, 9054 Dunedin, New Zealand, [2]Sunway University, Psychology, 5 Jalan Universiti, Bandar Sunway, 47500 Selangor, Malaysia, [3]University of Nottingham Malaysia Campus, Psychology, Jalan Broga, Semenyih, 43500 Selangor, Malaysia; mhyong@sunway.edu.my

Recognising emotions is considered a step towards developing empathy, and subsequently altruism. This ability has been reported in non-human species (primates and companion animals). However, it is unknown whether dogs and orangutans can match emotional faces to voices in an intermodal matching task or whether they show preferences for looking at certain emotional facial expressions over others, similar to human infants. We presented 52 domestic dogs (32 females, M=5.47 years, SD=4.01), 6 orangutans (4 females, M=18.17 years, SD=5.85), and 24 7-month-old human infants (10 females, M=7.16 months, SD=0.85) with two different human emotional facial expressions of the same gender simultaneously, while listening to a human voice expressing an emotion that matched one of them. Results showed that neither dogs or human infants looked longer at the matching emotional stimuli. But orangutans looked longer at the concordant angry face when listening to an angry voice, P<0.01. Both dogs and infants demonstrate the same pattern of preferential looking across the three pairings of emotion faces, both P<0.01. Dogs looked less at the sad face compared to both the happy and angry faces, Z=2.66, P<0.01, r=0.38 and Z=4.16, P<0.01, r=0.59, respectively, with similar findings for infants, Z=4.14, P<0.01, r=0.85 and Z=3.66, P<0.01, r=0.75 respectively. There was no preference in looking when the angry and happy faces were paired for dogs, Z=1.70, P=0.09, r=0.24, or infants, Z=0.89, P=0.38, r=0.18. Our results are consistent with most matching studies – both dogs and infants demonstrated an identical pattern of looking less at sad faces when paired with happy or angry faces (irrespective of the vocal stimulus), with no preference for happy vs angry faces. Orangutans did not demostrate any preference. Discussion focuses on why dogs and infants might have an aversion to sad faces, or alternatively, heightened interest in angry and happy faces. Preference for emotions might be a result of domestication, rather than an evolutionary trait.

Perception of emotional valence in goats

Luigi Baciadonna[1], Elodie F. Briefer[2], Livio Favaro[3] and Alan G. Mcelligott[1]
[1]Queen Mary University of London, Mile End Rd, E1 4NS London, United Kingdom, [2]Institute of Agricultural Sciences, Universitätstrasse 2, 8092 Zürich, Switzerland, [3]University of Turin, Via Accademia Albertina 13, 10123 Turin, Italy; luigi.baciadonna@qmul.ac.uk

Animals can potentially transmit information about emotional states through their vocalisations. We investigated whether goats can discriminate conspecific calls with different emotional valence (positive or negative). We used a habituation-dishabituation-rehabituation paradigm. Subjects were initially habituated to a stimulus by repeated exposures to 9 calls from the same animal, which was previously recorded either positive or negative situations. After the habituation to the stimulus, a stimulus of opposite valence was presented (n=3 calls from the same animal with different valence compared to the habituation phase). Finally, the stimulus that the subject was habituated to was presented again (n=1 call from the same animal) to check whether the animal showed a new shift in attention. Twenty-four goats (12 females, 12 males) were tested. The time spent looking at the speaker and the physiological reactions to the calls were measured. During the habituation phase, goats reduced the rate of looking towards the speaker (linear mixed-effect model (LMM); $\chi^2_{(1)}$=30.01, P<0.001), indicating the expected habituation effect. The rate of looking towards the speaker increased when the second call of dishabituation was played (LMM; $\chi^2_{(1)}$=5.58, P=0.01). During rehabituation, the rate of looking towards the speaker was higher when a negative call was played compared to a positive call (LMM; $\chi^2_{(1)}$=5.57, P=0.01). Heart rate decreased during habituation (LMM; $\chi^2_{(1)}$=30.01, P<0.001, mean first call: 123.85±3.71 BPM, mean last call: 108.59±3.48 BPM), regardless of the valence, and did not change in either the dishabituation or rehabituation phases. Heart-rate variability (RMSSD) during the habituation was affected by the valence (LMM; $\chi^2_{(1)}$=4.66, P=0.030); it was generally lower when a positive call was played (mean: 53.55±2.39 ms) compared to a negative call (mean: 57.21±2.01 ms). When the first call of dishabituation was played, heart-rate variability was affected by the valence (LMM; $\chi^2_{(1)}$=4.50, P=0.033); it was higher when a positive call was played (mean: 60.71±4.04 ms) compared to a negative call (mean: 49.64±4.09 ms). When the rehabituation call was played, an effect of valence was found (LMM; $\chi^2_{(1)}$=6.52, P=0.010); heart-rate variability was higher when a positive call was played (mean: 71.30±3.61 ms) compared to a negative call (51.14±7.33 ms). Our results indicate that goats discriminate between calls of different valence. Investigating the perception of emotion-linked calls in livestock is important for evaluating their potential role in emotional contagion.

Vocal expression of emotions in pigs

Elodie F. Briefer[1], Lorenz Gygax[2] and Edna Hillmann[1]
[1]ETH Zürich, Institute of Agricultural Sciences, Universitätstrasse 2, 8092 Zuich, Switzerland,
[2]Federal Veterinary Office, Agroscope Reckenholz-Tänikon Research Station ART, Centre for Proper Housing of Ruminants and Pigs, Tänikon, 8356 Ettenhausen, Switzerland; elodie.briefer@usys.ethz.ch

Emotional indicators are important to provide accurate assessment of non-human animal emotions, as well as a better understanding of the evolution of emotions through cross-species comparisons. Investigating indicators of both emotional arousal (intensity) and valence (negative or positive) simultaneously in the same species is important in order to obtain reliable indicators of these two dimensions. We investigated indicators of arousal and valence in pig vocalization, and hypothesized that the two dimensions would be associated with specific vocal profiles. We focused on grunts, as they are the most common pig call type and are produced in both negative and positive situations. We subjected 32 weaned pigs (33-38 days old), in a randomized order, to a negative situation (3 min of social isolation) and to a positive situation (5 min in pairs with food, water and toys) triggering grunt production. Emotional arousal was assessed using the heart rate of the pigs, measured during the situations using a non-invasive monitor. We analyzed frequency, duration and amplitude parameters in 571 good-quality grunts (7-10 grunts per pig and per situation). We found several reliable vocal indicators of the two emotional dimensions. For example, with increasing arousal, the peak frequency decreased (linear mixed-effect model (LMM): $F_{1,50.74}$=6.97, P=0.011), while the harmonicity increased (LMM: $F_{1,55.91}$=6.46, P=0.014). Grunts produced in the positive situation had a shorter duration (LMM: $F_{1,18.87}$=100.78, P<0.0001; mean±SD duration, log-transformed = -1.30±0.30 s), lower fundamental frequency ('F0', LMM: $F_{1,26.24}$=11.58, P=0.002; F0mean=62.36±5.07 Hz) and higher formants (e.g. F1; LMM: $F_{1,27.24}$=30.82, P<0.0001; F1mean, log-transformed = 6.20±0.12 Hz) than those produced in the negative situation (log(duration) = -0.82±0.32 Hz; F0mean=65.22±7.25 Hz; log(F1mean) = 6.12±0.12 Hz). These indicators could serve as very useful non-invasive indicators to assess both negative and positive emotions on-farm.

Exploring whether ear postures are a reliable indicator of emotional state in dairy cows

Helen Proctor and Gemma Carder

World Animal Protection, 5th Floor, 222 Grays Inn Rd, WC1X 8HB, United Kingdom; gemmacarder@worldanimalprotection.org

We have built upon our previous research to further test the suitability of ear postures as a measure of emotional state in cattle. In this current study, we used a positive and negative contrast paradigm to elicit the high arousal emotional states of excitement and frustration. We conditioned 22 dairy cows to anticipate the delivery of standard feed when a bell was rung, whilst they were being held in a stall. Each cow underwent four standard feed trials, during this time we believed that they would learn to expect the standard feed, and this represented a neutral stimulus as they have continuous access to this feed in their pens. We then changed the stimulus to concentrates feed, a high energy pellet feed. This was intended to induce a positive state of excitement in the cows as they had been expecting standard feed, and access to concentrates is normally highly restricted. Following five trials with the concentrates we changed the feed to inedible woodchip for one trial to elicit a frustrated emotional state. During all the trials, we observed the cow's ear postures and mean heart rate (bpm). Both the concentrates and woodchip elicited a high arousal state as the mean heart rate was significantly higher than in the standard feed treatment (F $_{(1.89, 357.29)}$=125.70, P<0.001) Furthermore, we found that the treatments had a significant effect on the types of ear posture performed. During feeding, an upright posture tended to be performed significantly longer when the cows were positively 'excited' (M=4:00), compared with being 'neutral' (M=3:41) or 'frustrated' (M=2:20) (F $_{(2, 40)}$=19.75, P<0.001). Whereas a forward posture was performed for significantly longer when the cows were considered to be frustrated (M=2:35) (F $_{(2, 36)}$=16.07, P<0.001) compared to when given the 'positive' (M=0:48) and 'neutral' (M=1:05) stimulus. These findings complement our previous work which found that a held back ear posture and a relaxed, drooping ear posture was associated with a positive, low arousal emotional state. As a result of this research we can now begin to attribute certain ear postures to emotional states, and we are one step closer to understanding and measuring the emotional lives of dairy cows. Ear postures represent a quick and objective measure that can be used in practice by farmers as well as in welfare assessments to improve the welfare of cows.

Characterising fear in dogs

Björn Forkman
University of Copenhagen, Department of Large Animal Sciences, Groennegaardsvej 8, 1870
Frederiksberg C, Denmark; bjf@sund.ku.dk

It is often suggested that fear is one of the basic emotions. When comparing the results of different fear tests the results seem to measure very different aspects of fear however. This even to the extent that it can be questioned whether they are testing the same affective state. The purpose of this study was to investigate the relationship between different types of fear tests. Database material from the Swedish Dog Mentality Assessment, (n=70,875, 34 breeds) was used. The assessment consists of a behavioural test battery. There are four sub-tests that are directly relevant for the current study: three of the tests contain an element of surprise ('sudden appearance', 'sudden noise', gunshot) with two of these containing an auditory stimulus ('sudden noise', gunshot), finally a fourth test measures the response to two slowly moving, unknown and probably threatening beings (test persons draped in white sheets with buckets with eyes). The tests are always done in the same order. The relationship between them is tested using a rank test (Spearmann correlation). The strongest overall correlation was found between the more long term reactions of the sudden noise and sudden appearance (R_s median for all breeds = 0.4, varying between 0.48 for rough-haired collie (n=3,335, P<0.001) to 0.02 for golden retriever (n=4,059, NS)) as well as the startle reaction for these (R_s median for all breeds = 0.36, P<0.001). Both the metallic noise and the gunshot are sudden and auditory but the correlation is comparatively weak (R_s median for all breeds = 0.14, P<0.001). The slow threat correlated only weakly with the other tests (R_s median for all breeds, between 0.05 and 0.13, P<0.001). In conclusion gun-shyness seems to be a specific type of fear that is not related to other types of fear. Fear of a more long term threat (several seconds) likewise seems to be of a different type than that shown in a startle reaction to a surprise. The results emphasises the need for a wide variety of fear tests with differing fear evoking stimuli to give a complete picture of the fear reactions in the domestic dog.

Behavioural assessment of the habituation of captured feral goats to an intensive farming system

David Miller, Patricia Fleming, Anne Barnes, Sarah Wickham, Teresa Collins and Catherine Stockman
Murdoch University, School of Veterinary and Life Sciences, 90 South St, Murdoch 6150 WA, Australia; d.miller@murdoch.edu.au

Introduced (feral) goats are an invasive animal species in Australia. They can be legally trapped by licensed operators for the domestic and export meat markets. Using quantitative and qualitative behavioural assessment (QBA), the habituation of captured feral goats to an intensive farming system (feedlot) was assessed. The goats were initially captured by a licensed feedlot operator on an extensive rangeland property and immediately transported to an intensive feedlot. Male goats (n=120) with a similar body mass and similar age were separated into two treatment groups: high interaction (HI, n=60) and low interaction (LI, n=60). Within each treatment group they were assigned to 3 separate pens containing 20 goats. In the HI group, a human entered the pens twice daily and calmly walked amongst the goats for 40 minutes. In the LI group, a human only briefly entered the pens to fill up feed bins (weekly). At the end of each week the goats were weighed and drafted into the 12 subgroups of 10 animals distinguished by differing ear-tag colours (i.e. 6 sub-groups per treatment). Each group was then moved into a small holding pen for an agonistic and flight response test. The goats were held in the holding pen for 2 minutes during which time video footage was taken and later analysed for number of agonistic contacts. Video footage collected at this time was also used for QBA analysis. The pen was then opened into a 3 m wide laneway the speed at which they exited was recorded. The videos of each group taken each week were shown in random order to 16 observers who used their own descriptive terms to score the animal using QBA. There was a high level of consensus between observers (42.2% variation explained, P<0.001). A generalised procrustes analysis (GPA) was used to identify the principle dimensions of consensus and variation explained between goats. Two main GPA dimensions of behavioural expression were identified. GPA dimension 1 (57.0% variation explained) differed between treatments (P<0.05); HI goats scored higher on GPA dimension 1 (more calm/content/at ease) compared to LI goats (more agitated/nervous/scared). GPA dimension 2 (13.1% variation explained) scores were not significantly different between treatments. There was also a significant effect of time on GPA dimension 1 (P<0.001) but not dimension 2, with goats becoming more 'calm/content/at ease' over the three weeks. The QBA data were in agreement with the quantitative measures of habituation (number of agonistic contacts, r^2=0.76, P<0.01; flight speed, r^2=0.63, P<0.05), thus supporting the usefulness of QBA as a tool for habituation and welfare assessment.

Welfare indicators for sheep: relationship between QBA and behavioural measures

Susan Emily Richmond, Francoise Wemelsfelder and Cathy M Dwyer
SRUC, Animal & Veterinary Sciences, King's Building, West Mains Rd, Edinburgh EH9 3JG, United Kingdom; susanrichmond@hsa.org.uk

The majority of sheep in the EU are managed extensively yet these systems can pose unique and complex problems for the animals' welfare. The AWIN project developed a novel animal-based welfare assessment protocol to gauge the physical and mental wellbeing of such sheep. Our main aim was to understand the relationship between the indicators assessing affective state and behaviour during a longitudinal study. This would help identify the best parameters for inclusion in the AWIN protocol. Ewes representative of the breeding flock's age and body condition scores were used (n=49). Ewes were marked to aid recognition but remained with the flock. On 13 visits spanning six months (pre-lambing to post-weaning) an observer performed qualitative behavioural assessments (QBA) on the individual ewes using a fixed-list of 21 terms previously developed by a two stage process: focus group and expert opinion. Four meaningful PCs were identified explaining 60.6% of the variation between sheep. Only the first two are discussed here due to space constraints. PC1 (22.6%) ranging from positive emotional states (content, bright) to negative states (apathetic, subdued) was summarised as 'General Mood'. PC2 (17.3%), ranged from low energy (relaxed, calm) to high energy (tense, agitated) and relates to 'Arousal'. The effect of reproductive state (pregnant, lactating, weaned) on QBA scores was investigated using a one-way repeated measures ANOVA followed by a Tukey test. Significant increases in General Mood occurred between successive reproductive states (F=177.15, P<0.001. Pre-lambing= -1.83±0.08, post-lambing=0.36±0.11, post-weaning=2.15±0.16). Arousal significantly increased in the post-lambing period but was not significantly different post-weaning (F=10.68, P<0.001. Pre-lambing= -0.39±0.09, post-lambing= -0.18±0.13, post-weaning=0.54±0.14). On five occasions (QBA assessments 9-13) data on nearest neighbour distance and response to a human approach test were collected along with group composition and vocalisation. Relationships between these behavioural indicators and QBA PC scores were assessed using Linear Mixed Models. Ewes with higher General Mood were nearer to their neighbours (W=4.61, d.f.=1, P=0.02) and were in groups with higher numbers of ewe vocalisations (W=5.51, d.f.=1, P=0.02). Ewes in a higher Arousal state fled when the human was at a greater distance (W=4.28, d.f.=1, P=0.04) and were in more vocal groups (W=3.51, d.f.=1, P<0.001). In conclusion, the affective state of sheep is associated with aspects of social behaviour and perceived predation threat. These findings support the use of the indicators assessed here as they provide complementary information on the welfare of sheep.

Qualitative Behaviour Assessment as a tool for monitoring cat welfare in vet nurse practice

Francoise Wemelsfelder[1], Kirsty Young[2] and Karen Martyniuk[2]
[1]*SRUC, Easter Bush, EH25 9RG Midlothian, United Kingdom,* [2]*SRUC Barony Campus, Parkgate, DG1 3NE Dumfries, United Kingdom; francoise.wemelsfelder@sruc.ac.uk*

Qualitative Behaviour Assessment (QBA) asks people to judge and then quantify the expressive quality of animal demeanour (e.g. as relaxed, agitated or lethargic). Such assessments take into account subtle dynamic details of posture and behaviour, and have the potential to contribute finely-judged, context-relevant information to the monitoring of animal health and welfare. Research has shown such information to be useful in animal welfare inspection protocols, but its practical relevance can be extended to animal care situations where identification of shifts in animal expression, as and when they happen, is the main concern. The present study aimed to develop QBA for such use, in a participative project with 7 experienced vet nursing professionals, and domestic cats as focal species. The main objectives were: (1) to generate video footage showing the expressive range of cats held for treatment in vet nurse practices; and (2) based on this footage, to generate QBA terms suitable for cat welfare monitoring, and test participants' level of agreement in using this terminology. The 7 participants between them generated 22 1-minute video clips of individual cats showing a range of pre- and post-operative expressions. Given participants' limited time availability, a focus group approach rather than Free Choice Profiling was used to generate a common terminology. At a first meeting, through initial individual assessment and subsequent collective discussion of the 22 video clips, participants generated and selected 14 QBA descriptors (relaxed, bright, confident, engaged, interactive, tense, aggressive, disorientated, frustrated, frightened, withdrawn, agitated, miserable, wary). At a second meeting, they each individually scored the same 22 clips on each of these terms, using a QBA app installed on electronic tablets. Principal Component Analysis (correlation matrix, no rotation) of scores generated 4 dimensions of cat expression: PC1 (38% of variation): confident/relaxed-tense/frightened; PC2 (16%): agitated-withdrawn; PC3 (11%): frustrated-wary, and PC4 (8%): aggressive-miserable. Kendall's coefficient of concordance showed inter-observer agreement for the ranking of clips on each of the 4 dimensions to be 0.76, 0.65, 0.78 and 0.60 (n=22, df=21, P<0.001) respectively. Given the participative character of video and term selection procedures, this was not an agreement based on entirely independent assessments, but it nevertheless indicates a meaningful shared judgement of cat expressivity, which participants can further validate for their practical use.

Lateralization during agonistic behaviour in pigs

Sophie Menneson[1], Simon P Turner[1], Gareth Arnott[2] and Irene Camerlink[1]
[1]Scotland's Rural College (SRUC), Animal Behaviour & Welfare, Animal Veterinary Sciences Research Group, West Mains Rd., EH9 3JG Edinburgh, United Kingdom, [2]Queen's University Belfast, Institute for Global Food Security, School of Biological Sciences, University Road, BT7 1NN Belfast, Ireland; g.arnott@qub.ac.uk

Cerebral lateralization has been increasingly studied in animal sciences as it can provide information on cognitive processing and emotions, and thereby may contribute to animal welfare. Cerebral lateralization can be observed by behavioural side preferences and has been studied for its strength and for its direction, e.g. across species attacks are mostly made from the left side of the body. Stronger lateralization has been related to increased cognitive abilities and may result in more efficient information gathering and processing. We investigated lateralization during agonistic behaviour of pigs. We hypothesized that pigs would attack from their left side and more lateralized pigs would be better information processors and therefore sooner establish dominance relationships. Weight matched pigs (n=136), unfamiliar to each other, were staged into dyadic contests in a novel arena without resources. Tests were ended when a winner was apparent or after 30 minutes. Tests were approved by the ethical committee and had end-points preventing injury other than skin lesions. Lateralization was recorded by scan sampling the pigs' position towards its opponent every 10 s during agonistic behaviour, and in addition recording the direction of each attack and retreat. Lateralization indexes (LI) were calculated for the direction (-1: left to +1: right) and the strength (0: weakly lateralized to +1: strongly lateralized). Lateralization indexes for direction showed a normal distribution for right and left side use, indicating no population bias. Pigs showed no side preference for attacking the opponent (59% attacked from the left; P=0.45) or retreating (P=0.21). The direction did not influence the contest duration (P=0.51).The strength of lateralization ranged between 0 and 0.76 (average 0.25±0.02). Contests took on average 5 min. Pigs that were strongly lateralized had a shorter contest duration (b=-277±101 s/0.1 LI; P=0.01), which is in line with our hypothesis. This suggests that the duration of aggression between pigs in commercial environments could be minimized by providing an environment in which individuals have sufficient space to properly execute lateralized behaviour.

What is the emotional state of dairy calves and young stock?

Marta Brscic[1], Nina Dam Otten[2] and Marlene Katharina Kirchner[2]
[1]University of Padova, Animal Medicine, Production & Health, Viale dell'Università 16, 35020
Legnaro (PD), Italy, [2]University of Copenhagen, Large Animal Sciences, Groennegaardsvej 8,
1870 Frederiksberg C, Denmark; mk@sund.ku.dk

Indicators of positive welfare states are still lacking in the animal welfare assessment schemes. qualitative behavioural assessment (QBA) using fixed-terms is, however, included in the welfare quality protocol (WQ) for cattle to assess the positive emotional states. Currently there is no standardized QBA protocol available for dairy calves. Therefore the study aimed at using 20 terms (Active-Relaxed-Uncomfortable-Calm-Content-Tense-Enjoying-Indifferent-Frustrated-Friendly-Bored-Positively occupied-Inquisitive-Irritable-Nervous-Boisterous-Uneasy-Sociable-Happy-Distressed) for investigating the dimensions achieved in the principal component analysis (PCA) and standardize QBA in dairy calves. Further, we collected on-farm factors to investigate potential influences on the explaining dimensions of the QBA-PCA. Four veterinarians previously trained for standardized QBA in cattle collected the data on dairy farms in Denmark (40) and in Italy (9); 31 conventional and 18 organic (FARMSTYLE). QBA, applied as first measure on-farm, was recorded on group level (Ø48.2 calves/farm; CALVES_NO). Breeds present on-farm were either Holstein (25), other milk types (13) or mixed (11) (BREED) and out/indoor housing was recorded for different ages (HOUSING). PCA based on correlation matrix was computed to evaluate QBA dimensions (PC). The obtained PC1-score was further used as dependent variable in a linear model (backward/forward procedure based on AIC) explained by the recorded farm factors. PCA summarized QBA descriptors on two main components, PC1 and PC2 (eigenvalues 8.0 and 2.7, explained variance 40.0 and 13.0%, respectively). Descriptors with the highest positive values on PC1 (Valence) were: Content (0.91) and Happy (0.92), whereas Frustrated (-0.74) and Uneasy (-0.72) loaded the opposite direction. PC2 (Arousal), revealed Tense (0.68) and Nervous (0.72) as the most positive terms and Indifferent (-0.36) and Distressed (-0.41) on the negative dimension. The farm factors CALVES_NO, FARMSTYLE, BREED and HOUSING stayed in the model ($P<0.05$), explaining ~20% of PC1's variance. Farms had higher estimates if following the organic disciplinary (+0.66; $P<0.05$), with Holstein (+0.41; $P=0.2534$) and older calves outdoors (+0.32; $P=0.3207$) as well as an increasing number of calves (+0.005; $P<0.05$). We conclude that the 20 terms achieved a relatively high portion of explained variance (54.0%) on PC1 (Valence) and PC2 (Arousal), providing a differentiated view on the emotional state of calves. Moreover, the explained variance by some farm factors indicated a need for further studies on potential influences.

Use of conditioned place preference to investigate emotion transfer in domestic hens

Joanne Edgar, Suzanne Held, Christine Nicol and Elizabeth Paul
University of Bristol, Dolberry Building, Clinical Veterinary Science, Langford, Bristol, BS40 5DU, United Kingdom; j.edgar@bris.ac.uk

Emotional empathy is a multifaceted and multilayered phenomenon in which the emotional state of a subject ('demonstrator') is reflected in the emotional state of an observer. To determine the capacity for emotional empathy, studies must demonstrate existence of emotional responses in both observer and subject. As valence is a defining feature of emotion, we used a conditioned place preference paradigm to investigate the existence of valenced, and therefore emotional, responses in observer and subject chickens. Subjects were a brood of domestic chicks with their mother as an observer. Subjects and observers were conditioned to associate a particular coloured box (red/yellow) with either a control condition or the chicks receiving air puffs. Then, in three separate replicates, subjects and observers were individually allowed to move freely between the boxes and their first choice along with time spent in each box used as an indicator of their preference, and hence a valenced response. Mann-Whitney U tests revealed that the chicks spent less time in the section of the apparatus in which they had received an air puff (AP 37.7±8.4%; z=1.958, P=0.05, r=0.38). Additionally, the chicks' first choice for the control side of the apparatus was chosen significantly more often than by chance in the first two replicates (Binomial: replicate 1: observed 0.74, P=0.02; replicate 2 observed 0.70, P=0.05). This response pattern indicates a conditioned aversion associated with experience of the air puff stimulus. Despite the clear place preference by the chicks, Mann-Whitney U tests showed that observer hens did not spend significantly more time in the control section of the apparatus (AP 44±6.9). However, during the first replicate, their first choice for the control section was significan%tly higher than expected by chance (binomial: observed 0.70, P=0.05). For future studies, although the CPP test appears to be an appropriate and valuable paradigm which could be utilised to measure direct aversion to stimuli, further research is needed to assess its suitability for measuring empathic processes. This paper will address issues of interpretation of CPP, including its relevance as a test of social learning and/or emotion transfer.

Evidence for emotional-lateralization in zebrafish

Mari Kondoh, Daniel M. Weary and Becca Franks
University of British Columbia, Animal Welfare Program, 180-2357 Main Mall, Vancouver, BC, V6T 1Z6, Canada; marikondoh@gmail.com

Zebrafish have become one of the most heavily used animals in scientific research and efforts to understand their welfare are increasingly important. We explored the utility of behavioral-lateralization – the tendency to favour one side of the body over the other – as an indicator of emotional state in zebrafish. Previous research in a wide range of species (from dogs to cichlids) has revealed evidence for emotional-lateralization: using one side of the body to interact with potentially dangerous stimuli and the other side to interact with positive stimuli. While studies with zebrafish have indicated a preferential left-eye use for inspecting novel stimuli, it is unclear whether this tendency is driven by emotional-lateralization. One possibility is that zebrafish use their left-eye for all information gathering regardless of emotional context. Alternatively, as the majority of the previous tests were conducted under potentially stressful conditions (in isolation and in aversive environments), these results could indicate that zebrafish only rely on their left-eye in negative (e.g. fear- or anxiety-inducing) emotional contexts, but would show no preferential use in more emotionally-neutral contexts. To separate negative or risky emotional contexts from more emotionally neutral contexts, we presented novel objects to fish under minimally stressful conditions (within stable social groups and in their hometank), gave them the opportunity to inspect the objects at a distance (less risky) or at close range (more risky), and measured their behavioural lateralization over time. We hypothesized that if zebrafish are emotionally lateralized, they would only preferentially use their left-eye in the most risky contexts – when the objects were first placed in the tank and when performing inspections at close range – and that they would show no lateralization in other contexts. Observing the fish at the tank level (n=6, 10 fish/tank), we tallied the number of left-eye vs right-eye inspections every 10 seconds during the first 100 seconds for each of 9 novel object presentations (generating a total of 51 observations). While object investigation remained high across the 100 second observation period (2.97/10 seconds), zebrafish only favoured their left-eye during close inspections in the first 10 seconds (73%, SE=7; crossed-random effects model to account for repeated sampling of tanks and objects: z=3.34, P<0.01) and showed no evidence of behavioural lateralization for other inspections (49%, SE=5; z=0.25, P>0.8). These results are consistent with an emotional-lateralization pattern of behaviour and suggest that lateralization measures may prove useful in assessing emotional state in zebrafish.

Vocal expression of emotions in European wild boars

Anne-Laure Maigrot[1,2], Edna Hillmann[2] and Elodie F. Briefer[2]
[1]University of Bern, Vetsuisse faculty, Division of Animal Welfare, Länggassstrasse 120, 3012 Bern, Switzerland, [2]ETH Zürich, Institute for Agricultural Sciences, Ethology and Animal Welfare Unit, Universitätstrasse 2, 8092 Zürich, Switzerland; anne.maigrot@usys.ethz.ch

Expression of emotions plays an important role in social species because it regulates social interactions. However, studies testing a direct link between emotions and vocal structure in animals, and especially wild animals, are rare. In particular, little is known about how animals encode in their vocalizations, information about the emotion they are experiencing. In this study, we combined methods developed to study animal emotions (dimensional framework) and vocalizations (source-filter theory), in order to investigate how European wild boars encode information about emotional valence in their vocalizations. To this aim, we recorded vocalizations produced by the animals in naturally occurring situations characterized by different emotional valence. The valence of the situations was inferred from knowledge of the function of emotions and of wild boar behaviour. Negative situations consisted in agonistic interactions and positive situations in affiliative interactions and feed anticipation. As these situations might also differ in their emotional arousal, we assessed arousal levels using movement as a behavioural indicator. We extracted and analysed 256 calls produced by 19 individuals and divided them into four categories (Grunt; Scream, Squeal and Mixed calls). We only considered the calls produced when the calling animal could be easily identified. We then carried out linear mixed effects models (LMM) to test the effect of emotional valence on 19 vocal parameters, including call duration, fundamental frequency and formant contours. P-values were assessed using likelihood-ratio tests. We found several differences between the positive and negative situations. For example, calls produced during the positive situations were shorter (LMM: χ^2=9.91, P=0.002; Positive: 0.4±0.2 s; Negative: 0.9±0.6 s) and less modulated in amplitude (LMM: χ^2=16.02, P<0.0001; Positive: 10.7±8.6 dB/s; Negative: 26.8±15.7 dB/s) than during negative ones, and the harmonicity of calls decreased from positive to negative valence (LMM: χ^2=14.93, P=0.0001; Positive: 4.5±3.2; Negative: 2.6±2.3). The results of these experiments provide better knowledge of the way in which emotions are encoded in wild mammal vocalizations. The vocal indicators of emotions that we found would be particularly useful for welfare assessment of wild species that cannot be easily approached and manipulated. Finally, comparisons of our results with those of similar studies in domestic species could reveal interesting findings about the effect of domestication on the structure of animal vocalizations.

Is mood state reflected in the acoustic parameters of pig vocalisations?

Mary Friel[1], Hansjoerg P Kunc[1], Kym Griffin[2], Lucy Asher[3] and Lisa M Collins[4]
[1]Queen's University Belfast, School of Biological Sciences, 97 Lisburn Road, Belfast BT9 7BL, United Kingdom, [2]Nottingham Trent University, School of Animal Rural & Environmental Sciences, Burton Street, Nottingham NG1 4BU, United Kingdom, [3]Newcastle University, Centre for Behaviour and Evolution, Henry Wellcome Building, Framlington Place, Newcastle NE2 4HH, United Kingdom, [4]University of Lincoln, School of Life Sciences College of Science, Brayford Pool, Lincoln, Lincolnshire LN6 7TS, United Kingdom; mfriel06@qub.ac.uk

The cognitive bias test has emerged as a valuable technique for investigating the valence of underlying emotional state in non-human animals. Optimistic responses to ambiguous stimuli are believed to be due to a positive underlying emotional state, while pessimistic responses to ambiguous stimuli are thought to be caused by a negative underlying emotional state. Vocalisations are known to reflect the emotional state of the caller and they have the advantage of being relatively easy to record and measure. There is increasing interest in exploring whether emotional valence is reflected in the acoustic parameters of vocalisations, yet the relationship between acoustic parameters and optimistic/pessimistic responses in the cognitive bias test has yet to be investigated. Here we assessed the relationship between the acoustic parameters of vocalisations produced during the cognitive bias test and optimistic/pessimistic responses to the ambiguous stimuli. Twenty seven 9 week old pigs were trained and tested in a location based cognitive bias test and vocalisations were recorded during the test then analysed using the acoustic analysis software PRAAT. Linear mixed effects models were used to investigate the relationships between the acoustic parameters of the vocalisations and response type (optimistic or pessimistic). Call duration was found to be significantly predicted by response type with pessimistic responses being associated with longer calls (means±S.E.: pessimistic = 0.42±0.04 sec; optimistic = 0.33±0.02 sec, $\chi^2(1)=6.416$, P=0.0113). These results link two methods for measuring emotions and provide the basis for future research to test predictions on the effect of emotional valence on acoustic parameters in different contexts. If the effects of valence on acoustic parameters can be reliably identified then measuring vocalisations in practical settings will enable quick and reliable assessment of the emotional status of vocalising animals. This will be invaluable for animal welfare science, where reliable, valid and practical welfare indicators are urgently needed.

Eye preferences in response to emotional stimuli in captive capuchin monkeys (Sapajus apella)

Duncan Andrew Wilson[1], Sarah-Jane Vick[2] and Masaki Tomonaga[1]
[1]Kyoto University Primate Research Institute, Language and Intelligence Section, Kanrin-41-2Inuyama City, 484-8506 Aichi Prefecture, Japan, [2]The University of Stirling, Division of Psychology, Stirling, FK9 4LA Stirling, United Kingdom; duncan.wilson.76x@st.kyoto-u.ac.jp

Eye preferences may provide a window into emotional responses in animals, which are important for predicting behaviour. The study investigated whether captive capuchin monkeys show eye preferences consistent with the Valence hypothesis of emotional processing; a right eye preference (left brain hemisphere dominance) for viewing positively valenced stimuli, and a left eye preference (right brain hemisphere dominance) for viewing negatively valenced stimuli was predicted. Eleven monkeys were presented with four images of different emotional valence (an egg and a capuchin monkey raised eyebrow face were categorised as positively valenced, and a harpy eagle face and a capuchin monkey threat face as negatively valenced) and social relevance (consisting of a capuchin monkey face or not) in a within-subjects design during 22 hours of research sessions. During each two-hour research session several monkeys were individually presented with all four images (30 seconds presentation of each image) on a computer screen in a 4×4 Latin square design. Eye preferences and arousal behaviours for viewing the images through a monocular viewing hole were recorded. Strong individual-level eye preferences were found; seven monkeys had a left eye preference and four monkeys had a right eye preference for viewing all the images (binomial tests, P<0.05 for all). However, eye preferences did not differ with the emotional valence of the images (t(10)=1.27, P=0.23), and so the results do not support the Valence hypothesis. As the stimuli were presented as images, rather than real objects, this may have reduced their emotional salience. However, the monkeys showed more arousal behaviours for viewing the negatively valenced images (M=0.66 (SEM=0.13) behaviours per session) than positively valenced images (M=0.40 (SEM=0.09) behaviours per session), F(1,10)=5.74, P=0.04, suggesting the stimuli had a degree of emotional salience. The strong individual-level eye preferences found may reflect the sighting-dominant eye (the eye used for monocular viewing tasks) and the constraint of the monocular viewing task itself, independently of hemispheric specialisations of emotional processing. Further research is needed to test this hypothesis.

Comparing positive and negative welfare indicators and explanatory factors across age groups in Danish dairy cow herds

Nina Otten and Marlene Kirchner
University of Copenhagen, Large Animal Sciences, Grønnegårdsvej 8, 1870 Frederiksberg C, Denmark; nio@sund.ku.dk

Positive welfare consequences can be assessed using behavioural observations alike the Qualitative Behaviour Assessment (QBA) while negative welfare consequences mostly are asessed as clinical measures. The ultimate negative welfare consequence, however, is death. The present study aimed at investigating patterns of predictive factors for the positive and negative welfare consequences for cows and calves (0-180 days). A study population of 25 Danish dairy herds (>100 cows, loose housed) was evaluated. On-farm recordings of positive welfare were performed by means of QBA developed for dairy cows and an adapted version for dairy calves. Outcome scores for positive welfare were analysed according to the Welfare Quality protocol yielding a criterion score for cows ($S_{QBA\ COWS}$) and calves ($S_{QBA\ CALVES}$). Register data on production type, herd size and proportions of mortality were obtained from national databases. Four outcome variables were assessed: $S_{QBA\ COWS}$, $S_{QBA\ CALVES}$, cow mortality and calf mortality for correlations (Spearmans rho) and potential predictive factors in multivariable models. Mean scores for positive welfare differed between age groups: $S_{QBA\ COWS}$ 71 points (SD=19.5, min-max: 36-96), $S_{QBA\ CALVES}$ 53 points (SD=7, min-max: 38-60), while medians for negative welfare consequences were equal: mortality cows 4.9% (min-max:1.7-9.8) and mortality cows 3.4% (min-max: 0.6-22.5). There were no significant correlations between the positive and negative welfare outcomes neither within or between the two age groups. The final linear models showed tendencies towards lower $S_{QBA\ CALVES}$ in organic herds (estimate=-8.1) together with increased calf mortality (estimate=0.6) (P=0.1, R^2=0.12); cow mortality tended to be lower in small herds (P=0.07). Results showed good positive welfare for both cows and calves in Danish dairy herds, however, the missing significant associations highlight the missing consistency in patterns for predicting welfare outcomes in dairy cows and calves.

Inter-observer reliability in applying Qualitative Behaviour Assessment to dairy goats: the importance of assessor training

Monica Battini[1], Sara Barbieri[1], Ana Vieira[2], Edna Can[2], George Stilwell[2], Françoise Wemelsfelder[3] and Silvana Mattiello[1]
[1]*Università degli Studi di Milano, Via Celoria 10, 20133 Milan, Italy,* [2]*Universidade de Lisboa, Avenida da Universidade Técnica, 1300-477 Lisbon, Portugal,* [3]*SRUC, Roslin Institute Building, Easter Bush, Midlothian EH25 9RG, United Kingdom; monica.battini@unimi.it*

This study examines the effectiveness of assessor training on the on-farm inter-observer reliability of 13 fixed-list Qualitative Behaviour Assessment (QBA) descriptors for goats. The study was conducted during the prototype testing of the AWIN welfare assessment protocol. Two assessors per country, for a total of four assessors, performed the observations independently of each other on 20 intensive dairy farms (10 in Italy and 10 in Portugal). None of the assessors had previous experience with QBA. The two Italians and one Portuguese together received the same training on QBA including both a theoretical (e.g. video clips representing different goat emotional states, later discussion) and a practical (e.g. scoring simulation on farm) session. The second Portuguese assessor only received a short explanation of the QBA principles. All assessors had previous experience with farm animals, but only two of them, one for each country, had sound experience with dairy goats. In Portugal, the assessor with experience on dairy goats was also the same who received the complete training on QBA. Data from each country were analysed separately with Principal Component Analysis (PCA – correlation matrix, no rotation). In the Italian analysis, PC1 was labelled as relaxed – aggressive, and PC2 as content – bored, explaining 28.9 and 26.5% of variance respectively. For Portugal, PC1 was labelled as content – agitated and PC2 as curious – bored, explaining 33.3 and 15.7% respectively. Depending on the scores' distribution, assessors' agreement for farm scores on PCs was investigated using either Pearson (r) or Spearman (ρ) correlations. The two Italian assessors showed excellent agreement on both PC1 (r=0.910, P=0.001) and PC2 (r=0.906, P=0.001), while the two Portuguese assessors showed no agreement on either PC1 (ρ=0.539, P=0.108) or PC2 (ρ=0.224, P=0.533). Due to logistic restrictions, sample size in this study was low; however these results suggest that adequate training can contribute to good assessor agreement, and also that absence of experience with a particular species can be overcome by such training, as was the case for the Italian assessors.

Equine welfare assessment: the use of QBA to evaluate positive emotional state

Francesca Dai[1], Emanuela Dalla Costa[1], Elisabetta Canali[1], Leigh Murray[2], Philipp Scholz[3], Francoise Wemelsfelder[4] and Michela Minero[1]
[1]Università degli Studi di Milano, DIVET, Via Celoria 10, 20137 Milano, Italy, [2]British Society of Animal Science, Penicuik, Penikuik, United Kingdom, [3]Polish Academy of Sciences, WólKa Kosowska, WólKa Kosowska, Poland, [4]Scotland's Rural College, West Mains Road, Edinburgh EH9 3JG., United Kingdom; francesca.dai@unimi.it

The assessment of equine welfare should include not simply the absence of negative experiences, but also the presence of positive emotional states. Qualitative Behavioural Assessment (QBA) has previously been successfully used to quantify positive aspects of animal welfare on farm. However, using QBA on equine facilities poses some difficulties: for instance, in horses the barren environment of single boxes might limit the expression of affective states. Also, the assessment of donkeys at group level was reported as difficult by assessors. The EU-FP7 AWIN project developed welfare indicators for horses and donkeys, including QBA. The aim of this study was to apply QBA to the assessment of positive welfare in equines. Two fixed QBA rating scales, one for horses (13 terms) and one for donkeys (16 terms), were developed by experts during the AWIN project and then used on 40 horse facilities (n=355) and 20 donkey facilities (n=20). In horses, the assessor first observed subjects undisturbed in their box for 30 seconds, and then for a further 30 seconds while grooming their withers. In donkeys, the assessment of each group took place outside of the paddocks for 10-15 min without disturbing the animals. Assessors had never entered the farms before and performed QBA assessments first, before any of the other indicators were measured. At the end of observation, assessors scored the individual horses or donkey groups on each of the QBA terms for their respective species. Principal Component Analysis (PCA – correlation matrix, no rotation) and ANOVA were used to analyse data. In horses, PC1 (44% of total variance) ranged from friendly/relaxed to annoyed/uneasy, and appears to reflect general mood. Horses that were pre-identified as 'positive' by having obtained only positive evaluations on all other welfare indicators (n=20), were characterised by significantly higher QBA scores on PC1 (P<0.01) than horses with negative evaluations (n=15). PC2 (17%) ranged from apathetic to aggressive and appears to reflect level of arousal. In donkeys, PC1 (33%) ranged from at-ease/relaxed to agitated/distressed reflecting aspects of both general mood and level of arousal, while PC2 (19%) ranged from happy/friendly to anxious/fearful, reflecting mostly a shift in general mood. In conclusion, the application of QBA to equines allowed identification of positive emotional expressions in both individuals and groups of animals.

The long reach of early life: how developmental experience shapes adult behaviour in the European starling

Daniel Nettle
Newcastle University, Henry Wellcome Building, Newcastle NE2 4HH, United Kingdom;
daniel.nettle@ncl.ac.uk

In many long-lived species, experience during early life can have effects on adult health and behaviour. Moreover, it is not just early trauma that can do this, but subtle variations of experience within the normal range. We can conceive of early-life inputs as fulfilling two roles in development. First, they provide raw materials, most obviously in the form of food, but also other kinds of necessary resources. Second, early-life inputs provide information about the state of the world and the individual's likely prospects within it. Some of the effects of early-life conditions, particularly the health effects, simply represent variation in the raw materials: you can't build as robust a phenotype if you are given less to work with. Others may represent adaptive calibration to future needs or environments. I will review our recent work on early-life effects in the European starling. Birds have a number of advantages over mammals as systems for the study of early life. They grow to maturity rapidly, and parental investment is direct rather than via lactation, making it easier to record and manipulate. Moreover, starlings readily allow for hand-rearing and cross-fostering. This has allowed us to conduct a series of experiments where siblings have been exposed to subtly different early-life period, and their behaviour in adulthood. I will present two sets of results. The first shows that restricting resources or increasing stress during early life leads to subtle deficits: for example, in the form of reduced flight performance and accelerated cellular ageing. These results are readily interpreted as consequences of having worse raw materials to work with in building a phenotype. I will then present results showing apparently strategic differences in the behaviour of birds as adults, particularly in respect of foraging, according to the developmental histories they have had. These results are more readily interpreted as adaptive shifts in priorities given early experience. The remarkable thing about these shifts is how persistent they are: our early-life manipulations often last just ten or twelve days, whereas the consequences are detectable over a year later.

Effects of prenatal stress and postnatal enrichment on exploration and social behaviour of suckling piglets

Sophie Brajon[1], Nadine Ringgenberg[1,2], Stephanie Torrey[1,2], Renée Bergeron[2] and Nicolas Devillers[1]
[1]*Sherbrooke R & D Centre, 2000 College st., Sherbrooke, QC, J1M 0C8, Canada,* [2]*Campbell Centre for the Study of Animal Welfare, 247 ANNU Building, University of Guelph, Guelph, ON, N1G 2W1, Canada; nicolas.devillers@agr.gc.ca*

Prenatal stress can affect behaviour of progeny, such as increasing emotional reactivity, anxiety and depressive-like behaviours. We hypothesised that enrichment can counteract negative effects and improve welfare of prenatally stressed animals. The effects of prenatal stress and enriched housing on the behaviour of suckling piglets were investigated in 41 litters. During gestation, sows had been assigned to either a mixing stress (T) or a control treatment (C). At lactation, half of the T and C sows were housed with their litter in straw enriched pens (E) while the others were housed in standard farrowing crates (S). Piglet exploration (5-min scan sampling) was recorded at d20 while piglet posture was recorded at d6, d12 and d20. One minute one-zero sampling was used to record locomotor and fighting play on the same days. Piglets were also individually subjected to a 5-min social isolation on d17. Data were analysed per day using mixed models with mixing stress, enrichment and interactions between factors as fixed effects. Less T piglets were laying down at d6 (mean±SEM: 64.1±1.0 vs 67.3±1.0% of piglets per scan, P=0.03), compared to C piglets. More E piglets were laying down than S piglets on d12 (61.6±1.8 vs 54.1±1.8% of piglets, P=0.006) and d20 (56.8±2.3 vs 48.8±2.4% of piglets, P=0.02) and they explored more their environment on d20 (16.0±1.1 vs 11.0±1.1% of piglets, P=0.004). However, E piglets performed less locomotor play (d12: 22.3±1.9 vs 29.0±1.9% of samplings, P=0.02; d20: 27.2±2.7 vs 36.2±2.8% of samplings, P=0.03) and fighting play (d12: 24.1±2.0 vs 30.6±2.0% of samplings, P=0.03; d20: 25.6±2.8 vs 35.1±2.9% of samplings, P=0.03) than S piglets. During the social isolation, E piglets attempted to escape more often (mean[CI]: 6.2[4.7;8.3] vs 2.7[1.8;4.0] times, P=0.001) and they exhibited high vocalisations more often (16.7 [11.5;24.1] vs 8.44 [5.7;12.4] times, P=0.01) and for longer durations (0.52±0.02 vs 0.44±0.02 s, P=0.008) than S piglets. Finally, the mean central frequencies of low calls were 466.5±18.6 vs 530.8±18.2 Hz (P=0.02) for T and C piglets, respectively. To conclude, effects of prenatal stress on behaviour were not counteracted by the enrichment (no interaction between factors), but they did not appear detrimental to welfare. Although the housing enrichment decreased piglet social interaction in the pen and was associated with greater emotional reactivity during social isolation, it may still improve welfare since it increased comfort (i.e. laying down) and promoted foraging behaviour.

Cognition of dairy calves exposed to nutritional enrichments during the milk-feeding stage

Kelsey Horvath and Emily Miller-Cushon
University of Florida, Department of Animal Sciences, 2250 Shealy Drive, Gainesville, FL, 32611, USA; kchorvath@ufl.edu

Environmental complexity is known to affect cognitive development, but it is common to house dairy calves individually in limited environments. The hypothesis of this study was that providing calves with simple nutritional enrichments may improve their success in a cognitive task and reduce reactivity to novelty. Individually-housed Holstein heifer calves were assigned at birth to conventional management (C; n=10), with access to milk via bucket (6 l/d) and grain concentrate *ad libitum*, or enriched feeding (E; n=10), with access to milk (6 l/d) via an artificial teat to allow natural suckling, and access to both chopped hay and concentrate *ad libitum*. Sample size was estimated to provide 80% power to detect treatment differences. At week 5 of age, calves were tested in a T-maze with a reward (0.2 l milk; fed according to familiar method) to assess: initial spatial learning, reversal learning (reward location was changed to the opposite arm), and response to an intramaze change (a novel object, a colored ball, was placed in the maze). Calves were tested in 5 sessions/d for 5 days or until they met the learning criteria (moving directly to correct side in 3 consecutive sessions). The time to complete test, movement in maze, kicks, and non-nutritive licking were recorded from video. Behavioral data were analyzed separately by stage in a general linear mixed model with session as a repeated measure, and sessions required to meet criteria were analyzed using the Wilcoxan signed rank test. In initial learning, the number of sessions required to meet criteria was similar between calves (P=0.12). However, E calves took longer to complete the task in early sessions (treatment by session interaction; P=0.02), due to increased time spent on the correct side of maze before obtaining the reward (26.72 vs 7.48 s; SE=3.4; P=0.005). In the reversal learning stage, there was no overall difference in the number of sessions to meet criteria (P=0.20), but E calves completed the task faster (19.84 vs 27.22 s; SE=2.10; P=0.03), and C calves spent 1.5× longer on the incorrect side of the maze than E calves (P=0.04), suggesting that C calves struggled to relearn the task. In both initial and reversal learning, C calves kicked more frequently (P<0.04). During the novel object session, E calves found the reward faster (6.11 vs 20.6 s; SE=4.06; P=0.01), whereas C calves spent longer in the middle of the maze where the novel object was located (2.08 vs 13.4 s; SE=5.33; P=0.04), and spent 5× more time licking (P=0.02). These results suggest that providing simple feeding enrichments or nutritional changes during the milk feeding stage may alter calf cognition and influence responses to environmental changes.

Impact of perinatal nutrition on spatial cognitive performance of pigs later in life

Caroline Clouard[1], Bas Kemp[1], Walter J. J. Gerrits[1], David Val-Laillet[2] and J. Elizabeth Bolhuis[1]
[1]Wageningen University, Animal Sciences, P.O. Box 338, 6700 AH Wageningen, the Netherlands,
[2]INRA, UR1341 ADNC, Domaine de la Prise, 35590 Saint Gilles, France; caroline.clouard@wur.nl

Perinatal nutrition can program cognitive functions in mammals. Accordingly, excessive intake of dietary fat and/or sugar during pregnancy and/or lactation has been found to alter learning and memory of offspring in rodents. Little is known, however, on the impact of perinatal nutrition, and the most sensitive period (prenatal or postnatal) for dietary programming of cognition in the domestic pig. We investigated the effects of prenatal vs postnatal exposure to a high-fat high-sugar (HFHS) diet on cognition of pigs. Thirty-two sows and their litters were allocated to 1 of 4 dietary treatments in a 2×2 factorial design, with 8-week prenatal (gestation) and 8-week postnatal (lactation and post-weaning) exposure to a HFHS diet or control (standard commercial) diets as factors. From weaning onwards (4 weeks of age), 3 piglets per litter were selected and housed in pens of 3 littermates. Starting 3 weeks after the dietary intervention, piglets were subjected to a spatial cognitive hole-board task, in which they had to discriminate 4 baited buckets out of 16 in a fixed configuration. After 28 acquisition trials, piglets were subjected to 16 reversal trials, in which the configuration was changed. Reference memory (RM) score was calculated as the ratio between the number of visits to the set of baited buckets and the number of visits to all buckets, and working memory (WM) score as the ratio between the number of different buckets visited and the total number of (re) visits. Data were analysed using a mixed model with pen as experimental unit. In both phases of the task, WM and RM scores increased over trials (P<0.001). Compared to controls, piglets exposed to the prenatal HFHS diet showed higher WM scores (0.79±0.01 vs 0.77±0.01, P=0.05) in the acquisition phase, and a trend for higher WM (0.83±0.01 vs 0.81±0.01, P=0.08) and RM (0.39±0.01 vs 0.37±0.01, P=0.06) scores in the reversal phase. In both phases, piglets exposed to the prenatal HFHS diet made fewer visits in total (acquisition: 14.9±0.36 vs 16.1±0.31, P=0.02 and reversal: 14.0±0.48 vs 15.4±0.46, P=0.04) and tended to have longer inter-visit intervals than controls (acquisition: 6.03±0.66 vs 4.74±0.40 sec, P=0.07 and reversal: 4.80±0.58 vs 3.64±0.19 sec, P=0.05), suggesting that enhanced motivation did not explain their improved cognitive performance. No effect of the postnatal diet or its interaction with the prenatal diet was found. Our study highlights the key role of late prenatal, but not postnatal, nutrition for programming of spatial memory in pigs, which may impact their ability to cope with later life challenges.

Effects of prenatal stress (PNS) on lamb behaviour, perinatal temperature and weight change

Leonor Valente[1,2], Kenny Rutherford[1], Carl W Stevenson[2], Nadiah Yusof[1], Jo Dobnavand[1], Kevin Sinclair[2] and Cathy Dwyer[1]
[1]*Scotland's Rural College, Animal Behaviour and Welfare, West Mains Road, Edinburgh, EH9 3JG, United Kingdom,* [2]*University of Nottingham, School of Biosciences, Sutton Bonington, LE12 5RD, United Kingdom; leonor.valente@sruc.ac.uk*

Stressful conditions, experienced in pregnancy, can have a detrimental effect on the development of offspring. However, to what extent husbandry conditions experienced by housed pregnant sheep have similar impacts is not known. We hypothesised that conditions such as high stocking density and social instability during pregnancy will have a negative effect on lamb survival, growth and behaviour. At week 11 of gestation we allocated 77 twin-bearing Scottish Mule ewes to either Control (C, n=42) or Stress (S, n=35) groups of 7 ewes balanced for parity. Groups differed in lying area (C: 2.70 m^2/ewe, S: 1.19 m^2/ewe), feed space (C: 71 cm/ewe, S: 36 cm/ewe) and social stability (C: remained undisturbed, S: underwent two social mixing events). All ewes were released into Control conditions one week prior to the start of lambing. Lamb vocalizations were counted for 30 min after birth. Body temperature was measured at 30 min, 2, 24, and 48 h after birth. Lambs were weighed at 24 h, at weaning and at slaughter. At 24 h the ability of lambs to recognise their dam was tested. There was an interaction (P=0.003) between PNS and lamb sex as C females vocalised more than C males (44.2 vs 24.1) whereas S females vocalised less than S males (28.04 vs 43.03; GLMM F=9.22). PNS also interacted (P=0.044) with lamb sex for temperature at 24 h as C females had lower temperatures than C males (39.3 vs 39.4 °C) but S females had higher temperatures than S males (39.4 vs 39.2; GLM, F=4.16). There was an interaction (P=0.017) between PNS and dam parity on lamb temperature at 30 mins with C lambs from multiparous ewes having higher temperatures than those of primiparous ewes (40.1 vs 39.4 °C), whereas S lambs had similar temperatures regardless of ewe parity (GLM F=6.05). PNS had no effect on lamb weight gain. In the choice test lambs spent more time close to their dam than to the alien ewe (83.5 vs 29.5 secs; Z=201, P<0.001). PNS did not affect latency to reach the dam or relative amounts of time spent close to the dam or the alien. Maternal stocking density and social instability can affect lamb body temperature and neonatal behaviour but effects are dependent on the parity of the ewe and sex of the lamb which might mean that some groups are more affected than others.

Development of fearfulness in horses: modulated by maternal care?

Janne Winther Christensen, Line Peerstrup Ahrendt and Jens Malmkvist
Aarhus University, Animal Science, Blichers Allé 20, 8830 Tjele, Denmark; jwc@anis.au.dk

Naturally occurring variations in the level of active maternal care have been shown to influence behavioural development and stress sensitivity in rodent offspring. This has not yet been investigated in horses, although emotional responses are of primary importance for both horse welfare and safety in the horse-human relationship. This study was part of a larger project investigating the development of fearfulness, stress sensitivity and learning capacity in foals and the potential association to variations in maternal care. Preliminary results from 44 mare-foal pairs on one stud are presented here. The horses were kept 24 h on pasture (min 2 pairs/pasture). Mare-foal interactions (e.g. suckling, vocalisations, aggression, grooming) and pair-distance were registered at pasture 2×2 h per week, during weeks 1-8 and 17-20 after birth. Fecal samples for analysis of fecal cortisol metabolites (FCM) were collected weekly from the mares and foals (foals only weeks 17-20, plus before and after abrupt weaning at 7-9 months of age). Fear tests (exposure to novel objects) were conducted when the foals were 20 weeks old, i.e. before weaning, and again after weaning at 12-14 months of age. The fear tests at 20 weeks were designed to measure fear reactions in the foal and mare independently, i.e. in the foal fear test only the foal was exposed to the novel objects while the mare was kept outside the test arena. Mares and foals were habituated to the test procedures before the tests. Behavioural (e.g. latency to pass the novel object) and heart rate (HR) reactions were measured. Behaviour and HR reactions correlated within tests at 20 weeks but were in opposite directions in mares and foals; foals with high HR had short latencies to pass the objects whereas mares with high HR had long latencies (e.g. foal fear test, latency to pass and HR (Spearman): r_s=-0.35, P=0.025 and mare fear test: r_s=0.60, P<0.001). This likely relates to the test set-up where the mare was used to motivate the foal to pass the objects, whereas food was used to motivate the mares. There were modest positive correlations in HR between mares and foals in the different tests (e.g. foal HR in foal fear test and mare HR in mare fear test (r_s=0.33, P=0.04), suggesting a link between mare and foal fearfulness. FCM values of the foals increased significantly in response to abrupt weaning and had returned to baseline on day 10 after weaning (RM ANOVA; ng/g (mean±se): Day 0: 19.5±1.8, Day 1: 30.9±1.9, Day 2: 27.6±1.2, Day 3: 27.3±1.3, Day 10: 19.5±0.9, F=15.4, P<0.001). The relationship between fearfulness, baseline and acute stress responses, and mare-foal interactions in the early post-natal period remain to be explored and will be presented.

Rearing complexity affects fearfulness and use of vertical space in adult laying hens

Margrethe Brantsaeter[1], Janicke Nordgreen[1], T. Bas Rodenburg[2], Fernanda M. Tahamtani[1], Anastasija Popova[1] and Andrew M. Janczak[1]
[1]*Norwegian University of Life Sciences, Prod. med., Ullevålsveien 72, 0454 Oslo, Norway,* [2]*Wageningen University, Behavioural Ecology Group, P.O. Box 338, 6700 AH Wageningen, the Netherlands; margrethe.brantsaeter@nmbu.no*

The complexity of the rearing environment is important for behavioral development in laying hens. This study aimed to test the hypothesis that laying hens reared in a complex aviary system would be less fearful and less sensitive to stress as adults than hens reared in barren cage environments. Laying hens (n=160) were reared in the same house, either kept inside the aviary rows to imitate cages (n=80) or in the conventional aviary system (n=80). The birds were then transported to the experimental facilities at 16 weeks of age and mix-housed in custom built pens. Two birds per treatment per pen where tested in human approach and novel objct tests at 19 and 23 weeks of age. The results were analyzed by ANOVA on scores for a fearfulness-related principal component, generated using principal component analysis. Flight responses and the relationship between flight response and fearfulness score were analyzed by logistic regression. Use of vertical space was analyzed by Wilcoxon matched pairs signed rank test. Aviary-reared (AV) birds had lower levels of fearfulness compared with cage-reared (C) birds both at 19 weeks (mean±std AV: 0.24±1.56; C: 0.74±1.72; $F_{(1,19)}$=5.66; P=0.03) and at 23 weeks (AV: -0.69±1.62; C: 0.19±2.11; $F_{(1,19)}$=4.49; P=0.05) of age. At 19 weeks of age, more aviary-reared birds tended to show a flight response compared to cage-reared birds (OR=2.4; P=0.09) when initially exposed to a novel object. Rearing complexity did not affect the flight response at 23 weeks of age (OR=1.4; P=0.6). The odds ratio of showing a flight response when exposed to the novel object was not influenced by the principal component fearfulness score (OR=0.89; P=0.2). At 19 weeks of age, more aviary-reared birds spent time on the low perch (median, 25th-75th percentile AV: 20.8, 14.4-23.0; C: 12.5, 10.4-18.4; P=0.01), the elevated platform (AV: 19.2, 13.0-25; C: 12.1, 6.3-22.1; P=0.01) and upper perch (AV: 16.7, 10.4-23.0; C: 5.2, 2.1-14.1; P<0.0001) compared to the cage-reared birds. However, at 23 weeks of age, these differences were not detected (P>0.0001). Conclusively, increased environmental complexity during rearing reduced fearfulness at 19 and 23 weeks of age and increased vertical space use at 19 weeks of age in laying hens.

Prenatal learning: a forgotten, but essential driver for behavioural development

Peter Hepper and Deborah Wells
Queens University Belfast, Animal Behaviour Centre, Psychology, Belfast, BT7 1NN, United Kingdom; p.hepper@qub.ac.uk

When Tinbergen first posed his question regarding how did / does behaviour develop, consideration was largely given to genetic influences before birth and experiential influences after birth. One area that may have significant influence for the development of behaviour and subsequent well-being is the prenatal period (conception to birth). Recently, studies have focused on the prenatal period largely with a view to determining how the prenatal environment may influence the structural and functional development of the individual's organs and physiology, e.g. maternal stress. The 'Barker hypothesis' demonstrated that experiential influences before birth may set the life-long functioning of certain organs. Where the programmed expectations for functioning in a particular environment in adult life are not met, the result can be ill-health and poor welfare. Comparatively less interest has examined the individual's ability to learn before birth. However, the ability to learn provides the organism with a much more focused and flexible ability to respond to its individual circumstances. Such an ability will be highly adaptive as the individual can tailor its response to the specific circumstances present in its environment at that time. Recent evidence suggests that prenatal learning may profoundly influence subsequent behaviour and behavioural preferences exerting a lifelong effect on the animal's behaviour. In this presentation I explore how, in the normal course of prenatal development, the individual will learn features of its environment, which effect its behaviour and preferences after birth and the survival advantages this confers upon the individual. Evidence is presented from all the major vertebrate groups: fish, amphibian, reptile, bird (domestic chicken) and mammal that natural 'prenatal' learning of stimuli influences subsequent preferences for that stimulus and this preference is retained into adulthood. Prenatal learning appears a ubiquitous phenomenon in vertebrates, following a similar developmental path and influencing behaviour over the individual's lifespan. Further, the stimuli that the individual is exposed to are not random but rather have been tailored by the evolutionary process to ensure that it is stimuli 'essential' for survival that have preferential access to the individual before birth and can be learned. Prenatal learning confers advantages in dietary selection, suckling in mammals and recognition of caregivers and kin. The talk concludes that prenatal learning enables the individual to adapt uniquely to its environment before and after birth and is a major driver for successful development. Thus, prenatal learning must be accounted for in any attempt to address Tinbergen's question of how does behaviour develop.

Ontogeny of coping styles in commercial sows: from birth to adulthood
Kristina Horback and Thomas Parsons
University of Pennsylvania School of Veterinary Medicine, 382 W Street Road, Kennett Square, PA 19348, USA; khorback@vet.upenn.edu

A coping style can be defined as a set of behavioral and physiological stress responses which are consistent over time and across contexts. Proactive individuals are reported to be more aggressive and active while reactive individuals are less aggressive and more cautious. Knowledge of coping styles among gestating sows could help to improve individual welfare by identifying those best suited for living in the complex environment of group housing. In order to assess the existence and temporal consistency of coping styles in sows, 36 genetically similar sows were observed from birth through fourth parity. The weight, response to human handling (duration vocalize/struggle), and behaviors during an open field test (OFT) were recorded when each sow was 3 and 5 weeks old. One year later, each sow was evaluated for her reaction toward a human handling her piglets 24 h after first farrowing. After weaning, aggressive and exploratory behaviors were recorded for 1 h during the sow's introduction to a large group pen. To investigate the relationship between coping styles and affective state, cognitive bias testing was carried out during the second gestation period. Spearman rank-order correlations were calculated among physical and behavioral variables. Sows which strongly resisted being handed at 3 weeks old displayed lower rates of retreating from a fight during pen mixing at 1 year old (r_s=-0.5, P<0.05), while those which strongly resisted to being handled at 5 weeks old engaged in more fights (r_s=0.5, P<0.01) and initiated more bites (r_s=0.5, P<0.05) during pen mixing at 1 year old. Sows which jumped at the walls during OFT at 3 weeks old displaced others more often (r_s=0.5, P<0.05) and nosed/sniffed others more often (r_s=0.4, P<0.05) during pen mixing at 1 year old. Sows which reacted aggressively toward a human when their first litter was handled weighed more at 3 and 5 weeks old (r_s=0.4, P<0.05; r_s=0.4, P<0.05), engaged in longer fights (r_s=0.5, P<0.01), initiated more bites (r_s=0.5, P<0.05), chased others more often (r_s=0.7, P<0.01) and displaced others more often (r_s=0.4, P<0.05) during pen mixing at 1 year old. Sows which displayed a more positive cognitive bias during their second gestation strongly resisted to being handled at 3 weeks old (r_s=0.4, P<0.05), initiated more bites (r_s=0.5, P<0.05) and chased others more often (r_s=0.7, P<0.01) during pen mixing at 1 year old. These results support the existence of coping styles in sows, with proactive copers displaying more aggressive and active behaviors throughout their lives. These results also suggest that coping styles may play a role in assessing affective state. Further research is needed to address whether proactive copers actually experience a more positive affective state.

Early life in a cage environment adversely affects spatial cognition in laying hens (*Gallus gallus domesticus*)

Fernanda M. Tahamtani[1], Janicke Nordgreen[1], Rebecca E. Nordquist[2] and Andrew M. Janczak[1]
[1]Norwegian University of Life Sciences, Animal Welfare Research Group, Department of Production Animal Clinical Science, Ullevålsveien 72, 0454 Oslo, Norway, [2]University of Utrecht, Emotion and Cognition Group, Department of Farm Animal Health, Yalelaan 7, 3584 CL Utrecht, the Netherlands; fernanda.machado.tahamtani@nmbu.no

Aviary and cage-rearing systems for laying hens differ from one another in environmental complexity. Spatial cognition, important for navigating the environment and finding resources, is adversely affected by a lack of environmental complexity during early life. However, to our knowledge, no previous studies have tested the effect of early exposure to varying degrees of environmental complexity on specific components of spatial cognition in chickens. The aim of the present study was to test the hypothesis that rearing in a cage environment relative to an aviary environment causes long-lasting deficits in the ability to perform spatial tasks. 52,000 white Dekalb laying hens were reared in separate corridors of a commercial rearing farm. Each corridor had either a cage- or an aviary-rearing system with densities of 25 and 12 birds/m^2 respectively. The cage system consisted of aviary-row cages that remained closed throughout the rearing period. All housing conditions used are consistent with current Norwegian legislation. At 16 weeks of age, 24 hens, 12 from each rearing system, were transported to experimental facilities, housed in furnished cages, and trained and tested in a holeboard task. The holeboard is an open-field arena with nine cups, three of which are baited with a food reward. Over time of exposure to the arena, the birds are expected to remember the position of the baited cups. Trial duration and working memory performance for each trial were recorded and tested in a repeated-measures-ANOVA. Aviary and cage-reared hens did not differ in trial duration and working memory during training, indicating that birds from both treatments were physically capable of performing the task. However, during the final test phase, the cage-reared hens required more time to locate rewards (Aviary: 59.95±7.02 s. Cages: 161.21±14.51 s. $F_{3,57}$=3.897; P=0.013; post hoc t=2.99; P=0.004) and had poorer levels of working memory (Aviary: 0.84±0.02. Cages: 0.65±0.04. $F_{3,57}$=2.75; P=0.051; post hoc t=-2.88; P=0.005). The findings thus support the hypothesis that rearing in a cage environment causes long-term impairment of short-term memory in chickens.

Maternal age affects leghorn chick response to the Anxiety Depression Model

Leanne Cooley[1], Simone Hendriksen[2] and Tina Widowski[1]
[1]*University of Guelph, Animal BioSciences, 50 Stone Road E, Guelph, ON, N1G 2W1, Canada,*
[2]*Wageningen University, Animal Sciences, P.O. Box 338, 6700 AH Wageningen, the Netherlands;*
lcooley@grayridge.com

Commercial strain leghorn breeders are reared and housed in a variety of cage and barn systems where they are exposed to various physiological and psychological stressors associated with housing type throughout rearing and lay. Epigenetic programming, as influenced by environment, can occur during two critical periods: gametogenesis when breeding stock are adolescents, and egg formation when egg composition affects the embryo. Previous research in birds has demonstrated behavioural adaptive responses from parental exposure to acute stressors and *in ovo* exposure to stress hormones. A knowledge gap exists as to whether maternal housing experience can affect offspring behaviour. The objective was to investigate the influence of maternal rearing and housing environments, and age on behavioural phenotype of progeny using the Chick Anxiety-Depression Model. Two cohorts of LSL-Lite leghorns were reared in aviary (AV) or conventional battery cages (CC) from day 1, then housed in AV (n=2), CC (n=12) or large furnished cages (FC) (n=12) from 16 wks of age. Hens from each housing environment (n=96) were inseminated with pooled semen at three ages: Young (25 wks of age), Ideal (44 wks of age) and Old (68 wks of age). Each cohort had four replicates of progeny (7 males and 7 females), identically reared in furnished floor pens, from the five rearing × housing treatments: Trt1 (AV × AV), Trt2 (CC × CC), Trt3 (AV × CC), Trt4 (CC × FC), and Trt5 (AV × FC). Two male and two female progeny from each replicate were individually isolated in a sound-proof enclosure and tested twice in random order, with and without a mirror, at either 5, 6 (n=480) or 32, 33 days of age (n=480). The mirror simulated the presence of a conspecific. Audio was recorded for 5 mins at 8000 Hz and analyzed for the mean number of total distress vocalizations (TDV). A repeated GLM procedure (SAS 9.4) (cohort=random) with LS-means was used for analysis. Test order was not significant, and TDV were lower ($F_{1,\,467}$=181.80, P<0.0001) when progeny were tested with (60.2±4.50) vs without the mirror (146.6±4.50). Sex, maternal rearing and maternal housing did not affect TDV, TDV from chicks tested at 5, 6 days of age were higher (126.2±4.50 than at 32, 33 days of age (81.3±4.50); ($F_{1,\,467}$=49.79, P<0.0001). Progeny of young hens had more TDV (124.8±5.51) than those from Ideal, (97.4±5.51) and Old (89.0±5.51; $F_{2,\,467}$=11.54, P≤0.0001). We conclude that progeny from the young hens exhibit greater stress reactivity to social isolation and that this decreases with both maternal age and age of progeny.

Does an active fetus become an active lamb?

Tamsin Coombs, Kirstin Mcilvaney and Cathy Dwyer
SRUC, West Mains Road, Edinburgh, EH9 3JG, Scotland, United Kingdom;
tamsin.coombs@sruc.ac.uk

In order to fully understand the origins of behaviour in mammals it may be important to understand fetal behaviour. With the advent of new technology it is now possible to directly observe behaviour in the fetus making it possible to evaluate its role in postnatal function. Therefore the aim of this study was to consider whether fetal activity and behaviour is a predictor of neonatal behaviour in sheep. Twenty-three multiparous singleton and twin bearing ewes of Blackface (BF) and Suffolk (S) breeds, (BF: Single n=7; Twin: n=6 litters; S: Single: n=5; Twin: n=5 litters) were selected for this study. Fetal activity data were collected by transabdominal ultrasonography during two 20 minute sessions at approximately gestational days 56, 77 and 98 (total gestation is 145 days). Fetuses were assigned an activity score every 5 minutes during scanning and each fetus, or pair of twin fetuses was scored 10 times in total as follows: (1) low activity: brief movements, mostly startles; (2) active: frequent movements, stretches and movement of head and limbs; (3) very active: vigorous continual activity including trunk rotations. Neonatal activity data consisted of detailed information regarding the postural changes made in the first 30 minutes of life obtained from analysis of digital video recordings. Fetuses that were most active at d56 (mean total activity: 15.11±0.5) spent longer lying laterally with head down following birth (mean: 159±38.6 s, r_p=0.53, P<0.05). BF fetuses that were most active at d56 (mean total activity: 15.69±0.8) had shorter kneeling and standing bouts (Kneeling: 5.26±0.81 s, r_p=-0.56; Standing: 19.55±3.74 s, r_p=-0.55, P<0.05) while S fetuses that were most active at d56 (mean total activity: 14.28±0.6) took longer to nose their mother (mean: 679.4±98.2 s, r_p=0.73, P<0.05). BF fetuses that were most active at d98 (mean total activity: 13.73±0.31) were quickest to reach the udder (mean: 1263±138 s, r_p=-0.96, P<0.05). However there was no direct relationship between fetal activity and latency to suck. The data presented here demonstrate that there is continuity in behaviour from the fetal to the neonatal period in sheep but it is not necessarily a linear relationship. Greater fetal activity early in gestation appears to be detrimental to neonatal behavioural progress while greater foetal activity later in gestation may result in more optimal neonatal behavioural performance. This is the first time that a relationship has been found between fetal and neonatal behaviour in sheep and although the sample size was small it provides a basis for a better understanding of the significance of fetal activity to behavioural development in the lamb.

Calf-cow-contact during rearing affects social competence and stress reactivity in dairy calves

Edna Hillmann[1], Cornelia Buchli[1,2], Alice Raselli[1], Rupert Bruckmaier[3] and Kerstin Barth[4]
[1]*ETH Zürich, Environmental System Sciences, Ethology and Animal Welfare, Universitätstrasse 2, 8092 Zurich, Switzerland,* [2]*University of Zurich, Vetsuisse-Faculty, Department of Farm Animals, Winterthurerstrasse 260, 8057 Zurich, Switzerland,* [3]*University of Berne, Vetsuisse-Faculty, Veterinary Physiology, Bremgartenstrasse 109a, 3001 Berne, Switzerland,* [4]*Johann Heinrich von Thünen-Institute, Federal Research Institute for Rural Areas, Forestry and Fisheries, Institute of Organic Farming, Trenthorst 32, 23847 Westerau, Germany; edna-hillmann@ethz.ch*

We investigated if rearing calves with contact to a dam or foster cow affects their social behaviour and stress reactivity compared to rearing calves without cow-contact, as it is usually done in dairy farming. Previous studies on dam rearing have been conducted on experimental farms, while this is the first study that used an on-farm approach to reflect the large variability between housing systems found in practice. 69 female calves (34 with cow-contact on 4 farms, 35 without cow-contact on 15 farms, 1-3 months of age) were subjected to a combined isolation/novel-object test, followed by a confrontation with a cow other than their dam or foster cow. Data were analysed using linear mixed effects model in R. Heart rate was increased during isolation and declined in the course of the trial, with higher isolation heart rate in the no-cow-contact compared to cow-contact calves: 160(111-203) vs 141(91-211) bpm (interaction test phase ' rearing system $F_{2,95}=5.7$, P=0.005). There was no effect of cow-contact on saliva cortisol concentration or behaviour during the isolation/novel-object test (0.58 vs 0.38 ng/ml). During confrontation, no-cow contact calves approached the cow more often than with-cow contact calves (6.0(0-16) vs 2.(0-6), $F_{1,27}=12.2$, P=0.002). Threatening behaviour of the cow led to a higher incidence of submissive behaviour in calves reared with cow-contact compared to no-cow-contact calves (2.0(0-8) vs 4.0(0-10), $F_{1,26}=16.9$, P<0.001). The rearing system had no effect on the number of socio-positive interactions during confrontation ($F_{1,27}=0.9$, P=0.3). We conclude that rearing with cow-contact reduced the cardiac but not cortisol stress response during isolation, and led to a more appropriate social behaviour compared to traditional rearing without cow-contact. Thus, contact to the dam or foster cows early in life may promote social learning in dairy calves that positively affects their integration into the dairy herd later in life.

Locomotor style of laying hens on inclined surfaces

Chantal Leblanc[1], Bret Tobalske[2], Stephen Bowley[3] and Alexandra Harlander[1]
[1]*University of Guelph, Animal Biosciences, 50 Stone Road East, Guelph, ON, N1G 2W1, Canada,*
[2]*University of Montana, Division of Biological Sciences, 32 Campus Drive, Missoula, MT 59812,*
USA, [3]*University of Guelph, Plant Agriculture, 50 Stone Road East, Guelph, ON, N1G 2W1,*
Canada; leblanc@uoguelph.ca

Housing laying hens in multi-tiered aviary systems allows for three-dimensional freedom of movement, which is important for the improvement of their welfare. However, birds are normally placed into these systems with no prior training or experience to gain the appropriate cognitive skills necessary. With this, comes increased injuries from collisions or falls within the system. For the first time, we aim to assess the fundamental locomotor strategies (hindlimbs and forelimbs) that are used in domestic fowl to reach elevated surfaces. The present study evaluates locomotor style of the hindlimbs and forelimbs in laying hens of four different strains (Lohmann Brown, Hyline Brown, Dekalb White, Lohmann LSL) in relation to the degree of incline (0-70°), platform height (70 and 160 cm), the age of the birds (8-36 wks) and the ramp surface substrate (sandpaper or wire grid). All birds were presented with all inclines, heights and surfaces in a systematic testing order on a weekly (8-9 wks), biweekly (11-21 wks) and five-week basis (26-36 wks). When birds were successful, the mode/modes of locomotion were observed and recorded for each test. Motivation to climb was provided by having 5 same-age birds in a crate on each platform height. Behavioural counts of wing-assisted incline running (WAIR) and aerial ascent were analyzed using the Glimmix procedure via SAS (version 9.4). Results indicate that birds only used their hindlimbs (no wing use) to master inclines up to 40°. The wire surface material improved traction by providing more grip and decreased wing use ($t_{265=}$5.24, 2.25±0.43, P<0.0001). White feathered strains performed more wing-associated locomotor behavior compared to brown feathered strains (HB: $t_{1481=}$2.28, 1.23±0.54, P=0.0229; LB: $t_{1481=}$2.87, 1.43±0.50, P=0.0042). Birds started to perform WAIR and aerial behaviour at 40° and continued to do so at 50, 60 and 70°. Younger birds (8-11 wks) performed more WAIR to successfully master steeper inclines (60-70 deg), but as they grew older (13-36 wks) this behaviour decreased ($t_{3180=}$4.15, 3.65±0.88, P<0.0001) and they performed more aerial ascent ($t_{3183=}$-2.42, -3.32±1.37, P=0.0157). These results suggest that providing 40 degree ramp inclines in the design of three-dimensional housing systems will allow for easy training and climbing (without wing use) to prevent injuries in adult domestic fowl.

How do group-housed dairy calves use automated grain feeders?

Catalina Medrano-Galarza[1], Alison Vaughan[2], Anne Marie De Passillé[2], Derek Haley[1] and Jeffrey Rushen[2]
[1]*University of Guelph, Department of Population Medicine, Guelph ON, N1G 2W1, Canada,* [2]*University of British Columbia, Faculty of Land & Food Systems, Agassiz BC, V0M 1A0, Canada; cmedrano@uoguelph.ca*

Our objectives were to determine (1) calves' latency to the first grain feeder visit, and (2) pre- and post-weaning grain feeding patterns (meal size, total daily grain intake [TDGI] and duration of visits). A total of 188 calves housed in groups of 9 from 6d of age were allowed 12 l/d/calf of whole milk, and grain *ad libitum* through automated feeders. Weaning started at a median of 51d-old, and lasted a median of 9d. The MIXED procedure of SAS was used to determine differences in TDGI and duration of visits to the grain feeder. Most calves (75%) visited the grain feeder within 48 h of introduction to the group pen; 95% had visited by day 5. During calves' first visit, 82% only entered the feeding stall, 12% triggered a feed-drop, and 6% consumed grain. A large variation was found in the time between the first feeder visit and the first visit with grain intake (a median of 12d; range: 0-78d). During calves' first month in the group pen, TDGI occurred mainly in meals <50 g (median: 100% of TDGI; 25th-75th percentile: 50-100). By the second month, meals <50 g were still the most common (67%; 25th-75th percentile: 46-100), but meals of 50-99 g had increased (24%, 25th-75th percentile: 0-40). After 11 wks, only 14% (25th-75th percentile: 4.5-22) of TDGI occurred in meals <50 g, while 58% (25th-75th percentile: 30-69) occurred in meals of 100-299 g. Only 25% of the calves had meals >300 g, accounting for 16-47% of their TDGI. Variation between calves in meal size decreased during post-weaning. One wk post-weaning, TDGI was mainly consumed in meals of 100-299 g (47%; 25th-75th percentile: 36-55) and 300-999 g (33%; 25th-75th percentile: 15-49). By wk 8 post-weaning, the percentage of TDGI occurring in meals of 100-299 g decreased to 26% (25th-75th percentile: 19-35), while meals of 300-999 g increased to 61% TDGI (25th-75th percentile: 52-71). TDGI on wk 1 in the group pen was lower than wk 4 (median: 3 vs 14 g), and significantly increased by wk 11 (137 g; P<0.001). TDGI increased (P<0.003) during the first 3 wks post-weaning (wk 1: 1,686 vs wk 2: 2,290 vs wk 3: 2,776 g); afterwards there were no differences. The duration of a grain feeder visit increased during the first 3 wks on the group pen from a median of 1.4 min to 2 min (P<0.001), then increased to 2.7 min by wk 11 and to 5 min one wk post-weaning (P<0.001), without differences thereafter. Meal size, total grain intake and duration of visits to the grain feeder increased with age. However, there were large differences between calves in how they interacted and consumed grain from an automated feeder, highlighting the individual variability of the adaptation to solid feed.

Affiliative relationships within a group of dairy calves

Joao H. C. Costa, Nicola A. Adderley, Marina A. G. Keyserlingk and Daniel M. Weary
Animal Welfare Program, University of British Columbia, Faculty of Land and Food Systems,
2357 Main Mall, Vancouver, BC, V6T 1Z4, Canada; jhcardosocosta@gmail.com

An affiliative relationship refers to a preferential bond between two individuals, irrespective of a familial relationship. The existence of affiliative relationships between cows has been suggested by the uneven distribution of socio-positive interactions. Social structure among dairy calves remains poorly understood and the stability of these relationships over time has not been characterized. The aim of this study was to investigate the development and stability of affiliative relationships between young dairy calves. Calves (n=8) were grouped when they were 7±3 d old and housed in a sawdust bedded pen (4.87×7.31 m). We measured and ranked affiliative relationships using 24 h continuous video recording of the group of 8 dairy at 30, 50, and 70 d of age. We scored the video for social resting (resting in close proximity to another calf), playing (social and locomotor play) and allogrooming (sniffing and licking one another). We ranked preferences for social partners based upon the frequency of socio-positive behaviours; preferred partners were assessed separately for each class of behaviour. Agreement across behaviours within age and vice versa were described using Spearman Rank correlation (only correlations with P<0.05 are reported here). Descriptively, the frequency of affiliative behaviour increased between 30 (257.8±24.5 social interactions) and 50 d (420.8±36.3 social interactions) but was similar between 50 d and 70 d of age (454.8±55.5 social interactions). Affiliative partnership ranks were moderately stable between 50 d and 70 d of age for allogroming (r=0.51), resting (r=0.49) and playing (r=0.42), but it was poorly correlated between 30 and 50 d of age for all behaviours. Also, the partnership rank established based upon one behaviour was well correlated with the rank based on another behaviour for calves at the same age (e.g. partnership rank on allogroming vs social resting at 30 d, r=0.56; at 50 d, r=0.60; and at 70 d, r=0.57). These results indicate that affiliative relationships become more stable as calves age. In summary, group housed dairy calves appear to develop affiliative relationships within 30 d of birth, and these relationships become more stable as the calves age.

Locomotion and spatial preferences of domestic chicks during rearing

M Kozak[1], B Tobalske[2], C Martins[1], A Harlander[1] and H Wuerbel[3]
[1]UoGuelph, 50 Stone Road E, ON N1G2W1, Canada, [2]UoMontana, 32 Campus Dr, Missoula, MT, USA, [3]UoBern, Hochschulstrasse 4, Bern, 3012, Switzerland; mkozak@uoguelph.ca

Aviaries, or flight cages, were probably first used to display wild birds in the 19th century. They contain vertical and horizontal elevated structures to simulate a natural environment. Aviaries for domestic birds kept for egg laying, a floor-housing system with extra tiers to increase vertical space utilization, were developed in the 20th century allowing for more birds to be housed per square meter of building space. Domestic birds perform terrestrial (walking, running) and aerial locomotion (controlled aerial ascent/descent and jumping). Their locomotion has multiple functions and is significantly affected by their immediate environment but is also influenced by age. The question of how the locomotor activities and environmental preferences of growing chicks change in complex aviaries has rarely been investigated. The present study measured how chicks utilized the ground, elevated horizontal surfaces and inclined surfaces (ramps/ladders) within an aviary from 1-9 weeks of age. One hundred and twenty chicks (Lohmann [LSL]-lite, Lohmann Brown [LB], Dekalb White [DW], Hyline Brown[HB]) were sorted by line (10 birds/pen) and placed into 12 aviaries ($182 \times 243 \times 280$ cm), which had 4 platforms at two different heights (70 and 160 cm) connected vertically by a ramp and ladder. The chicks' spatial use, aerial and perching behaviour were measured (30 min /pen/week). To analyze all occurrences of motor patterns and location, pens were visually divided into 4 sections: ground (S1), low perch to first platform (S2;15-69 cm), in-between first and second platform (S3;70-159 cm), and the second platforms and high perch (S4;160 cm+). The data were analyzed using Glimmix (SAS) with age and line as fixed effects. Results are presented as LSM±S.E.M. Chicks performed locomotive events on the ground (S1) significantly ($F_{1,72}=75.8$, P<0.0001) more often during their first week of life (33.7 ± 6.3) compared to wk 5 (13.3 ± 2.7, P=0.001), 6 (15.8 ± 3.1, P=0.006), 7 (12.9 ± 2.6, P=0.0007), 8 (8.2 ± 1.7, P<0.0001) and 9 (7.1 ± 1.5, P<0.0001). Locomotor and perching events on elevated surfaces increased over time ($F_{1,64}=9.4$, P=0.003) from 13.4 ± 2.5 in wk 2 to 21.1 ± 3.8 in wk 9 (P=0.08) where most behaviour was performed in S2 (45%). Chicks utilized incline surfaces during wks 2-5. LSL chicks (19.5 ± 2.5) performed locomotive and perching events on elevated surfaces (S2, S3, S4) significantly often ($F_{3,64}=2.7$, P=0.05) than LB (13.5 ± 1.9, P=0.04), DW (12.9 ± 1.7, P=0.02) and HB chicks (12.1 ± 1.7, P=0.01). Chicks almost exclusively locomote at ground level and on S2, and they began using S2 as early as 2 weeks of age. Therefore introducing lower platforms and ramps/ladders to rearing environments might improve space use.

Development of perching behaviour in 3 strains of pullets reared in furnished cages

Andrea M. Habinski, Linda J. Caston, Teresa M. Casey-Trott, Michelle E. Hunniford and Tina M. Widowski
University of Guelph, Campbell Centre for the Study of Animal Welfare, Animal Biosciences, 50 Stone Rd E, Guelph, ON, N1G 2W1, Canada; afinney@uoguelph.ca

Furnished rearing cages are becoming more widely available to replace conventional systems for pullets. To date, there is no information on how pullets develop perching behaviour in furnished cages or how this varies between strains. The objective of this study was to evaluate the use of perches and a platform in a commercial furnished rearing 'Combi-Cage' system by 3 pure-bred heritage strains of pullets (Rhode Island Red (RR), Columbian Rock (CR) and White Leghorn (WL)). We hypothesized that there would be strain differences in perch use, with RR expected to use vertical space less than white strains. Four cages per strain each housed 100 pullets up to 6 wks of age then 50 pullets until 14 wks. Each cage had a platform (36 cm high) and three perches, which ran the length of the cage: one above the feeder (43 cm high), one fixed above an adjustable drinker, and one located above the platform (61 cm off cage floor). The length of the cage was visually split into 4 sections. The number of pullets on each section of each location was recorded one d per wk at 12:00 h from 1-14 wks of age and at 16:00 h (lights out) from 4-14 wks of age. Mixed model repeated analyses were used to test the effects of strain, age and their interaction on the use (% of pullets in cage) of the platform, the 3 perches and all 4 locations combined (vertical space use) for both time periods. General linear models were used to determine the use of the locations and sections based on strain and type of location. Overall, pullets used vertical space more during the day (24.04±2.32%) than lights out (14.01±1.35%). Age ($P<0.05$), strain ($P<0.05$) and strain×age ($P<0.05$) affected vertical space use at both time periods. Perch use generally increased with age. CR pullets used the perches the most (overall average use during day; 26.70±2.51%), followed by the WL (12.94±1.80%) and then the RR pullets (9.99±2.15%). Across all strains, perch use differed according to location at both time periods ($P<0.0001$). Use of all locations remained low at lights out (combined overall mean of 3.45±0.67%). During the day, the perch above the feeder was used most (overall mean of 12.5±4.30%) and the perch above the platform was used least (overall mean of 0.37±0.047%). There was also a preference for the sections of the platform and perches that were next to the cage walls ($P<0.05$). Differences in strain perch use indicate that current cage design may not be ideal for all strains. Overall, these results suggest that the design of furnished rearing cages requires improvements in order to ensure the furnishings are used by pullets to their full potential.

The effect of alternative feeding strategies for broiler breeders in feeding motivation during rearing

Aitor Arrazola[1,2], Elyse Mosco[1,2], Tina Widowski[1,2], Michele Guerin[2,3] and Stephanie Torrey[1,2]
[1]*Animal Biosciences, University of Guelph, Guelph, ON, N1G 2W1, Canada,* [2]*Campbell Centre for the Study of Animal Welfare, University of Guelph, Guelph, ON, N1G 2W1, Canada,* [3]*Population Medicine, University of Guelph, Guelph, ON, N1G 2W1, Canada; aarrazol@uoguelph.ca*

Since broiler breeders have the same genetic potential for fast growth as broilers, they are feed restricted to reduce obesity-related problems in fertility. Pullets are feed restricted up to 43% of *ad libitum* feed intake for the same body weight during the rearing phase. In North America, producers frequently use 'skip-a-day' feeding programs that eliminate one or more non-consecutive feeding days per week in exchange for feeding larger quantities of feed on on-feed days. This chronic feed restriction leads to welfare concerns over the birds' lack of satiety and the impact of fasting periods on feeding frustration. The objective was to examine the effect of rationed alternative diets and non-daily feeding schedules on a feed intake test of broiler breeder pullets. At 3 weeks of age, 1,680 Ross 308 pullets were conveniently allocated into 24 pens fed with one of four isocaloric treatments: (1) daily commercial diet (control); (2) daily alternative diet; (3) skip-a-day commercial diet; (4) graduated commercial diet with varying on-feed days per week. The alternative diet had an inclusion of 40% soybean hulls and 1-5% calcium propionate, increasing with time. Pullets fed with the alternative diet were fed an additional 50% feed to compensate for dietary energy dilution. A feed intake test was performed at 4, 8, 12, 16 and 20 weeks with a subsample of 10 birds per pen, 5 on on-feed and 5 on off-feed days. At feeding time, birds were moved to test pens within the same room and fed *ad libitum* for 20 min with the home pen diet. Data were analyzed with a mixed model procedure, with week as a repeated measure and pen as a random variable. There was no overall effect of treatment on feed intake relative to body weight ($F_{3,17}$=2.49, P=0.095). Treatments differed in feed intake relative to body weight over weeks ($F_{12,1140}$=13.29, P<0.001) and depending on feeding day ($F_{3,1140}$=8.44, P<0.001). Pullets on a non-daily feeding scheme had received a larger feed allotment the previous day. But, only those under a fixed skip-a-day treatment had a lower feed intake compared to control on off-feed days (difference: 0.14±0.04 g/g BW). Pullets fed on a daily basis (2.13±0.04 g/g BW) were as feed motivated as those fasted for 40 hours (2.11±0.04 g/g BW and 2.08±0.04 g/g BW, respectively for skip-a-day and graduated) for the same diet. Pullets on a fixed skip-a-day treatment can habituate to their feeding scheme but not those under a graduated treatment.

A nest by any other name: pre-laying behaviour of hens in furnished cages

Michelle E. Hunniford and Tina M. Widowski
University of Guelph, Animal Biosciences, 50 Stone Rd. E., N1G 2W1, Canada;
mhunnifo@uoguelph.ca

Furnished cages (FC) are supposed to facilitate the performance of motivated behaviours by providing adequate resources. However, many factors may influence how hens perceive nests spaces. The objective of this research was to investigate the impact of cage size, design, space allowance and rearing on nesting behaviour. We hypothesized that nest use would be associated with indicators of positive hen welfare and settled nesting: less aggression, calmer searching, and more sitting. Three experiments used LSL-Lite hens housed in two sizes of FC (12 large: 41,296 cm^2; 12 small: 20,880 cm^2). Each FC had perches, a curtained nest with plastic mesh flooring, and a smooth plastic scratch mat. Cage size, room and tier were constant factors throughout; a linear mixed model (SAS) was used to analyze dependent variables. First, the effect of space allowance (low: 520 cm^2/bird; high: 750 cm^2/bird) and cage size was investigated using three measurements: egg location, egg timing and sitting/aggression. Hens in large FC laid more eggs in the scratch area (large: 21.9±0.5 vs small: 7.2±0.2%; P<0.01); aggression peaked with the peak laying period; and hens in small FC with low space allowance were the most aggressive (P=0.04) despite laying most eggs in the nest. Nest use, therefore, did not necessarily indicate positive welfare. Second, we examined the impact of rearing environment (cage vs aviary) on nesting. Focal observations were conducted at 22 and 32 wks of age. Individually marked hens were observed from lights-on until oviposition; behaviour patterns in the last hour were analyzed. Aviary-reared hens (n=50) laid more eggs outside of the nest (17.6±0.6 vs 9.6±0.4%; P<0.0001) and, especially at week 22, were less active (10.0±1.6 vs 18.1±2.3 min; P=0.0007) than cage-reared hens (n=52). Rearing seemed to affect how hens perceived potential nest sites. Finally, we used small groups (n=6) to investigate whether adding a wire partition in the scratch area would affect nest choice and pre-laying behaviour. Twelve groups of hens were transferred from conventional cages into FC for one week. Nests were the same in each FC, but the scratch area either had a wire partition (W) or not (NW). Egg location and individual hen behaviour were measured. Hens in the W treatment entered (W: 38.7±5.0 vs NW: 19.4±3.1 bouts; P=0.002), searched (W: 15.1±1.6 vs NW: 6.1±1.2 min; P=0.009), and laid more eggs (W: 18.3±1.8 vs NW: 7.3±1.3%; P<0.0001) in the scratch area, and entered the nest less frequently (W: 18.6±1.7 vs NW: 26.1±1.6 bouts; P=0.008), than NW hens. The wire partition seemed to create another nest site, which affected both nest use and pre-laying behaviour. Overall, a variety of factors influenced how hens perceived potential nest sites in furnished cages.

The behaviour of beef cattle unloaded for feed, water, and rest en route during long distance transportation

Derek Haley[1,2], Michael Ross[2,3], Ray Stortz[3], Hannah Flint[2], Ken Bateman[1,2], Tina Widowski[1,3] and Karen Schwartzkopf-Genswein[4]
[1]Campbell Centre for the Study of Animal Welfare, University of Guelph, Guelph, ON, N1G 2W1, Canada, [2]Ontario Veterinary College, University of Guelph, Population Medicine, Guelph, ON, N1G 2W1, Canada, [3]Ontario Agricultural College, University of Guelph, Animal Biosciences, Guelph, ON, N1G 2W1, Canada, [4]Agriculture & Agri-Food Canada, Lethbridge, AB, T1J 4B1, Canada; dhaley@uoguelph.ca

Transport regulations in Canada require that cattle on a journey expected to exceed a total of 52 h, be unloaded for feed, water, and rest, for a minimum of 5 h. We observed cattle during these rest stops to document their behavioural priorities for insight into how they may be feeling, and how well they were coping with these long journeys. We recorded behaviour live during the first 5 h at the rest station. Every 5 min we counted the number of animals eating, drinking, lying, and 'other'. Data from 87 pens of cattle (from 53 loads) were examined, using mixed linear regression models, to test for associations between different behaviour and possible explanatory variables, with load as a random effect. Behaviour was not strongly associated with the time cattle had spent in transit before arriving at the rest station. Behaviour changed over the first 2 h after they were unloaded (P<0.01), but then was consistent over the subsequent 3 h (P<0.25). During the first 2 h, on average, 51.6% of the cattle in each pen were eating, 2.7% drinking, and 26.7% lying and during the subsequent 3 h the averages were 19.2, 1.1, and 63.6%, respectively. Cattle arriving in the afternoon had higher odds of being observed lying (P<0.02). The same was true for pens of cattle that contained only some of the animals from a load, vs pens containing the entire load, and most differences in behaviour were associated with cattle being grouped together in the pen by mixing animals from different truck compartments. In a second study observing 24 loads of cattle, we tested the effect of doubling the normally available feeding space vs normal (one circular metal hay feeder, 7.5 m circumference, with 18 slanted openings, each 0.40 m wide). Each load was split in two with half the animals assigned to each treatment condition, in otherwise identical pens. These data were analyzed using the PROC MIXED procedure with SAS. Doubling feeding space resulted in cattle spending more time eating than with the normal space provided (88.2 vs 68.1 min, P<0.01), with fewer interruptions (P<0.05), with no effects on drinking or lying. These results suggest systematic evaluations of the way cattle are managed and cared for at rest stations can have important impacts on their behaviour, which might ultimately affect their welfare during long-distance transport.

The effect of alternative feeding strategies on feather condition and corticosterone levels in broiler breeder pullets

Elyse Mosco[1,2], Aitor Arrazola[1,2], Tina Widowski[1,2], Margaret Quinton[2], Michele Guerin[1,3] and Stephanie Torrey[1,2]
[1]*University of Guelph, Campbell Centre for the Study of Animal Welfare, Guelph, ON, N1G 2W1, Canada,* [2]*University of Guelph, Animal Biosciences, Guelph, ON, N1G 2W1, Canada,* [3]*University of Guelph, Population Medicine, Guelph, ON, N1G 2W1, Canada; elyse.mosco@ucdconnect.ie*

To limit problems such as obesity and lameness, broiler breeders are severely feed restricted during the rearing phase. This restriction results in signs of chronic hunger and frustration such as redirection of foraging behaviour and high arousal. We examined whether alternative feeding strategies improved feather condition and decreased corticosterone (CORT) levels during rearing under simulated commercial conditions. At 3 wk, 1,680 Ross 308 pullets were allocated to 24 pens and fed one of four treatments: (1) daily commercial diet (C); (2) daily alternative diet (A); (3) skip-a-day commercial diet (S); (4) graduated commercial diet with varying feed days per week (G). G schedule was 5/2 (5 days on-feed, 2 non-consecutive days off-feed) from wk 3-4, 4/3 from wk 5-11, 5/2 from wk 12-18, and daily from wk 19-22. The A diet was supplemented with 40% soybean hulls and 1-5% calcium propionate, increasing over time. Blood samples were collected after feeding for 6 birds/pen at 5, 11, 18 and 21 wk, 3 on both on- and off-feed days. CORT samples were extracted from plasma and measured via EIA. Biweekly, 10 birds/pen were scored for feather coverage (FC). A score of 0 to 5 was assigned to 6 areas (head, neck, back, vent, wing, tail), with 5 being the worst for each area. Data were analyzed by mixed model, with week as a repeated measure and room, pen and bird as random variables. Sum of feather condition was log transformed for analyses. CORT decreased with age ($F_{3,179}$=199, P<0.001) but there was a trt×wk×feed day interaction ($F_{9,102}$=3.79, P=0.0004); CORT was higher for non-daily treatments, particularly on off-feed days (t_{133}=-4.1, P<0.001; daily: 1.53±0.10 ng/ml; non-daily: 2.17±0.15 ng/ml), although differences disappeared by wk 21. Most feather damage was to the neck, vent and tail, areas indicative of feather pecking. There was a trt×wk interaction ($F_{27,1196}$=1.86; P=0.005) on FC. FC for all birds worsened with time, but did so faster for C birds. Daily fed birds (C,A; score=2.10±0.05) had worse FC than non-daily fed birds (S,G: 1.61±0.05; t=2.30; P=0.03), and C (2.40±0.08) tended to have worse FC than A (1.80±0.07; t_{20}=1.98; P=0.06). CORT results indicate that non-daily fed birds had higher levels of arousal on non-feed days, which dissipated with increasing time on the feeding schedule. FC, indicative of feather pecking, declined slower with alternative feeding strategies, suggesting a decrease in redirected foraging behaviour.

Both tail docking and straw provision reduce the risk of tail damage outbreaks

Mona Lilian Vestbjerg Larsen[1], Heidi Mai-Lis Andersen[2] and Lene Juul Pedersen[1]
[1]Aarhus University, Department of Animal Science, Blichers Allé 20, 8830 Tjele, Denmark, [2]Aarhus University, Department of Agroecology, Blichers Allé 20, 8830 Tjele, Denmark; mona@anis.au.dk

The purpose of this ongoing study is to investigate whether tail docking and straw provision reduce the risk of tail damage outbreaks in slaughter pigs. The study is planned to include four batches of slaughter pigs to a total of 112 pens (range: 11 to 18 pigs per pen). Three batches have been completed (n=80 pens) and preliminary results based on these data will be presented. Each pen was randomly distributed to three treatments: (1) TAIL: pens with undocked (n=36) or docked pigs (n=44; docked to 50% of the original length); (2) STRAW: pens not provided with straw (n=40) or pens provided with 150 g of straw per pig per day (n=40); and (3) STOCK: same-size pens (13.10 m^2) with 18 (n=40; 0.73 m^2/pig) or 11 pigs (n=40; 1.19 m^2/pig). Tail damage for each individual pig was registered three times per week through the entire slaughter pig period, and a tail damage outbreak at pen level was registered when at least one pig in the pen had a bleeding tail wound. After the first tail damage outbreak, tail damage was no longer recorded in that particular pen; therefore, per definition only one tail damage outbreak could be registered per pen. Of the 80 pens, we recorded a tail damage outbreak in 42. The risk of a tail damage outbreak was analysed using the Cox Proportional-Hazards Regression for survival data in R including the covariates TAIL, STRAW, STOCK, batch number, the average pen start weight and the interaction between TAIL and STRAW. No interaction was found between TAIL and STRAW, and significant hazard rate ratios (HRR) were seen only for TAIL (P<0.0001) and STRAW (P<0.01). Incidence of tail damage outbreaks in pens with undocked and docked pigs were 69% (95% CI [50-81]) and 39% (95% CI [22-52]), respectively, and the risk of a tail damage outbreak was almost three-fold higher for pens with undocked pigs compared to pens with docked pigs (HRR=2.92; 95% CI [1.56-5.46]). The incidence of tail damage outbreaks in pens without and with straw provided were 68% (95% CI [49-79]) and 38% (95% CI [21-51]), respectively, and the risk of a tail damage outbreak was almost three-fold higher for pens without straw compared to pens with straw (HRR=2.81; 95% CI [1.47-5.37]). These preliminary results suggest that both tail docking and straw provision can reduce the risk of a tail damage outbreak on pen level, whereas this was not the case for a reduction in stocking density. These results provide important knowledge to the effectiveness of tail docking and straw provision to prevent tail damage outbreaks. The next step is to investigate the relative effectiveness of tail docking and straw provision.

Management factors associated with mortality of dairy calves

Leena Seppä-Lassila[1], Kristiina Sarjokari[1,2], Mari Hovinen[1], Timo Soveri[1] and Marianna Norring[1]
[1]*University of Helsinki, Department of Production Animal Medicine, P.O. Box 57, 00014 University Of Helsinki, Finland,* [2]*Valio Ltd, P.O. Box 10, 00039 Valio, Finland; marianna.norring@helsinki.fi*

Mortality of dairy calves can reflect suboptimal production environment or management, and can act as a crude indicator of overall calf welfare. We aimed to identify potential management factors affecting calf mortality in order to eventually help reducing calf mortality. Mortality data of 13580 calves from 82 farms with average (±standard dev.) herd size of 125±41 cows were acquired from Finnish Agricultural data processing center. Data on management practices were collected during farm visits by interviewing farmers with a structured questionnaire. The calf mortality data were analyzed using linear regression models. At the herd level, the mean (±sd) mortality of calves that were less than 7 days of age, was 5.2±2.3%. Every increase of 10 cows in herd size increased mortality rate by 0.13 percentage points (P=0.019). Separating sick calves from healthy decreased the mortality rate by 1.38 percentage point (P=0.005). In addition, higher mortality tended to be associated with longer latency from birth to colostrum intake (0.13 percentage point increase for every one hour delay; P=0.083), lower average parity (1.72 percentage point increase in mortality rate for every one parity decrease in the mean; P=0.097) and a smaller proportion of breeds other than Ayrshire or Holstein in the herd (0.29 percentage point decrease for every percentage point decrease in breed proportions; P=0.053). The mean (±sd) mortality of older calves aged 7 to 180 days was 5.7±6.2%. Mortality rate increased with a shorter whole milk feeding period (0.01 percentage point increase for every one day shorter period; P=0.008), longer period in the calving pen with the dam (0.19 percentage point increase for every one day; P=0.016), and smaller average herd production level (0.32 percentage point increase by every 1000 kg decrease in average milk yield; P=0.001). Mortality of calves was 0.39 percentage point lower on farms that used a veterinarian to disbud calves instead of layman (P=0.024). In addition, mortality tended to be lower on farms which sold lower proportion of calves (0.008 percentage point decrease by every percentage point increase of proportion of calves sold; P=0.065) and that had natural ventilation instead of forced (0.47 percentage point decrease, P=0.071). More attentive management policys; advancing health by colostrum feeding, separating sick calves, and using a veterinarian for disbudding can help reducing calf mortality rate on farm. Larger farms, farms with smaller average production and farms with lower average parity had higher calf mortality rates. Particular consideration should be used on these farms when tackling mortality issues.

Influence of milking-parlour size on the behaviour of dairy cows

Yamenah Gómez[1,2], Melissa Terranova[1,2], Michael Zähner[2], Edna Hillmann[1] and Pascal Savary[2]
[1]ETH Zurich, Department of Environmental System Sciences, Rämistrasse 101, 8092 Zurich, Switzerland, [2]Agroscope, Institute for Sustainability Sciences ISS, Tänikon 1, 8356 Ettenhausen, Switzerland; yamenah.gomez@agroscope.admin.ch

Over a period of ten years, breeding selection for higher milk yield has caused Brown Swiss and Holstein dairy cows to increase in size by 5 and 11 cm, respectively. Despite this, milking-parlour size has not yet been adjusted in Europe. As a result, milking-stall space per cow may be too small, which could have a negative impact on animal welfare during milking. Our goal was therefore to investigate the influence of different space allowances in the main types of milking parlours on the behaviour of cows. On 15 Swiss farms, we measured parlour and cow size and randomly selected 10 cows per farm for behavioural observation during milking (time taken to enter the milking parlour, hind-leg activity, excretion and rumination). Individual space ratios were calculated for each cow by dividing milking-stall length by cow length and stall width by cow width. Space ratios >1 indicate that cow length or width was less than stall length or width. We analysed space-ratio effects on behaviour using generalised linear mixed models. With an increasing space ratio ($\geq$1.2), cows exhibited significantly less hind-leg activity and excretion, and more rumination events (all $P<0.02$), and took less time to enter milking parlours ($P<0.1$). Based on these results, proper minimum requirements for milking-stall dimensions should be 1.2 times the cow size. In our study, the averaged ratio of 1.1 for group milking parlours did not meet these requirements. Adjustments based on our results could lead to an improvement in cow welfare during milking.

Effects of long-term exposure to an electronic containment system on the behaviour and welfare of domestic cats

N. Kasbaoui, D.S. Mills, J. Cooper and O. Burman
University of Lincoln, School of Life Science, Joseph Banks Laboratories Green Lane, LN6 7DL Lincoln, United Kingdom; nkasbaoui@lincoln.ac.uk

Free-roaming cats are exposed to several risks associated with their lifestyle, such as disease, poisoning and injury. One way of mitigating these risks is to contain cats, for example using an electronic boundary fence system that delivers an electric 'correction' via a collar if a cat ignores a warning sound and attempts to cross the boundary. There is no empirical evidence to indicate the welfare implications of these systems. The aim of the study was to determine the welfare impact on cats of long-term exposure to an electronic containment system. We compared 46 owned domestic cats, one group (n=23) having been exposed to the containment system for more than 12 months (AF-group) and one group (n=23) with free outdoor access and no containment system in place (C-group). We assessed the cats' welfare using two approaches. First, a 'cat-centred approach' involving behavioural tests: an unfamiliar person test, a novel object test, a sudden noise test and a cognitive bias test. Behaviours indicative of negative affective state or sensitivity to noise were analysed individually and as grouped factors following PCA. Secondly, an 'owner-centred approach' was used by asking them to complete a questionnaire focused on their cats' behaviour. There was no significant difference between groups of cats in the time spent outdoors. We found few significant differences between the populations. In the unfamiliar person test, C-group did significantly more 'lip licking' than AF-group (median licks/second ± SE; AF:0.0±0; C:0.005±0.038); Mann-Whitney P=0.029); AF-group looked at and explored the unfamiliar person significantly more (mean factor loading ± SE; AF:0.41±0.31; C:-0.39±0.12; t-test P=0.031); and interacted more with the unfamiliar person than C-group (mean factor loading ± SE; AF:0.4±0.29; C:-0.5±0.14; t-test; P=0.013). In the novel object test, AF-group looked at and explored the object significantly more than C-group (median factor loading ± IQR; AF:0.05±1.53; C:-0.54±0.61; Mann Whitney P=0.012). No significant differences were found between AF and C groups for either the sudden noise or cognitive bias test. Regarding the questionnaire, C-group reported that their cat showed more irritable behaviour than AF owners (median factor loading ± IQR; AF:-0.57±1.53; C:0.27±1.43; Mann Whitney P=0.022), and, compared to C owners, AF owners thought that their cats housesoiled more often (median ± IQR; AF:0.0±0.9; C:0.0±0 Mann Whitney P=0.048). Overall, AF cats were less neophobic than C cats. These findings indicate that an electronic boundary fence with clear pre-warning cues does not significantly impair the long-term quality of life of cats.

Feeder space relates to aggression, jostling, and feeder sharing in laying hens

Janja Širovnik Koščica, Michael Jeffrey Toscano and Hanno Würbel
University of Bern, ZTHZ, Division of animal welfare, Burgerweg 22, 3052, Switzerland; janja.
sirovnik@vetsuisse.unibe.ch

Information to determine minimum feeder space (cm per bird) for laying hens is lacking for modern systems (e.g. aviaries) and current requirements are primarily based on studies on hens in conventional cages. The lack of information likely contributed to the variation in recommended and/or regulated linear feeder space ranging from 3.8 cm/hen in the USA to 10.0 cm/hen in the EU. The aim of the present study was to determine the effects of different feeder spaces on behaviour and production of laying hens kept in multi-tier systems, though the present work focuses on behavioural results. The experiment was conducted in a commercial laying hen barn that contained 20 pens with 200 LSL hens housed in each. Following an initial feeder space of 6 cm/hen for all pens at 18 weeks of age that served as a baseline, at 20 weeks of age all five treatments (3.8, 6.0, 8.0, 9.0, and 10.0 cm/hen) were introduced to pens in a stratified way (n=4 pens/feeder space). Body mass was assessed regularly (eight times), crops were palpated and the feeder was monitored closely to assure constant feed availability and consumption. Behaviour was assessed by video at 19, 21, 32, 39 and 45 weeks of age and included: aggression, jostling with and without feeding attempt, duration of feeding bouts, number of hens feeding from the perch placed above the feeder (abnormal feeding) and number of hens simultaneously feeding normally (feeder sharing). For statistical analysis, feeder space and age were considered fixed factors with feeder space treated as a continuous variable. Across surveyed feeder spaces, the results revealed that feeder space negatively related to aggression (n=1.45/min in 4.0 cm/hen vs n=0.4/min in 10.0 cm/hen treatment, P=0.002), jostling with feeding attempts (n=0.88/min in 4.0 cm/hen vs n=0.12/min in 10.0 cm/hen, P=0.038) and without feeding attempts (n=0.13/min in 4.0 cm/hen vs 0.05/min in 10.0 cm/hen, P=0.027), feeder sharing (n=3.64 hens in 4.0 cm/hen vs n=1.88 hens in 10.0 cm/hen, P<0.001) and abnormal feeding position at 21 weeks of age (n=0.33/min in 4.0 cm/hen vs 0.03/min in 10.0 cm/hen, P=0.013). Our findings suggest that laying hens change their behaviour in relation to feeder space, though more information is needed to determine how these changes relate to animal welfare. Moreover, our effort provides a valuable tool to determine appropriate feeder space based on important behavioural responses.

Dairy cow preference for different types of outdoor access during the night

Anne-Marieke Cathérine Smid
University of British Columbia, Animal Welfare Program, 2357 Main Mall, V6T 1Z4 Vancouver,
Canada; amcsmid@gmail.com

When dairy cows are offered access to the outdoors, they display a partial preference for being outside. However, little is known about what aspects of the outdoor environment are attractive to dairy cattle. This study tested the preference of dairy cattle for pasture vs an outdoor sand paddock. Seventy-two lactating pregnant cows were assigned to 6 groups of 12 animals. After a baseline phase of 2 days in which cows were kept inside, cows were provided access in random order by group to either a pasture (FP; forced pasture phase) or sand paddock (FS; forced sand phase) for 2 nights each from 20:00 h until morning milking at approximately 08:00 h. During the next 3 nights cows were given a choice (C; choice phase) between the pasture and sand paddock. During all three phases cows were always allowed free access to the barn. Behaviours (feeding, standing with the 2 front hooves in the stall (perching) and standing with 4 feet in the stall) were recorded when cows were indoors during the day and night periods. Data were analyzed using a mixed model with group treated as the experimental unit. Cows spent $91.2\pm6.3\%$ of the time outside in the FP phase and $44.9\pm8.9\%$ in the FS phase. Time spent feeding indoors did not change ($P=0.88$), but cows perched less when indoors during both forced phases ($P<0.01$) compared to the baseline period. During the C phase, cows spent $11.1\pm2.3\%$ of the time indoors and spent more time on pasture than on the sand paddock (88.9 ± 2.7 vs $0.99\pm0.71\%$). In summary, cows spend a considerable amount of time outside when given the opportunity during nighttime and show a preference for pasture vs a sand paddock as an outdoor area.

The effects of nesting material on the toxicological assessment of cyclophosphamide

Brianna N. Gaskill[1], Catherine Brochu[2] and Christina Winnicker[3]
[1]*Purdue University, 625 Harrison St, West Lafayette, IN 47906, USA,* [2]*Charles River, 1580 Ida-Métivier, Sherbrooke, QC J1E0B5, Canada,* [3]*Columbia University, 615 W. 131st St., New York, NY 10027, USA; bgaskill@purdue.edu*

Nesting material for mice is an ethologically relevant environmental enrichment, which reduces cold stress, improves feed conversion, and increases litter size. Mice housed in typical laboratory temperatures have increased tumor growth and reduced adaptive immunity compared those housed in thermoneutrality, indicating the effect of cold stress on immune function. However, the field of toxicology is hesitant to introduce any new items into the cage because of possible interference with results. Therefore, the goal of this experiment was to determine whether provision of nesting material would affect results of a known toxicological agent. We hypothesized that nesting material would reduce study related stress but would not alter immunological parameters from a well-studied toxicological drug, cyclophosphamide (CY). A 90 day toxicity study was performed to assess the following treatments in a balanced factorial design in both sexes: nesting (0 or 10 g of crinkled nesting material); drug (50 mg/kg CY or 10 ml/kg saline IP weekly). Mice were housed in groups of 3 and data was averaged per cage (n=4; 32 cages total). Detailed health examinations and body weights were performed weekly, clinical pathology and immune parameters were collected at the termination of the study. Fecal pellets were collected at 0, 4, 6, and 12 weeks to analyze corticosterone metabolites as a measure of stress. Piloerection was observed less frequently during health examinations in nested mice (P<0.05). Male, but not female, body weights were highest in the nested mice injected with saline and lowest in those injected with CY (P=0.05). There were no significant differences in hematology, clinical chemistry, or relative % of T lymphocytes between nested and non-nested groups. Relative % of B lymphocytes was increased in groups provided nesting material (P=0.045), regardless of drug treatment. Corticosterone metabolite levels were not different within saline treated mice (P>0.05) but in CY groups, the nested mice had significantly lower levels than those without (P<0.05). All mice that received nesting material, regardless of drug treatment, had metabolite levels similar to their respective baseline levels (P>0.05), while mice with 0 g had increased levels from baseline (P<0.05). This study illustrates that access to 10 g of nesting material does not interfere with clinical pathology parameters on a standard toxicology study, but may support immune function and buffer stress, improving mouse welfare.

Enrichment and management influence the emotional state of farm mink

Anne Sandgrav Bak and Jens Malmkvist
Aarhus University, Department of Animal Science, Blichers Allé 20, 8830 Tjele, Denmark;
annes.bak@anis.au.dk

The mink producers aim to decrease fearfulness and increase exploration in mink in order to improve welfare and to make handling easier. But whether the emotional state and temperament are influenced by cage environment and management procedures is not clear. The purpose of this study was to examine how cage enrichment and management influence mink's reaction in the stick test. The stick test is a simple voluntary approach-avoidance test used to categorize the temperament of mink in selection of breeding animals. Six hundred young farm mink were included in the experiment. The animals were kept in pairs with one female and one male and housed in standard production cages with a nest box topped with straw at the AU research farm. We used a factorial design to investigate the influence of enrichment (A. control vs B. shelf and tube in the cage) for a 4-week period. Additionally, effects of different type of short-term management events were tested equally distributed within each of the main factor group: a. no treatment, b. provision of canned cat food treat for 3 days, c. being caught and kept in a small trap for 15 min. In the following 1.5 week, all animals were caught and pelt evaluated by the farm personnel as part of the yearly sorting and selection of breeding animals before pelting. We tested stick test reaction (mink latency to react, number and duration of explorative contact, minimum distance towards test spatula, categorization as: exploratory, fearful, aggressive or uncertain) the day after the short-term treatments and again the day after the farm sorting procedure. Preliminary results show that mink with cage enrichment were more explorative during the stick test compared to animals without enrichment (P<0.001). Besides, there was a higher number of animals given the score uncertain in the group without enrichment (P<0.001). An effect of the short-term treatment of the mink was also evident. Trapped mink kept a longer distance during the stick test compared to the mink given cat food (P=0.001) and the animals not exposed to any short-term treatment (P=0.017). Finally, results of the stick test after sorting suggest that the animals are reacting more fearfully after being sorted compared to before regarding explorative distance during the test as well as score given (P<0.001). Overall, these results suggest that the environment as well as management practices affect the behaviour of mink during the stick test with increased explorative behaviour following enriched housing and decrease in exploration following short-term 'negative' event such as capture/involuntary handling.

Using animal protein and extra enrichments to reduce injurious pecking in furnished cage housed laying hens

Krysta Morrissey[1,2,3], Sarah Brocklehurst[4], Tina Widowski[1,3] and Victoria Sandilands[2]
[1]*Animal Biosciences, University of Guelph, Guelph, ON, N1G 2W1, Canada,* [2]*Monogastric Science Research Centre, SRUC, Auchincruive, KA6 5HN, United Kingdom,* [3]*CCSAW, University of Guelph, Guelph, ON, N1G 2W1, Canada,* [4]*Biomathematics and Statistics Scotland, James Clerk Maxwell Building, King's Buildings, Edinburgh, EH9 3FD, United Kingdom; kmorriss@uoguelph.ca*

Currently, infra-red beak treatment (BT) is permitted in the UK. Non-BT flocks have increased risk of feather damage and mortality due to injurious pecking (IP). Though BT reduces symptoms of IP, it does not alleviate the causal factors. Insufficient foraging opportunities can increase IP, which is especially important for caged hens as foraging substrates are usually inadequate. Adding enrichments or increasing time spent feeding (e.g. by increasing dietary fibre) may satisfy foraging motivation in cages. Other aspects of the diet may also affect behaviour. We used a factorial design with three factors, each with two levels (protein source: plant based 'P' or 5% animal protein inclusion 'A', dietary fibre: control 'C' or 10% dilution with oat hulls 'F', and environment: control 'N' or extra enrichment (ropes, pecking mats, and beak blunting boards) 'E') to assess their effects on hen behaviour and welfare from 16-35 wk of age. A total of 1,344 non-BT Hyline Brown hens were housed in 64 21-hen furnished cages (754.3 cm^2/hen). One focal hen/cage was observed for beak related behaviours over 3 d every 2 wk. On the fourth day, cages were scanned (6 times/d) and all hens performing oral behaviour were recorded. This experiment was approved by SRUC's animal ethics committee. Data were analysed in Genstat (16th ed.) using generalised mixed models with significance at α=0.05. There were no overall treatment effects on IP or pecks at the extra enrichments. At certain ages, F hens pecked at the feed more often than C hens (P<0.002) but performed less spot pecking (P<0.001). E hens performed less spot pecking than N hens (P<0.001, SED=0.109, back transformed mean and estimated mean±SE (logit scale) for proportion of hens spot pecking, N: 0.025 (-3.65±0.07), E: 0.014 (-4.28±0.08)), and this trend differed over time (P<0.001). Neither protein source nor fibre inclusion affected feather cover, but there was an improvement for E hens over N hens (P=0.022, SED=0.130, back transformed mean and estimated mean±SE (logit scale) for proportion of feather scores >0, N: 0.133 (-1.88±0.34), E: 0.096 (-2.24±0.34). Though effects were slight, E appeared to improve hen welfare by reducing feather damage and spot pecking, which may be indicative of reduced foraging frustration. Extra fibre reduced spot pecking as well, though overall the flock performed quite well. Treatment differences may have been more apparent with higher baseline IP.

Shelter 'runways' to improve bird distribution in the outdoor area of free-range laying farms

Isabelle Pettersson, Claire Weeks and Christine Nicol
University of Bristol, School of Veterinary Science, Langford House, Langford, Bristol, BS40 5DU, United Kingdom; i.pettersson@bristol.ac.uk

Free-range hens often cluster near the popholes on the range, poaching the ground and causing high stocking densities in this area. We aimed to improve hen distribution by encouraging birds to range further from the house through the use of specially designed shelters. Fourteen free-range flocks (average size 13,729) were visited when 40 weeks old. Birds were counted outside three times a day and averages taken. Overall range use was 9.74% (SD: 10.55) across all flocks with 33.21% (SD: 24.50) of these birds recorded within 2 m of the house, 18.46% (SD: 14.56) recorded at 2-10 m and 39.82% (SD: 24.89) at 10 m+. Long, narrow shelters were designed that spanned from within 2 m of the house and extended 10 m out on the range. The shelters had a metal frame with blue netting secured over the top (107 cm wide and 50 cm tall). These shelters were installed at the start of the next flock cycle for the fourteen previously studied farms. The percentage of birds in the 2-10 m area increased significantly to 30.29% (SD: 13.29) at 40 weeks in the second flock cycle (Z=-2.417, P=0.016). This second round of 40 week observations took place at a different time of year from those in the baseline flocks due to typical 72 week commercial flock cycles. Scan counts of birds using the shelters (every minute for 5 minutes per shelter) were performed at 25 and 40 weeks. Counts for birds under and within 1 m of the shelter ('total') were taken as well as counts for those underneath ('under') and those perching on the shelter ('on'). All three counts increased significantly from 25 to 40 weeks of age (Total: t(13)=-2.727, P=0.017, Under: Z=-2.229, P=0.026, On: Z=-2.936, P=0.003). Focal observations of birds leaving the house via two popholes (one within 10 m of the shelter and one over 10 m away) were made per shelter. At 40 weeks 62.26% of focal birds used a shelter within 5 minutes. This was a significant increase from 48.56% at 25 weeks (t(12)=-2.656, P=0.021). Birds coming out of the closer pophole were significantly more likely to use the shelter than those from the further pophole at both 25 (χ^2(1)=21.013, P<0.001) and 40 weeks (χ^2(1)=12.083, P=0.001). In conclusion, shelter runways can improve bird distribution on the range and were used more by older birds.

Cow comfort, evaluated by lameness, leg injuries, and lying time, after facility and management changes in dairy barns

Emily Ann Morabito, Herman W. Barkema, Ed Pajor, Laura Solano, Doris Pellerin and Karin Orsel

University of Calgary, Production Animal Health, 3330 Hospital Dr NW, Calgary, AB, T2N4M1, Canada; emily.morabito@ucalgary.ca

Cow comfort is a key component to maintaining animal welfare, decreasing lameness, and decreasing injuries. In 2011, 81 freestall dairy barns in Alberta, Canada, were evaluated on animal based measures (leg scores, injury scores, lying time, production information) and environmental measures (barn design, stall measurements, management practices). The results of these evaluations were benchmarked, allowing producers to assess their farm's cow comfort in regards to the Canadian Dairy Code of Practice, other participating farms in Alberta, as well as averages across Canada. Besides the benchmarked results of the study, no farm-specific recommendations were made. The purpose of the current cross-sectional reevaluation trial (2015) was to determine what farmers have changed over the past 5 years, and to quantify the impact of these changes in regards to the same evaluations of cow comfort. In 2015, 15 farms were selected that made changes to their freestalls between 2011 and 2015, and were compared to 15 other farms did not make any changes, or made changes to areas not related to the freestalls between 2011 and 2015. Data were collected between April and December 2015. A random selection of farms from 2011 were utilized to compare to the 2015 data. Changes in environmental and animal-based measures were analyzed using descriptive, univariate and multivariate analyses. Multilevel mixed logistic regression models have been constructed (including farm as a random effect.) Lying time increased on the farms that made changes between 2011 and 2015 from 10.80±0.65 to 11.48±1.24 h/d (P<0.05), whereas lying time did not change on farms that did not make changes (10.15±0.61 in 2011 vs 10.22±0.62 h/d in 2015; P>0.05). In farms that made changes to freestalls, lameness prevalence changed between 2011 and 2015 from 20.33±3.96 to 11.07±1.29% (P<0.05). Knee injuries, and hock injuries in farms that made changes between 2011 and 2015 were 8.56±4.02 and 6.83±2.41% to 5.56±2.25 and 4.02±1.96%, respectively (P>0.05). On farms that did not make changes, the 2011 and 2015 prevalence of lameness, hock injuries and knee injuries were 16.41±3.24, 5.59±1.79 and 4.17±1.59 vs 14.29±2.29, 5.18±2.05 and 4.92±1.24%, respectively (P>0.05). Farms that made changes to the freestalls improved in some areas of cow comfort and animal welfare when compared to farms that made no changes, or changes not related to the freestall area.

Using elevated platforms to improve broiler leg health on commercial broiler farms
Eija Kaukonen, Marianna Norring and Anna Valros
University of Helsinki, Department of Production Animal Medicine, PL 57, 00014 Helsingin
Yliopisto, Finland; eija.kaukonen@helsinki.fi

Broilers are much less active than laying hens of the same age, and have a high incidence of poor gait and reduced leg health. Facilitating more activity by providing a more complex environment might be one method to improve leg health. Because broilers are not eager to use perches, they are ineffective at increasing their movements, leading us to examine the use of elevated platforms. Altogether 18 flocks with platforms and 18 control flocks were included. Six consecutive batches on three commercial broiler farms were analyzed. On each farm, two houses were included: one contained 30-cm high plastic platforms with ramp access, and the other house served as a control. Farmers recorded the use of platforms twice a week with a 5-point scale. Leg health was assessed by gait scoring, as the presence and severity of tibial dyschondroplasia (TD) and footpad lesions. Gait was evaluated before slaughter using a 6-point scale according to the Welfare Quality® Assessment (WQ) protocol for poultry. The severity of TD was determined at the slaughterhouse from 200 birds per flock using a 4-point scale, and footpad condition was assessed at slaughter following WQ protocol with a 5-point scale. Effects of age and platform on gait were analysed using separate general linear univariate models. The data of TD and footpad lesons was analysed with nonparametric tests. Platform use was between 50 and 100% throughout the whole growing period. Overall 29% of the tested birds had gait score ≥3. Younger scoring age was associated with lower mean gait score (P=0.001) and lower percentage of birds with gait scores 3 (P=0.001) and 4-5 (P=0.007). Leg health was better in birds with access to platforms: Access to platforms resulted in lower mean gait score (P=0.001), lower percentage of birds scoring 3 (P=0.001) and lower TD percentage (P=0.038). 72%±3.7 of the assessed birds had healthy footpads. The presence of platforms had no effect on footpad condition. Offering broilers modified perching items (platforms) and additional possibilities for locomotion seems to improve the leg health of broilers.

Behaviour of Danish weaner and grower pigs is affected by the type and quantity of enrichment material provided

Franziska Hakansson, Marlene K. Kirchner, Vibe Pedersen Lund, Anne Marie Michelsen and Nina Otten
University of Copenhagen, Department of Large Animal Sciences, Groennegaardsvej 8, 1870, Denmark; fh@sund.ku.dk

Inappropriate behaviour is known to reduce the welfare of pigs and therefore, determining factors influencing the quality of pig behaviour in commercial systems is of importance. As part of a larger project, this study investigated the effect of selected management parameters on different aspects of the behaviour of weaner- grower (w/g) pigs at 19 integrated Danish farms. Applying the Welfare Quality® protocol (WQ) for growing and finishing pigs, a human-animal relationship (HAR) test, scan samplings of social and exploratory behaviour (SB and EB) and a Qualitative Behaviour Assessment (QBA), both for sows and w/g pigs, were performed at each farm. Additionally, space allowance (WQ), tail biting (WQ), percentage of nursing sows, breed, weaning-age, type and amount of rooting material were collected. From the single measurements, WQ-criteria scores and the corresponding principle score for 'Appropriate Behaviour' were calculated according to the latest published version of WQ. The relation between selected management factors and the aggregated behaviour scores was tested with the help of Pearson correlations (*/ ** = significance at 0.05/ 0.01 level). The results of this study indicate an effect of rooting material and the percentage of nursing sows on w/g pig's behaviour. The amount of rooting material correlated with results of the HAR test (r_p=-0.481*), EB (r_p=0.616**) and the principal score 'Appropriate behaviour' (r_p=0.489*). Thus, a higher amount of rooting material present at observation led to fewer occurrences of fear of humans, higher incidences of exploratory behaviour and consequently, a higher principal farm score. However, the type of rooting material rather than the amount was related to tail biting (r_p=-0.675**). Hence, at farms with 'high-value' rooting material (e.g. straw) a lower amount and severity of tail biting was registered. Pigs EB further increased with available pen space (r_p=0.584**). The percentage of nursing sows was positively correlated to the principal score (r_p=0.63**), but also to single behaviour measures. At farms where the sows scored high in the QBA, the w/g pigs did as well (r_p=0.598**). Breed and age of weaning had no statistical effect on the assessed behaviour of w/g pigs. In conclusion, this study found a relationship between certain management factors and the behaviour of w/g pigs and further underlines the importance of rooting material in pig production. This project was financed by the Danish Ministry of Environment and Food.

Measures of cow welfare in a hybrid pasture dairy system

Cheryl O'connor, Suzanne Dowling and Jim Webster
AgResearch Ltd, Ruakura Research Centre, Private Bag 3123, Hamilton, New Zealand;
cheryl.o'connor@agresearch.co.nz

In the New Zealand dairy industry around 12% of dairy farmers use off-paddock facilities, particularly during the winter. Lying time is a key indicator of whether or not these hybrid pasture systems are meeting welfare requirements. Minimum lying times of 8 h/d are recommended to farmers in the New Zealand Dairy Cattle Code of Welfare and by the dairy industry. While lying time is relatively easy to measure in a research setting using loggers, it is more difficult to measure on-farm. The relationship of bedding quality, cow dirtiness and gait score with lying time were therefore assessed over 5 weeks in a hybrid pasture system to determine their utility as farmer friendly indicators of cow welfare. Two hundred non-lactating dairy cows were divided into 2 replicates of 100 balanced for breed, body weight and expected calving date. Cows were kept on a stand-off pad of woodchips for 18 h/day at a space allowance of 5.4 m^2/cow, with 6 h/day on pasture for 5 weeks. Lying times were recorded continuously for 60 cows per replicate using Onset Pendant G data loggers. Gait score of the same 60 cows was recorded once a week on a 5-point scale. At the same time each cow was scored for dirtiness of their left side ranging from clean to dirty on a 5-point scale for each of five body areas. One person recorded the gait score and one person the cleanliness score throughout the trial. Bedding moisture content was measured weekly in 10 stratified 25×40 cm quadrat samples across the stand-off pad. Data were analysed by repeated measures (REML) following smoothing (Bayesian or Spline). Lying time declined over the five weeks from 10.5 h/day during the first week to 6.7 h/day during the fifth week (SE=0.36; P<0.05). On 17/35 days the herd daily lying time was less than the minimum recommended requirement of 8 hours, including the last 10 days of the trial. There was a significant increase in overall dirtiness score from mean 7.9 to 22.7 (SED=0.35; P<0.001) from week 1 to 5, but there was no effect of time on stand-off on gait score. Over the 5 week period the moisture content of the bedding increased from 62 to 78% (SED=1.9; P<0.001). For comparison a small group of 16 cows kept under the same conditions, with the exception that they had fresh woodchip bedding throughout, maintained a consistent daily lying time of 10.5±1.2 h throughout the five weeks. This study has demonstrated that during 5 weeks in a hybrid pasture system, bedding quality deteriorated, cows got dirtier and lay down less to such an extent that welfare was compromised at both a herd and individual cow level. Bedding moisture and cow dirtiness may therefore be useful farmer-friendly indicators of cow welfare due to their relationship with reduced lying times in hybrid pasture systems.

Influence of milk feeding method on dietary selection of dairy calves

Kelsey Horvath and Emily Miller-Cushon
University of Florida, Department of Animal Sciences, 2250 Shealy Drive, Gainesville, FL 32611,
USA; emillerc@ufl.edu

Dairy calves that consume hay perform less non-nutritive oral behavior, suggesting that forage may satisfy a need for oral manipulation. The hypothesis of this study was that providing an outlet for natural sucking behavior in the form of an artificial teat would reduce dietary selection of hay. Individually-housed Holstein heifer calves were randomly assigned at birth to receive their milk (6 l/d) via either bucket (Bucket; n=10) or artificial teat (Teat; n=10). All calves were provided chopped hay and grain concentrate *ad libitum*. Calves were weaned by reducing milk allotment to 3 l/d from 42-49 d of age. Dietary selection of concentrate and hay was measured from birth until 7 d post-weaning (d 56), and calves were weighed weekly to assess growth. Data were summarized by week and analyzed over the course of the milk-feeding stage in a general linear mixed model with week as a repeated measure. Changes in dietary selection due to milk-weaning were analyzed with time period (7 d pre-weaning, 7 d weaning, 7 d post-weaning) as a repeated measure. During the milk-feeding stage, method of milk delivery did not affect selection of concentrate (138.3 g/d; SE=39.0; P=0.41), but hay intake was subject to a treatment by week interaction (P=0.031), with Teat calves consuming less hay than Bucket calves in earlier weeks of the milk-feeding stage (wk 4; 14.8 vs 25.9 g/d; SE=3.4; P=0.035), but increasing intake of hay to a greater extent over time. Through the weaning transition, Teat calves had consistently greater hay intake (102.3 vs 78.0 g/d; SE=14.2; P=0.047) and increased concentrate intake through weaning to a greater extent than Bucket calves (treatment by time period interaction; P=0.019). Consequently, concentrate intake, as a % of total intake, evolved differently between treatments (P=0.040), with Bucket calves decreasing the proportion of concentrate in their diet from pre-weaning to post-weaning (90.2 vs 86.7% of total intake) and Teat calves increasing their dietary proportion of concentrate (77.7 vs 83.4% of total intake). Average daily gain tended to be consistently greater for Teat calves through the weaning transition (0.65 vs 0.46 kg/d; SE=0.086; P=0.064). These results suggest that milk feeding method has potential to influence dietary selection of hay and concentrate in dairy calves, but that the effects of milk feeding method vary over time and in response to weaning.

Use of selected welfare quality® indicators to investigate nutritional conditioning and deficiency in broilers

Emma Fàbrega[1], Joaquim Pallisera[1], Eva Mainau[2], Insaf Riahi[1], David Torrallardona[1], Toni Dalmau[1], Antonio Velarde[1] and Maria Francesch[1]
[1]IRTA, Animal Nutrition and Welfare, Veïnat de Sies, 17121 Monells, Spain, [2]UAB, School of Veterinary Science, Department of Animal and Food Science, Facultat Veterinària, 08193 Bellaterra, Spain; emma.fabrega@irta.cat

Broilers have been said to adapt to early feed deficiencies by increasing their absorptive capacity at the intestinal level. The main objective of these studies was to investigate nutritional conditioning as a feeding regime strategy to improve the lifetime performance, nutrient utilisation and environmental impact in broiler chickens. Several Welfare Quality (WQ) indicators were assessed at individual and group level to evaluate the effect of nutritional conditioning on welfare. Two trials (n=600 per trial) of male broilers housed in groups of 25 in 24 pens were subjected to either a phosphorus/calcium (P/Ca, trial 1) or methionine (Met, trial 2) nutritional conditioning, by offering a diet deficient in the target nutrient during the first week of life (6.9 vs 4.5 g P/kg or 4.4 vs 3.0 g digestible Met/kg, respectively). Conditioned (C) and non-conditioned (NC) chickens were split in two groups, so that at 21 d half of the animals were offered a balanced diet (B; 5.8 g P/kg or 4.3 g digestible Met/kg). Aiming to highlight the possible advantages of conditioning, the other half were fed a diet deficient in the nutrient under study (D; 3.5 g P/kg or 2.8 g digestible Met/kg). This resulted in four treatments (NC-B, C-B, NC-D and C-D). The effects of conditioning and late deficiency on welfare were evaluated by means of group indicators (percentage of panting and huddling and number of broilers in locomotion or standing) three times (day 11, 21 and 39) and individual gait score (score 0-5, from normal to totally impaired) on day 39. In trial 1, a significantly higher percentage of broilers subjected to late P deficiency presented locomotion problems (% of broilers with gait score 4-5 was: 3.6, 3.6, 21.4 and 25.0, and with gait 0-1: 39.3, 42.9, 17.9, and 7.1, in treatments NC-B, C-B, NC-D and C-D, respectively, P<0.01). In trial 2, no differences in gait scores were observed for methionine conditioning or deficiency. The group indicators evaluated were not significantly affected in any of the trials, although a significant decrease over time on the number of animals standing and moving was observed (P<0.001). Thus, in the present trials, gait score was found to be a useful indicator to detect effects on welfare when broilers were subjected to a nutrient deficiency clearly associated with locomotion (P/Ca), but not when Met was depleted. Therefore, when simplifying the WQ protocol the selection of the indicators is crucial and should be tailored to the expected effects.

Tail biting: prevalence among docked and undocked pigs from weaning to slaughter

Helle Pelant Lahrmann[1], Marie Erika Busch[1], Rick D'eath[2], Bjørn Forkman[3] and Christian Fink Hansen[3]
[1]*SEGES P/S, Pig Research Centre, Axeltorv 3, 1609 Copenhagen, Denmark,* [2]*SRUC, Animal & Veterinary Sciences, West Mains Road, EH9 3JG, United Kingdom,* [3]*University of Copenhagen, Department of Large Animal Sciences, Grønnegaardsvej 2, 1670 Frederiksberg C, Denmark; hla@seges.dk*

The vast majority of piglets reared in the EU and worldwide is tail docked to reduce tail biting even though the EU animal welfare legislation bans routine tail docking. This study compared the prevalence of tail lesions among docked and undocked pigs at a conventional piggery in Denmark, that usually tail-docked all pigs, and in which the incidence of tail injuries were low. A total of 1,927 DanAvl Duroc × (Landrace × Large White) pigs (964 docked and 963 undocked) were housed in pens measuring 2.4×5 m pens with 20-22 pigs/pen from weaning at 4 weeks of age till slaughter at 110 kg body weight. Pigs were given an *ad lib* access to commercial diets, water and straw (10 g per pig per day) and had permanent gain to two vertical soft wood sticks. No tail injuries were recorded among the tail docked pigs. Amongst the undocked pigs a total of 22% had a tail lesion recorded at least once during the study period. On average 3.6% (CI: 2.31; 4.82) of the pigs had a tail lesion on the days of tail inspection. Tail lesions were more often present in pens with pigs weighing 30-60 kg (34.3% [CI: 25.6-44.3]; F=12.0; P<0.05) than in pens with pigs weighing 7-30 kg (12.7% [CI: 8.29-18.9]) and 60-90 kg (11.0% [CI: 6.85-17.3]). On average more pigs with undocked tails (0.81 pig/ pen) than with docked tails (0.31 pig/ pen) were moved to hospital pens (t=2.36; P<0.05) but tail docking did not influence mortality. At the abattoir more undocked pigs (2.00%) compared with tail docked pigs (0.32%) got a tail lesion remark (χ^2=11.2; P<0.001). Surprisingly some docked pigs got a tail lesion remark at the abattoir, even though no docked pigs were observed with a tail lesion on the farm. Generally, tail wounds healed before slaughter. Pigs that normally get a tail biting remark during meat inspection are cases were the wound has got infected resulting in a local or systemic abscess. In conclusion, this study suggest that housing pigs with intact tails under conventional Danish conditions will increase the prevalence of damaging tail biting behaviour and the need for hospital pens even in well managed piggeries. Furthermore, the result showed that abattoir tail lesion remarks severely underestimate the prevalence of tail bitten pigs from weaning to slaughter on the farm.

The behaviour of dairy cows with artificial shade in a temperate region of Mexico

Ricardo Améndola-Massiotti[1], Norma Castillo[1], Germain Martínez[1], Lucía Améndola[2] and Juan Burgueño-Ferreira[3]
[1]*Universidad Autónoma Chapingo, Zootecnia, km 38.5 Carr Méx-Texcoco, 56230 Chapingo, Mexico,* [2]*University of British Columbia, Animal Welfare, 2357 Main Mall, V6T 1Z4 Vancouver, Canada,* [3]*CIMMYT, Biometría, Km. 45, Carr Méx-Veracruz, 56237 Texcoco, Mexico; r.amendola51@gmail.com*

Shade is a valuable resource for grazing dairy cows, but little is known on its effects in Mexican temperate regions where most animals graze without shade. We aimed to evaluate the impact of artificial shade on the behaviour of grazing dairy cows, their body temperature and herbage intake, in the rainy and cloudy season. Grazing without shade (NS) was compared to shade (AS) provided by 80% shade polypropylene cloth (6×4 m; 4.8 m^2/cow). Four experimental units of five Holstein-Friesian cows in milk of 520±15.7 (standard error) kg liveweight (LW) grazed on five 0.5 ha paddocks divided in four quarts (one quart per group) in a block design with two replicates per paddock. We evaluated two grazing cycles of five days per paddock. Body temperatures were registered five times daily with an infrared thermometer. Average herbage intake was measured as the difference between pre-and post- grazing herbage dry matter (DM), grazed herbage composition was estimated from hand-plucked samples. Behaviours of cows (grazing, ruminating, lying, affiliative and agonistic interactions) were measured through 10 min-interval scans during 24 h once per paddock-cycle (10 events). Data were analysed using Generalised Linear Mixed Models; means were compared with the Pdiff procedure (SAS). Average highest daily Temperature Humidity Index was 79±2; body temperature was lower ($P<0.05$) in AS than in NS cows (37.7±1.13 vs 38.3±1.39 °C). Average herbage intake tended ($P=0.08$) to be 18% higher in AS than in NS (3.93±0.47 vs 3.23±0.46 kg DM/100 kg LW/d), though no difference ($P>0.05$) in grazing time was detected. The Neutral Detergent Fiber content of consumed herbage was lower ($P<0.05$) in AS than in NS (50.7±1.1 vs 53.3±1.1%), related with 9% lower ($P<0.05$) ruminating time while lying in AS than in NS. There was no effect ($P>0.05$) on the frequency of affiliative interactions; agonistic interactions tended to be less frequent ($P=0.07$) in AS than in NS (1.69±0.22 vs 1.89±0.33 interactions/cow/day). Cows which had access to shade had lower body temperatures; the associated greater comfort was the probable cause of higher grazing efficiency (trend to higher intake in the same grazing time and more selective grazing), and trend to lower frequency of agonistic interactions. Access to shade had slight beneficial effects on cows in spite of the temperate and cloudy weather conditions; hence further assessment is justified.

Comparison of activity levels among different housing conditions and profiles in cats and their correlations with cortisol

Toshiki Maezawa, Miho Takeuchi, Ayuhiko Hanada, Ayaka Shimizu, Natsumi Sega, Haruna Kubota and Yoshie Kakuma
Teikyo University of Science, Department of Animal Sciences, 2-2-1 Senjusakuragi, Adachi-ku, Tokyo, 1200045, Japan; maroni.2008.0822@gmail.com

Many cats are kept not only at home but also in various housing environments such as institutions, sanctuaries, or cat cafés. In the present study, we compared the activity levels in cats among different husbandry systems and various profiles using bio-logging technique. We also measured urinary cortisol as a physiological parameter of stress and correlation with the activity levels was examined. We used six teaching resident cats kept at university (4 males; 2 females, 3-7 yrs, 3-4 kg), 46 cats kept at cat cafés (26 males; 20 females, 2-17 yrs, 3-9 kg), and 30 cats kept at home as pets (16 males; 14 females, 2-17 yrs, 3-7 kg, 10 allowed outside). All but six female café cats were neutered. An accelerometer was attached to the collar of each cat and acceleration was recorded every 2 sec. The activity level was calculated as the sum for 'Active' per day. The 24-hour activity level was measured for 22 days for university cats, once for café cats, and twice for pet cats. Urine of university cats (n=132) and pet cats (n=22) were collected on the following day and that of café cats (n=13) was collected within a month of the activity measurement. Urine was collected non-invasively using two-layer litter box with non-absorbent litter which all cats usually used. Cortisol and creatinine were measured to calculate urinary cortisol/ creatinine ratio (UCCR). We found that pet cats were significantly less active than university cats (P<0.001) and café cats (P<0.001) by one-way ANOVA ad Tukey's pots-hoc test. The activity level of neutered male cats was significantly greater than that of neutered female cats (P<0.05 by t-test), whereas there were no significant differences depending on body weight, age and going outdoors. No significant Pearson's correlations were found between the activity level and UCCR in university cats, those in café cats, and those in pet cats (P>0.05, respectively). Our findings suggested that pet cats are less active than those kept in other facilities because of less stimulations at home. Because UCCR was not affected by housing conditions or activity levels, less activity in pet cats may just induce obesity and other health concerns. We should examine the influence of other factors at home such as the number of family members and cats, or space allowance on the activity levels of cats in future studies. As few studies have been done on the activity patterns and behaviour in neutered cats kept freely at home and other facilities, it is important to collect basic information to help considering appropriate keeping condition for cats and to advise caretakers on behavioural needs in cats based on knowledge from these studies.

The effect of material composition of perches on bone mineralisation and keel integrity in caged White Leghorns

Patricia Y. Hester[1], Sydney L. Baker[1,2] and Maja M. Makagon[1,2]
[1]*Purdue University, Department of Animal Sciences, 125 South Russell St, West Lafayette, IN 47907, USA,* [2]*Current Address: University of California at Davis, Department of Animal Science, One Shields Ave, Davis, CA 95616, USA; phester@purdue.edu*

Perches in cages allows hens to express their natural roosting behavior, but a disadvantage is that metal perches as compared to hens with no perches increase keel damage. The objective of this study was to determine if alternative perching material such as AstroTurf or cork would help alleviate the deleterious effects that metal perches have on keel integrity. Forty cages each contained 4 White Leghorns, aged 16 wk. Eight cage replicates were assigned to 5 perch treatments of no perch, softwood, metal, metal perch covered with AstroTurf, or metal perch covered with cork. Hens, 52 wk of age, were sedated with pentobarbital and euthanized via cervical dislocation. The left wing, thigh, drum, and the breast were dissected from each hen and subjected to dual energy x-ray absorptiometry to determine bone mineralisation. The excised keel with muscle removed was examined for fresh and old healing fractures and scored for deformations. The MIXED model procedure of SAS was used to analyze data. Hens with access to a perch, regardless of its material composition, had bone mineral density values similar to control hens without perch access. The short duration of the study of 36 wk mostly likely contributed to the lack of an effect on bone mineralisation as previous studies that lasted 71 wk with metal perches showed an improvement in bone mineral density as compared to hens with no perch access. Likewise, bone mineral content was not affected by treatment with the exception that hens with access to the metal or the cork perch had higher keel ($P<0.05$) and humeral ($P<0.03$) bone mineral content, respectively, than controls but this did not improve keel integrity because bone mineral density is the more important trait contributing to improved bone quality. Perch presence caused a deterioration in keel quality. Hens with access to a cork (83%) or wooden (83%) perch had a greater proportion of fractures ($P<0.01$) than hens without a perch (43%). Hens with access to metal (76%) and AstroTurf (67%) perches were intermediate in values with respect to fracture incidences. Hens with access to a cork or wooden perch also had poorer keel scores ($P<0.0001$) than hens without a perch. The overall proportion of hens with keel fractures was 70%. Of these fractured keels, 21.4% were fresh, 59.5% were old, and 19.1% had both old and fresh keel fractures. In conclusion, softer perching materials such as AstroTurf and cork did not improve keel integrity as bone mineral density and keel fracture proportions were similar among perching material treatments.

The use of artificial grass to attract wild sika deer

Akitsu Tozawa and Ken-ichi Takeda
Shunshu University, 8304 Minami-Minowa-mura, Kamiina-gun Nagano-ken, 399-4598, Japan;
akitsu@shinshu-u.ac.jp

The Ministry of the Environment in Japan estimated that the number of wild sika deer (*Cervus nippon*) had reached almost 2.5 million in 2012, double the number 10 years ago. Furthermore, the increasing damage to crops and forests has meant that deer population control is necessary. However, the number of hunters has fallen to 185,000, half the number 30 years ago, and the average age of hunters has increased with 66% of hunters now being over 60 years old. In addition, predators, such as wolves, have become extinct in Japan. Therefore, efficient and labour-saving methods of capturing deer are requred. Deer are herbivores, which means that they are likely to distinguish green colours and attracted to green spaces. Using these characteristics, we experimentally evaluated the use of artificial grass to attract wild sika deer. The patch of artificial grass used was 3.5 cm high and 2×4 m wide in area. No odorants were added to the artificial grass. Four areas were used: three areas were corral traps in the forest (Areas 1-3), which were not in use at the time, and one was near an animal trail (Area 4) that showed evidence of deer use. The corral traps (Areas 1-3) were located in the forest where human disturbance was low. However, people occasionally passed by Area 4. All areas were within a 10-km radius. The artificial grass was placed 4-5 m away from the entrance of the corral trap or the animal trail. Two motion sensor cameras were set and the numbers of deer recorded by the cameras were counted. The study periods took place between February (Winter) and May (Spring), 2015 and the data were collected over 4 week periods, both before and after setting the artificial grass, except in Areas 1 and 4. In these two areas, data were collected for 3 weeks before grass laying because of overlaps with the hunting season. The deer numbers were converted by logarithmic transformation, and a paired t-test was used for statistical analysis. The deer numbers had no significant difference between before and after laying the grass (Area 1. 3.1 vs 5.0 heads/10 camera days; Area 2. 0 vs 6.8; Area 3. 7.7 vs 12.3; Area 4. 23.8 vs 25.7; $P=0.106$). The results suggested that artificial grass had no influence to attract wild deer. However, deer numbers did increase in all areas after laying the grass. Since a deer's field of view is limited, the effective range of visual stimuli that will attract deer is also reflected to a small area. We need further studies to attract wild deer using a combination of stimuli such as visual and olfactory stimuli.

Behavioural expression of oestrus in cattle varies with stocking densities

Kathrin Schirmann, Artur C. C. Fernandes, Marina A. G. Von Keyserlingk, Daniel M. Weary and Ronaldo L. A. Cerri
The University of British Columbia, Faculty of Land and Food Systems, 2357 Main Mall, Vancouver, BC, V6T 1Z4, Canada; kathrinschirmann@gmail.com

Behaviour is often used to identify oestrus on dairy farms, but behaviour can be greatly affected by housing conditions. The objective of this study was to assess the effect of stocking density on standing behaviour at the time of oestrus. Cows were moved into one of two stocking densities (67 or 133%; cow:stall ratio of 1:1.5 and 1:0.75) at 40±3 days in milk (DIM) and monitored for 34 d. Cows were fitted with data loggers to record standing time. All animals underwent a pre-synchronization program consisting of two PGF_{2a} injections (PG; Estrumate) 14-d apart starting at 47±3 DIM. The presence of oestrus was monitored beginning the day after the second PG application at 61±3 DIM for 5 consecutive days using per rectum palpations with ultrasonography (US). Ovulation was confirmed by the disappearance of the largest follicle during US examinations by the 5th day after PG and the appearance of a corpus luteum 12 d after PG. Data were analysed using t-test and ANOVA. A total of 101 animals had a confirmed ovulation. Overstocked cows spent on average 1 h/d ± 11 min (P<0.01) more time standing and had fewer (P<0.01) and longer (P<0.01) standing bouts than understocked animals. On the day of oestrus, compared with baseline values, both groups had greater standing times (+152±36 min and +145±30 min for over- and understocked cows respectively; P<0.01) and an increased longest standing bout of the day (+142±30 min and +135±24 min for over- and understocked cows respectively; P<0.01). Compared with baseline understocked cows also showed an increase of 1.4±0.5 h in standing time on the day before oestrus (P<0.01). We conclude that overstocking affects overall standing time as well as changes in standing behaviour associated with oestrus.

Behaviour of broilers in semi-commercial organic rearing – behaviour and mortality of hybrids with rapid or slow growth rate

Jenny Yngvesson, Anna Wallenbeck, Lotta Jönsson and Stefan Gunnarsson
Swedish University of Agricultural Sciences, Animal Environment & Health, P.O. Box 234, 53223 Skara, Sweden; jenny.yngvesson@slu.se

Regulations of organic broiler production within the EU demand long rearing periods and this is an incentive for using slower growing genotypes, as conventional broilers grow extremely fast and thereby risk growth related welfare problems, including abnormal behaviour. Earlier evaluations of hybrids have, to our knowledge, not resembled the commercial situation. We compared two slow growing hybrids (Rowan Ranger, Hubbard CYJA57) with one fast growing (Ross 308), under semi-commercial organic conditions. One hundered chicks of each hybrid were reared toghether in one large group on wood shaving litter, with *ad libitum* organic food, water, roughage and perches. Focal, 1-0 observations of behaviour were carried out on 10 birds per hybrid, during 10 min/bird. Scan sampling of all birds within 10 2×2 m^2 squares. Focal and scan observations were repeated four times during rearing (day 15, 44, 60 and 78 after hatching). Birds were slaughtered at 82 days of age. Individual birds were the statistical unit and hybrids were compared using Mood Median test. Differences between hybrids were small and individual variaton in behaviour was large. Ross birds perched (mean ±SEM) less (0.3±0.03) than the slow growing birds (0.8±0.2 and 0.7±0.3) (Mood Median χ^2=2.6, Df=2, P=0.002) but even Ross birds were seen perchng on the last observation at 78 days of age. Ross had 20% mortality and the slow growing birds 2% each. Frequent behaviours in all birds, using both observational methods, were stting/lying and sleeping (lying with eyes closed and head resting). Foraging behaviour, a very frequent behaviour in jungle fowl *Gallus gallus* and laying hens, was only rarely observed (14 times and mainly by Hubbard birds). Dusbating was seen in all hybrids, also during the later parts of the rearing period, however not in all individual. Compared to natural behaviour in *Gallus gallus*, both the conventional fast growing and the slow growing broilers showed a deviating behavioural pattern. Though behaviour did not differ to a great extent between hybrids, mortality did, raising ethical concerns for using fast growing birds in organic production.

The influence of music on the behaviour of canines in a daycare facility

Melissa L.A. Jollimore[1], Catherine Payzant[2] and Miriam B. Gordon[1]
[1]Dalhousie University, Plant and Animal Sciences, Haley Institute, 58 River Rd., Truro, NS, B2N 5E3, Canada, [2]Dalhousie University, Mathematics and Statistics, 316 Coburg Road, Halifax, NS, B3H 4R2, Canada; miriam.gordon@dal.ca

Dog daycare facilities have become increasingly popular as a means to look after dogs with busy owners. Auditory enrichment has been shown to be beneficial in creating a positive environment and decreasing behavioural indicators of stress for dogs in a rescue kennel environment, but this has yet to be studied in a daycare facility. This study investigated the impact of audio stimulation on the behaviour of 6 regular dog daycare attendees during a one-hour kenneled period. Experimental music treatments (classical, country, rock and a control (no audio music)) were played at the daycare facility for four hours before the dogs were placed in kennels for 'nap time', where behaviours were recorded. Dogs were exposed to each treatment at least 4 days prior to behavioural recording. Behavioural observations for each treatment were collected on four separate days. Data was recorded at five minute intervals, for one hour, and the behaviours shown at that instant were counted and recorded. The behaviours included activity, vocalization and location in kennel. Dogs slept more (P=0.002) during classical music (63.5±5.8%) than any other musical selection (control=27.8±6.3%, country = 16.4±4.5%, rock = 16.7±2.8%). There was a tendency (P=0.10) for the dogs to locate themselves at the front of the kennel more during classical music (60.9±8.8 vs 50.9±8.1, 21.3±5.4, and 42.6±6.7% for control, country, and rock, respectively). The time dogs spent vocalizing was non-significant (P=0.40). This study suggests that playing classical music during a kenneled nap time at a daycare facility can influence the dogs to rest. The animal's psychological and physiological well-being at the daycare is instrumentally important for the business' success.

Behaviour and preferences of fattening lambs exposed to two different size of wheat straw

L Aguayo-Ulloa[1,2], C González-Lagos[3], M Pascual-Alonso[2], M Villarroel[4], M Martínez-Novoa[2] and G María[2]
[1]CORPOICA, Km13, Vía M-C, 0, Colombia, [2]University of Zaragoza, M. Servet 177, 0, Spain, [3]Pontificia Universidad Católica, Av. LBO340, 0, Chile, [4]Polytechnic University of Madrid, RdM7, 0, Spain; levrino@unizar.es

In Spain, intensive sheep production systems use concentrate and straw to feed traditional light lambs. We aimed to evaluate two different sizes of straw on the behaviour and physiology of fattening lambs. Eight Rasa Aragonesa entire weaned male lambs (65d, 17.1±0.79 kg) were housed individually (150×120 cm, 1.80 m^2 per lamb) for 5 weeks to complete the commercial fattening period. They were all given grass hay during 1st week. After, 4 lambs were given long wheat straw (LST) and another 4 lambs received chopped wheat straw (CST), for one week. Then the groups were switched for another week, the first group receiving CST and the second LST (two-period crossover design), to avoid the possible effects of exposure order. After that, all straw was removed for another week (NST). Feeding (concentrate and straw) and water consumption were *ad libitum*. Consumption and lamb weights were recorded weekly. Behaviour was recorded for 3 days at the end of each week using video (8 am-12 pm and 6 pm-8 pm) by scan sampling (1′ every 5′). The choice tests were performed during the 5th week. Blood samples were taken immediately after the removal period (NST) and after the choice test week to assess stress response indicators. After the 1st week, lambs consumed more concentrate and had a higher average daily gain. However, forage consumption decreased in both exposure periods (-45.4% and -66.7% for LST and CST, respectively). The proportions of elicited behaviours changed as a function of straw size and supply (interaction term P<0.001). All behaviours, except eating concentrate and moving, varied as function of straw source (P<0.001). Interestingly, lambs ate and played with fiber source more often during the 1st week and for LST, than for CST (P<0.001 for all contrast). Also, CST lambs stood up more (U) and had more stereotypic behaviours (S=licking bars, walls, feeders) (P<0.001). Moreover, when the provision of straw was interrupted, resting behaviour (R) decreased (P<0.001) and U and S increased, reaching maximum levels (P<0.001). No significant differences were found for physiological stress variables, with the exception of non-esterified fatty acid, which doubled by the second sampling. During the choice test, lambs preferred (P<0.05) LST compared with CST. Although straw consumption was low, its presence influenced an important part of the behavioural budget. The NST period emphasized the behavioural pattern already shown during the CST period. It is concluded that the role of straw in this type of lamb may be more related to sensory stimulation than to feeding.

The effect of Pet Remedy on the behaviour of the domestic dog

Sienna Taylor[1] and Joah Madden[2]
[1]Hartpury College, Hartpury House, Gloucester, GL19 3BE, United Kingdom, [2]University of Exeter, Washington Singer, Exeter, Devon, EX4 4QG, United Kingdom; sienna.taylor@hartpury.ac.uk

Stress affecting behaviour in companion animals can have an adverse effect on animal health and welfare and their relationships with humans. This stress can be addressed using chemical treatments, often in conjunction with behavioural therapies. Pet Remedy, a natural valerian based stress relief product for mammals is commercially available on the market but its efficacy has yet to be scientifically tested in dogs using a placebo-controlled trial. The aim of the study was to explore whether Pet Remedy lowered anxious behaviour in dogs placed in a novel environment compared to a placebo product. The behavioural response of 28 mixed breed dogs placed in a novel 'stressful' environment was tested under a repeated measures, randomized order, double blind, placebo controlled trial. The dogs were placed in a 3×3 m enclosed room and exposed to both a placebo and Pet Remedy plug-in diffuser for 30 minutes (necessitated by study design) with an intervening period of approximately 7 days between treatments. The dogs' behaviour was video recorded during the course of exposure without the presence of the owner. The observer was screened behind the holding pen at all times to reduce observer presence on the dogs' behaviour. Behaviours were categorised into anxious and relaxed behaviour categories with number of occurrences and durations recorded. Data were analysed by principal component analysis (PCA) to reduce behaviour states into a smaller group of linear combinations. Paired t-tests were used to compare overall levels and levels of change in the composite and individual behavioural measures. Dogs exposed to Pet Remedy exhibited significantly higher overall levels of a composite measure of locomotory activity (M=0.18, SD=0.98) including locomotion, standing hind paws, rear and wall bounce than when exposed to the placebo across the entire 30 minute period; $t(27)=2.16$, P=0.041. Significantly lower occurrences of event-based yawning behaviour (M=0.03, SD=0.06) were also exhibited in the Pet Remedy condition in comparison to the placebo; $t(27)=2.44$, P=0.021. No test order effects were observed. Dogs that display suppressed behaviour when exposed to acute novel environments may in particular benefit from Pet Remedy with the product stimulating locomotory activity whilst also reducing anxious yawning behaviour. However, it is particularly important that treatment choice is selected on a case by case basis depending on how individual dogs react to stressful situations. In cases where stress manifests itself in lethargy, Pet Remedy may provide a useful tool for reducing the occurrence of some behaviours associated with anxiety.

The effect of roughage composition (rolled and baled hay) on repressing wool-biting behaviour in housed sheep

Chen-Yu Huang[1] and Ken-ichi Takeda[2]

[1]Shinshu University, Interdisciplinary Graduate School of Science and Technology, Department of Bioscience and Food Production Science, 8304 Minami-Minowa-mura, Kamiina-gun, Nagano-ken, 399-4598, Japan, [2]Shinshu University, Institute of Agriculture, Academic Assembly, 8304 Minami-Minowa-mura, Kamiina-gun, Nagano-ken, 399-4598, Japan; 12st553a@shinshu-u.ac.jp

Wool-biting is an abnormal behaviour in housed sheep, causing serious animal welfare problem in the sheep industry. Our previous study showed that roughage in the form of rolled hay could potentially provide sheep with appropriate oral stimulation by eliciting their normal foraging movements, and consequently, repressing the wool-biting behaviour upon feeding. In this study, we intended to verify this hypothesis further by mixing rolled and baled types of hay and quantifying the amount of rolled hay needed for sheep to repress the wool-biting behaviour. We used a 4×4 counterbalanced experimental design, consisting of 16 Friesland ewes divided into 4 groups (4 ewes / group) and provided with different proportions of rolled hay (100, 67, 33, and 0%) in roughage as treatments. Each sheep was fed daily 1 kg of roughage at 08:30 and 15:00 and 350 g of concentrate at 15:20. Behavioural observations were conducted from 16:00 to 18:00, two days a week, for four continuous weeks. During the observation period, the wool-biting bouts were recorded. Wool-biting frequency (times of biting bout per hour) showed a significant difference across treatments (Kruskal-Wallis test, χ^2=22.16, df=3, P<0.01, 1−β=0.92). However, only the 0% rolled hay treatment (6.92±1.09) had significantly lower wool-biting frequency than that by the other treatments (Dunn's test, P<0.01), and no difference was noted among the treatments with 100% (1.17±0.46), 67% (1.17±0.47), and 33% (1.67±0.48) rolled hay. This result indicates that the presence, and not the amount, of rolled hay influences the wool-biting behaviour. This also verifies our earlier observation that rolled hay has the effect of satisfying sheep with oral stimulation, further showing its efficiency in repressing wool-biting behaviour even at low quantities.

Laboratory rats housed in two-level cages are less anxious in a novel environment

SYLVIE Cloutier[1,2], Melissa Liu[1] and Wendy Ellison[1]
[1]*Washington State University, Center for the Study of Animal Well-being, Department of Integrative Physiology and Neuroscience, P.O. Box 647620, Pullman, WA 99164-7620, USA,* [2]*Canadian Council on Animal Care, 800-190 O'Connor Street, Ottawa, ON, K2P 2R3, Canada; scloutier@ccac.ca*

We hypothesized that laboratory rats raised in a complex environment show less signs of anxiety when placed in a novel environment. Pairs (n=17) of male and female Long-Evans rats were housed in one of two cage environments: (1) a one level cage (L1), or (2) a two-lcvel cage (L2) from birth to 30 days of age with their dam and littermates, and thereafter in same sex pairs. Production of 22 and 50 kHz ultrasonic vocalizations (USV), which can be reliably used to measure rat affective states and behaviour were assessed in an Emergence test when the rats were 51-68 days old. The Emergence test is used to assess anxiety (fear associated with novelty) by assessing the rat's preference for dark enclosed places, over bright exposed ones. A longer time engaged in full body emergence from the dark enclosed box and a shorter latency to full body emergence into the brighter exposed place, denote less anxiety. Rats were the unit of analysis. General linear model ANOVAs were used. During the test, L2 rats produced more 50-kHz USV (GLM ANOVA, $F(1,41)=7.82$, $P=0.008$; mean±SE, L1: 2±1.1, L2: 21±5.6 calls/5 min), took a shorter time to emerge from the dark box ($F(1,41)=17.15$, $P=0.0002$; L1:183±28.8, L2: 61±14.6 s), entered more frequently ($F(1,41)=21.38$, $P<0.0001$; L1: 1.8±0.42, L2: 4.2±0.33 entries/5 min) and spent more time ($F(1,41)=15.80$, $P=0.0003$; L1: 60±16.0, L2: 144±13.6 s) in the bright, exposed box. Thus L2 rats were in a more positive affective states based on call production, and were less anxious based on their shorter latency to emerge and the longer time spent in the bright exposed box compared to L1 rats. These findings show that allowing rats to grow up in a complex environment affects their response when faced with a novel situation later in life, allowing them to cope better with the unexpected. Our results have implications for the design of cage environment and the welfare of laboratory rats.

Does flooring substrate impact kennel and dog cleanliness and health in commercial breeding facilities?

Moriah Hurt[1], Judith Stella[2], Audrey Ruple-Czerniak[1], Paulo Gomes[3] and Candace Croney[1]
[1]Purdue University, 725 Harrison Street, West Lafayette, Indiana 47907, Veterinary Pathology Building room 135, USA, [2]USDA-APHIS, 725 Harrison Street, West Lafayette, Indiana 47907, Veterinary Pathology Building room 135, USA, [3]Purdue University, 625 Harrison StrFeet, West Lafayette, Indiana 47907, Veterinary Clinical Sciences, USA; judith.l.stella@aphis.usda.gov

Evaluation of dog kennel flooring surfaces is needed to understand their potential impacts on dog health and well-being. This study aimed to characterize kennel and dog body cleanliness on different flooring types common in US breeding kennels. Subjects were adult dogs (n=111) housed indoors on diamond coated expanded metal, polypropylene, or concrete flooring with access to outdoor concrete runs and grass yards at five commercial breeding facilities. Each indoor kennel and dog was visually assessed for cleanliness using a 1-5 scale (1 = <1% of dog, kennel floor covered in debris, 5 = >76% covered in debris). Fecal contamination of floors was characterized by swabbing kennels immediately after cleaning procedures were completed using electrostatic dry cloths, later cultured for *Escherichia coli*, a common species used to indicate fecal contamination. Because of high variation in breed, management and other factors, valid cross-facility comparisons could not be made. Descriptive statistics were therefore used to describe dog and kennel cleanliness scores and fecal contamination on each flooring type. Mean (±SD) kennel cleanliness scores at each facility were 1.81(±0.40), 1.10(±0.31), 1.52(±0.51), 1.00(±0.00), and 1.00(± 0.00). The mean (± SD) dog body cleanliness scores for each facility were 1.03(±0.18), 1.03(±0.18), 1.00(±0.00), 1.00(±0.00), and 1.05(±0.22). For both kennel and dog cleanliness a score of 1 was considered ideal. The number of kennels at each facility that were culture positive for *E. coli* from samples collected after cleaning ranged from 7 to 30%. Overall, dogs and kennels at these facilities appeared to be clean. Differences in the recovery of *E. coli* may reflect differences in management practices such as cleaning protocols or factors pertaining to the dogs. These findings indicate that despite differences in indoor flooring choices, routine cleaning procedures were effective at reducing fecal contamination and debris that could introduce health risks for dogs at these facilities.

Heat load affects measures of aversion in dairy cows

Jennifer M. Chen[1], Karin E. Schütz[2] and Cassandra B. Tucker[1]
[1]*University of California, Davis, CA, USA,* [2]*AgResearch, Ltd, Hamilton, New Zealand;* *jmchen@ucdavis.edu*

Spraying cows with water reduces heat load. Cows willingly use spray in some studies, but show reluctance to wet the head or entire body in others, which may reduce cooling. Spray flow rate may explain this reluctance, as higher flow rates generate greater spray impact, but it is unknown whether cows find higher-impact spray aversive. An aversion race can be used to evaluate the emotional valence of treatments, based on the prediction that animals will approach negative (vs positive) stimuli more slowly and require greater pressure from a handler. Our objective was to evaluate how heat load affects the degree to which dairy cows show reluctance to approach spray with different flow rates in an aversion race. We compared unsprayed treatments (n=8 with feed, n=16 without) to 0.4 l/min (n=14) and 4.5 l/min (n=17) spray, which differed 8-fold in impact (1.1 vs 8.9 kPa, respectively). Over 10 trials, cows were required to approach treatments in a covered, narrow raceway (air temperature outside ranged from 21.3-43.5 °C). We compared the pressure (0-6 scale) handlers applied, time to complete the race (transit time), and head posture using GLMMs with cow as the experimental unit. Overall, handlers applied pressure (binary measure, score >0) half as often when feed was offered (P=0.001), but there were no other treatment differences in handling pressure or transit time (P≥0.424). Thus, we excluded the feed treatment when examining interactions with heat load and flow rate. In warmer weather, transit time increased (by 13 s per 10 °C increase; P=0.043). Unsprayed cows moved more slowly as respiration rate increased (by 7 s per 10 breaths/min increase; P=0.017), but sprayed cows did not show this response. Handling pressure was not affected by heat load or treatment (P≥0.129), but head posture depended on flow rate. Cows lowered their heads nearly 5 times as often when approaching 4.5 l/min compared to 0.4 l/min or no spray (P<0.001). This may have been an attempt to reduce exposure of sensitive body parts to higher-impact spray. Indeed, when we applied von Frey monofilaments to the ear and shoulder (n=56), the ear responded at one-sixth the level of force, on average, compared to the shoulder (P<0.001), indicating greater sensitivity. In conclusion, each measure provided different insights about reluctance to approach treatments. Transit time increased in response to heat load, rather than aversion. Handling pressure reflected willingness, as pressure was needed less often when cows were offered a feed reward, but this measure did not differ among spray flow rates. Therefore, although cows lowered their heads to avoid exposing sensitive areas to higher-impact spray, they did not show any other indicators of aversion to this type of spray.

Behaviour of organic lactating sows given access to poplar trees

Sarah-Lina Aagaard Schild, Lena Rangstrup-Christensen and Lene Juul Pedersen
Aarhus University, Department of Animal Science, Blichers Allé 20, 8830 Tjele, Denmark;
sarah-lina.schild@anis.au.dk

Every third piglet born in Danish outdoor pig production dies before weaning. This fact is in conflict with both the idea of increased animal welfare in the organic production and the organic principles. A there is a need for studies identifying management procedures, which may be used to lower piglet mortality. Heat stress in the sows is one factor known to increase piglet mortality. And so the aim of the current study was to investigate how access to poplar trees (an alternative shaded area) affects signs of heat stress and paddock use in lactating sows. Seventy-two lactating sows (median parity 4; range 1-8) and their litters were included in the study. Fifty-eight test sows were housed in paddocks (490 m^2) with access to poplar trees and 14 control sows were housed in standard paddocks (278 m^2) without shade. Piglets were kept inside the huts for the first ten days pp (post partum). Sows were monitored using focal scan sampling on a daily basis from three days before until eight days after expected farrowing and at approximate day 14 and 28 after expected farrowing. The average total litter size was 17 piglets and approximately 31% of these died before weaning (day 49). The average ambient temperature was 16.5 °C. Data were analysed using the proc mixed procedure in SAS. The results showed that sows with access to poplar were more often outside the hut before farrowing and on d 0 and 1 pp (59.3% and 19%) compared to controls (42% and 6.4% respectively) ($P<0.05$). Sows with poplar also increased the use of the paddock at high compared to low temperature (at 25 vs 18 °C: 45 vs 30% of time) compared to controls (at 25 vs 18 °C: 30 vs 30% of time) ($P<0.05$). When in the paddock, sows with access to poplar were observed lying more (29%) compared to controls (18%) ($P<0.0005$). There was an interaction between temperature and observation day ($P<0.05$). All sows spent more time lying at high vs low temperature particularly before farrowing, on day 0 and 1 pp and again after d 11 pp. Sows with access to poplar, spent more time in the poplar during days before farrowing and after d 11 pp, than on other days ($P<0.0005$), and at high compared to low temperature ($P<0.05$). In conclusion access to poplar increased sows use of the paddock, particular at high temperatures. The use of the poplar trees was higher before farrowing and after d 11 pp; likely because the piglets then were able to leave the hut. In general there was a trend for all sows to lie down more when outside the hut at high temperature.

Effects of alternative deep bedding options on dairy cow preference, lying behaviour, injuries, and hygiene

Tania Brunette[1], Elsa Vasseur[1], Trevor Devries[2] and Renée Bergeron[2]
[1]McGill University, Ste-Anne-de-Bellevue, QC, Canada, [2]University of Guelph, Guelph, ON, Canada; rbergero@uoguelph.ca

Cows spend more time lying down when stalls are comfortable, and bedding plays a key role in the comfort of the lying surface. The cost associated with good-quality bedding forces farmers to look for alternatives. Switchgrass, a high-yielding perennial grass, could constitute a promising alternative. The first objective was, therefore, to compare cow preference for 2 types of deep beddings in a free-stall, switchgrass and switchgrass-lime mattress, using wheat straw on rubber mat as a control. Nine Holstein lactating cows were submitted to a 3 choice preference test. Cows were housed individually in pens containing 3 stalls with different lying surfaces: (1) rubber mat with 2 to 3 cm of chopped wheat straw (S); (2) deep bedded switchgrass (SG) and; (3) deep bedded switchgrass, water and lime mixture (SGL). Trios of animals were tested simultaneously, one in each test pen for 14 d periods consisting of 2 d of adaptation, 3 d of restriction to each stall, and 3 d of free access to all 3 stalls, during which cows were video recorded. The second objective was to test the effects of these 3 types of bedding on lying behaviour, injuries, cow cleanliness, and teat end bacterial contamination. The 3 bedding treatments were compared in a Latin square design using 24 cows, with beddings being switched every 4 wk. Lying behaviour was measured with data loggers. During free-choice periods, total lying time was higher (P<0.02) on SG (9.4±1.14 h/d) than SGL (1.0±1.14 h/d), and total standing time and the number of standing and lying bouts were higher (P<0.02) on SG (2.0±0.31 h/d, 10.1±1.08 bouts/d, 8.2±0.82 bouts/d) than on S (0.6±0.31 h/d, 3.5±1.08 bouts/d, 1.4±0.82 bouts/d) and SGL (0.6±0.31 h/d, 2.6±1.08 bouts/d, 0.9±0.82 bouts/d), suggesting a preference for SG. In experiment 2, total lying time and number of lying bouts were on average 9.6±0.90 h/d and 8.1±0.80/d and did not differ between treatments. No treatment effects were found for hock and knee injury scores, which were on average less than 1 (scale of 0 to 3), or for leg, flank and udder cleanliness scores, which were close to 0. Treatment differences (P<0.001) were found for bedding dry matter, which was highest for SG (74.1±1.10%), lowest for SGL (63.5±1.10%) and intermediate for S (68.5±1.10%). This may explain the higher (P<0.05) *Escherichia coli* count on teat ends of cows on SGL (0.92±0.165 $\log_{10}$ cfu/g) compared to SG (0.47±0.165 $\log_{10}$ cfu/g) and S (0.14±0.165 $\log_{10}$ cfu/g), although counts were low overall. In conclusion, cows preferred the deep-bedded switchgrass surface over the other 2 surfaces, perhaps because it was more compressible. Switchgrass appears to be a suitable bedding alternative for dairy cows.

Go slow feed bowls and the behaviour of the domestic dog during feeding: a desirable environmental modification?

Louise Anne Buckley and Josephine Lees

Harper Adams University, Department of Animal Production, Welfare and Veterinary Sciences, Newport, Shropshire TF10 8NB, United Kingdom; lbuckley@harper-adams.ac.uk

Go slow feeding bowls are marketed to dog owners as a way to slow down feeding rate and enrich the feeding experience of the domestic dog. However, is this a environmental modification that the dog wants? Eating more slowly is not necessarily synonymous with an enhanced feeding experience. This study investigates the effect of a popular, commercially available, go slow bowl on canine behaviour during feeding, the effect of experience on behaviours shown and the preference of the dogs for feed bowl type. Healthy adult pet dogs (n=10) of various types (mainly mongrels and cross breeds) and bodyweights (4.5-30 kg) were used as their own control in a randomised cross over study. The two treatments were: 1. Dog bowl (DB), and 2. Go slow feeder (GSF). Data were collected once daily for nine days per treatment, with each dog fed separately. Dogs were fed half their daily ration of a semi – moist diet during the test situation. Data were recorded using instantaneous sampling (1 sec intervals) until the ration was fully consumed or ten minutes elapsed. Behaviours recorded included: oral contact with food, oral contact with bowl, non – oral contact with bowl, contact with owner, and other. At the end of the study, each dog was tested for a further ten days in a simultaneous presentation choice test. First feeder approached and feeder selected to eat from were recorded. Behavioural data were analysed using the Wilcoxon Matched Pairs test and reported as a median percentage of test time budget to account for differences between dogs in latency to consume ration. Choice test data were converted to proportion of choices for DB and analysed via the sign test. Dogs spent most time directly in contact with food (GB: 97%; GSF: 86%). However, when fed from the GSF dogs were significantly more likely to perform non – oral ($W_{9}=0$, P=0.004; DB: 0%; GSF: 2%) and oral ($W_{10}=0$, P=0.002; DB: 0%, GSF: 2%) feeder – contact behaviours. No other categories were significant and experience with feeder type did not modify behavioural expression. When given a choice between feeder type, dogs were significantly more likely to approach the DB first ($T_{8}=0$, P=0.008), and were more likely to start eating from the DB ($T_{10}=0$, P=0.002). The median (± interquartile range) preference for approaching the DB first and feeding from the DB was 0.75 (0.6-0.8) and 0.85 (0.8-0.9) respectively. It is concluded that GSF bowls do modify the dog's feeding behaviour in a small but durable way. However, the preferences exhibited suggest that this environmental modification is not 'enriching' the dog's feeding experience. Thus, the use of GSF bowls for this purpose is not recommended.

A pecking device as an environmental enrichment for caged laying hens

Toshio Tanaka and Yuko Moroki
Azabu University, School of Veterinary Medicine, 1-17-71 Fuchinobe, Chuo-ku, Sagamihara-shi, 252-5201, Japan; tanakat@azabu-u.ac.jp

From the year 2012, battery cages for laying hens were banned in the EU and only alternative systems and enriched cages were allowed. In Japan, guidelines to promote welfare for laying hens were established in 2009 (Japan Livestock Technology Association). Cage systems are, however, predominant in the Japanese poultry industry, and it might be difficult to use other systems in the near future. Therefore, improving the welfare of caged hens is an urgent problem in Japan. So, to improve the welfare of conventional caged laying hens, a pecking device was introduced on the cage floor. Twenty-four White Leghorn hens aged 15 months were divided into four groups: single-housed hens with the device, single-housed control hens, pair-housed hens with the device, and pair-housed control hens. The pecking device was a wooden board (single: 18 cm wide, 5 cm deep, 0.2 cm high; pair: 36 cm wide, 5 cm deep, 0.2 cm high) with stones. Stones between 0.5 and 1.0 cm in diameter were attached to the top of the board with an adhesive. The number of stones (0.7 ± 0.2 cm/stone in diameter, 0.3 ± 0.1 g/stone) used for pecking devices was 43 for single-housed hens and 86 for pair-housed hens. The behaviors of hens were videotaped from 10:00 to 18:00 for 7 days total. Behavioral data was recorded by an instantaneous sampling method at 1-min intervals. Pecking data was collected during three periods of 1 h each (10:00 – 11:00; 13:30 – 14:30; 17:00 – 18:00). The number of pecks, time spent and number of pecking bouts at the device were also recorded by a behavior sampling method during three periods of 1 h each. Bouts were recorded as discrete events if they were separated by a minimum of 3 sec. Hens housed with the device pecked at various pecking objects (floor, cage wall, etc.) less often than control hens (6.6 vs 13.1%; $P<0.05$). Agonistic behavior was also lower in hens with the device than in hens without the device (0.2 vs 1.6%; $P<0.05$) implied that the possibility of improvement in quality of pecking stimuli with the device. Not only time spent pecking, but also quality of pecking might be important to fill their need for stimulation. Both single- and pair-housed hens more often pecked at device in the evening. Response to various pecking objects also showed that pecking behaviors were most frequently expressed in the evening ($P<0.01$). Increased foraging at dusk is well-known; therefore, the increase in pecking behavior in the evening might reflect the hens' general circadian rhythm. These results indicate that the device made of stones could promote some instinctive behavior. Enhancement of behavioral repertories and reduced agonistic behavior with a pecking device might improve the welfare of caged laying hens.

The effect of rubber mats on lying and feeding behavior of lame and non-lame sows housed in farrowing stalls

Magnus R Campler and Monique D Pairis-Garcia
The Ohio State University, Department of Animal Sciences, 2029 Fyffe Rd., Columbus, OH 43210, USA; campler.1@osu.edu

Lameness is the third most common reason for involuntary culling of sows and may negatively impact litter performance as a result of poor lactation due to decreased sow activity and feed intake. Thus, modifying environmental conditions to improve sow comfort in the farrowing stall is critical. The objective of this study was to evaluate the effect of rubber mats on lying and feeding behavior of lame and non-lame sows housed in farrowing stalls. A total of 197 multiparous sows were evaluated by one veterinarian and assigned as lame (score of 1, toe-tapping or non-weight bearing on one leg) or non-lame (score of 0) one week prior to farrowing. Non-ambulatory animals or animals that were considered in need of medical treatment, were not included in the study. Sows were randomly assigned to one of two treatments implemented in the farrowing stall: (1) rubber mat (non-lame: n=55; lame: n=45); and (2) control (non-lame: n=56; lame: n=41). Rubber mats were placed from the back of the stall up to shoulder height of the sow which ensured that both front and hind limbs were on the mat when standing. Lying and feeding behavior was recorded by live observation using a 15 min scan sampling method over a 2 hour period post-feeding. Behavioral observations and lameness scoring were conducted once a week for 4 weeks. Data were analyzed using Proc Glimmix in SAS. Lying behavior was not affected by week or lameness status prior to the study (P>0.05). Sows provided with a rubber mat performed fewer lying bouts compared to the control (5.6±0.1 vs 5.9±0.1 bouts/2 h, P=0.0366). Overall, the number of lying bouts decreased over time (7.1±0.1, 6.0±0.1, 5.5±0.1 and 4.5±0.1, bouts/2 h, P<0.0001, week 1-4, respectively). Feeding behavior was not affected by treatment or lameness status prior to the study (P>0.05). Feeding bouts decreased between week 1 and 2 and increased between week 2 and 4 (0.82±0.1, 0.6±0.1, 1.0±0.1 and 1.4±0.1, bouts/2 h, P<0.0001, week 1-4, respectively). The results of this study suggest that the presence of a rubber mat decreased lying bout number but the decrease was not associated with sow lameness status prior to the study. However, the lower number of lying bouts in sows housed on rubber mats may indicate a more comfortable stall flooring which may promote standing behavior around the time of feeding. The overall decrease in lying bouts over time corresponds to a generally increased sow activity levels post-farrowing. In addition, the presence of a rubber mat, regardless of lameness status did not influence feeding behavior. We argue that the motivation to feed may be so strong that lame sows will stand up to feed regardless of leg status.

Young farmed silver foxes (*Vulpes vulpes*) prefer meat bones when given access to alternative enrichment objects

Anne Lene Hovland[1], Anne Marit Rød[1], Tarja Koistinen[2] and Leena Ahola[2]
[1]Norwegian University of Life Sciences, Dept. of Animal and Aquacultural Sciences, P.O. Box 5003, 1432 Ås, Norway, [2]University of Eastern Finland, Department of Biology, P.O. Box 1627, 70211 Kuopio, Finland; anne.hovland@nmbu.no

Access to appropriate types of environmental enrichment is important for captive animals kept in uniform housing systems as a means to reduce monotony and to allow the expression of important behavioral needs. Relevant enrichment objects must reflect the species' biology and likings, and their preference should thus be examined prior to supplementing their housing environment. Here, we assessed 30 juvenile silver foxes short-term preference for 5 familiar objects: a mutual pulling device mounted between two cages, straw, rawhide bone, meat bone (cattle femur) and a plastic cube. The objects were made simultaneously available in the foxes' home cage after one week of deprivation and their behaviour was then video recorded. Measures of preference were latency to first contact and use, including object interaction time the 1st h after access. In addition, object interaction for 1 h was recorded after 2 and 4 days of use. Wear and tear, measured as weight reduction or gnawing marks was examined to estimate the objects' robustness and durability. The 1st h, foxes preferred the meat bone, followed by the rawhide, the pulling device, and straw and cube as the last and least used objects ($P \leq 0.05$). Foxes spent about 49% of the time interacting with the meat bone the 1st h after access. Preference for the bone and rawhide was still evident after 2 and 4 days of use ($P \leq 0.05$). All of the objects were, to a various degree, gnawed. After one week of use, about 30% of the cattle bone mass was reduced, compared to 88% of the rawhide bone mass, 13% of straw and 2% of the pulling device. Also, the cube was gnawed but remained 100% intact throughout the study. Although the plastic cubes were almost indestructible and therefore durable objects, their low ranking questions their relevance as suitable enrichment objects for foxes. The pulling device, initially intended to elicit social play, was only used 1.3% of the time and mostly alone (92%; $P \leq 0.05$). Meat bones, and also rawhide, may be particularly relevant enrichment objects for farmed foxes as they elicit gnawing and have sensory qualities that stimulate foxes' exploration and predatory motivations especially.

Perching behaviour in broiler chickens exposed to two different perch designs

Victoria Sandilands[1], Chang ping Wang[2] and Nicholas Sparks[1]
[1]SRUC, Monogastric Science Research Centre, Auchincruive Estate, KA6 5HW, United Kingdom,
[2]Jiamusi University, Xiangyang District,154007, Jiamusi, Heilongjiang Province, China, P.R.;
vicky.sandilands@sruc.ac.uk

Perching behaviour is a natural and highly-motivated behaviour in chickens, however, historically meat chickens (broilers) have been reared infrequently with perches in conventional housing systems. There is greater interest in meeting farm animals' needs which has resulted in an increase in (among other things) perch provision in broiler housing, but producers report that perches are infrequently used and that use declines with age. Perches provided on commercial farms are often of a similar design used for hens (e.g. 45° between perches) in which broilers must jump from perch to perch, which is possibly inadequate for broilers which are less mobile than hens. We compared the use of two different perch designs during both lights on and off over the production period. At 1 week of age, 80 broilers were distributed by weight into 8 pens of 10 chicks each (5 male, 5 female). Each pen contained two wooden A-frame perches placed at opposite ends of the pen, both made of 5 perch rails of 150 cm per rail: 'C' (42° between rails) and 'W' (12° between rails with wire adjoining the rails). With W perches, birds could stand on either the rails or the wire. Lights were on from 08:00 – 00:00. Birds were assessed for their location every 2 h for 48 h in each week from 1 to 5 weeks of age. Data were analysed using ANOVA in Genstat (15th edition). Mean % birds and standard errors of differences are shown. A large proportion of birds of all ages were observed on the litter during lights on (87.5%) and lights off (84.1%), but this was significantly affected by bird age: birds observed on the litter decreased during lights off from 98.8% at 1 week to 57.5% at 5 weeks (SED 2.6%, P<0.001). At all ages, a greater proportion of broilers were observed using the W perch (8.4-18.1%) compared to the C perch (0.8-2.5%) (SED 1.3%, P<0.001). Broilers used the C perch more during lights on (2.3%) than lights off (0.9%) (SED 3.8%, P<0.001). Use of the W perch was greater at lights on (11.1-12.4%) than lights off (1.3-6.8%) at 1-3 weeks of age, then higher at lights off (20.9-41.8%) than lights on (6.2-9.2%) at 4-5 weeks of age (SED 2.4%, P<0.001). Most of the birds observed on the W perch were attributed to birds standing on the rails (84%) and not the wire (16%). It appears that broilers will increase their perch use if the perches are easier to negotiate (by using shallower angles and wire to move from rail-to-rail) and that such designs will increase night-time use with bird age. The frequent use of well-designed perches by broiler chickens should therefore not be overlooked in commercial systems.

Up-regulation of IGF1 in frontal cortex of piglets exposed to an enriched arena

AB Lawrence[1], R Peters[2] and SM Brown[3]
[1]SRUC, West Mains Road, Edinburgh EH9 3JG, United Kingdom, [2]Aberdeen University, Kings College, Aberdeen AB24 3FX, United Kingdom, [3]University of Edinburgh, Roslin Institute, Penicuik EH25 9RG, United Kingdom; alistair.lawrence@sruc.ac.uk

This project aimed to study the effects of exposure to an environmentally enriched (EE) arena on piglet behaviour and brain gene expression, with a focus on IGF1 and related genes. IGF1 expression in the brain is known to correspond to regions experiencing synaptogenesis and axonal growth, and has been shown to be upregulated in the frontal cortex of rats after a play experience but not after social defeat, indicating a possible role in modulating positive affect. Eight litter groups were maintained post weaning. At 2 weeks post wean 2 pairs of 'test' piglets, plus 1 pair of companion piglets, per litter were separately given access to an enriched arena for 15 minutes per day on 5 consecutive days for habituation. Piglets were recorded in the arena for 15 minute periods on 3 consecutive days. On the final day one pair of test piglets plus companion piglets (per litter) were given access to the arena. One male piglet from each litter pair was culled 15 minutes after EE and brain tissue collected for PCR. One male piglet from the litter pair not run through the arena was also culled and tissue collected as a within litter control. RNA was extracted from frontal cortex and QRT-PCR for selected genes run on a Stratgene MX3000. In both the home pen and the EE arena litters spent the largest proportion of their active time in foraging behaviour (Home pen Mean=35.72%,SEM=3.87; Arena Mean 75.94%,SEM 1.90) with a significant increase in the enriched arena (t=12.38, P<0.001). A significant increase in running behaviour was observed in the enriched arena (Home pen mean=0.98%,SEM=0.25; Arena mean=2.93%,SEM=0.50); t=3.62, P=0.009) with significant decreases in non-locomotory play (Home pen mean=8.5%, SEM=1.62; Arena mean=1.21%, SEM=1.64; t=4.82, P=0.002) and inactivity (Home pen mean=19.38%, SEM=3.46; Arena mean=4.21%, SEM=1.25; t=4.6, P=0.002). A significant fold change (FC) increase (FC=1.07, t=4.42, P=0.002) was observed in IGF1 gene expression in the frontal cortex of piglets exposed to EE compared to those not exposed on the day of culling. No change in expression was observed in the IGF1 receptor gene nor in any of the binding proteins tested (IGFBP1-6). There was a weak tendency for increased expression of the neurotrophic factor BDNF1 (FC: 1.03; t=1.54, P=0.1). This work points to a potential and novel role for IGF1 in mediating brain effects of EE. We believe it is also the first to explore effects of EE on pig brain physiology and development. Future work will seek to better understand the role of EE-induced IGF1 on neuronal health and generation of positive affective states.

Dairy calf preference for enrichment items when housed in an individual hutch

Hannah Manning[1], Jessica Pempek[1], Maurice Eastridge[2], Emily Cosentino[1] and Katy Proudfoot[1]
[1]The Ohio State University, Veterinary Preventive Medicine, 1920 Coffey Road, Columbus, OH 43210, USA, [2]The Ohio State University, Animal Sciences, 2029 Fyffe Road, Columbus, OH 43210, USA; pempek.4@osu.edu

Housing pre-weaned dairy calves in individual pens or hutches is commonplace; however, this type of housing restricts social contact, and the barren environment may hinder the calves' behavioural repertoire. This study aimed to determine calf preference and use of environmental enrichment items added to a hutch. Jersey heifer calves (n=10) were housed in individual hutches on loose gravel and bedded with straw. The outdoor pen enclosure (1.2×2.7 m) contained: 2 artificial teats (1 attached perpendicular, 1 attached at a 45-degree angle), a stationary brush (L-shaped, two 46 cm push brooms), a calf 'lollie' (0.6×7.2 cm PVC pipe with 9.5 mm holes inserted for the throughput of dry molasses), and a rubber chain link (30.5 cm long). The location of each item was alternated per calf. To determine preference and use, calves were continuously video recorded twice weekly during wk 1, 3, and 5 of age from 08:00 to 20:00 h; the frequency of use of each item was collected using the Noldus Observer software program. Behaviour data were not normal, so each variable was log transformed before analysis. A paired t-test was used to determine preference for enrichment items (data were first averaged across time periods). A repeated measures ANOVA was used to determine if enrichment use changed over weeks (SAS, Version 9.4). Preliminary analysis revealed no difference between use of the straight or angled teat; thus, these variables were combined. Calves used the stationary brush most often, followed by the calf lollie, rubber chain, and artificial teat (mean frequency±SE: 16.9±1.2; 13.0±1.2; 8.9±1.2; 4.7±1.2 no./12 h, respectively; $P<0.01$). Brush usage increased ($P<0.05$) from wk 1 to 5 (10.0 to 25.8±1.3 no./12 h). Usage of the lollie also increased with time (wk 1 to 5: 7.6 to 18.5±1.3 no./12 h). Use of the artificial teat and rubber chain did not change over time ($P>0.05$). Results indicate that, out of the options provided, calves preferred the brush and lollie to the rubber chain and artificial teats. As these items likely elicited different behaviours (e.g. grooming, head-butting, and sucking), further analysis is needed to determine how the calves used each item. Additionally, further analysis to assess the impact of these items on play behaviour and abnormal sucking behaviour is still required.

Exploratory behaviour of male and female growing pigs

Lisa Mckenna and Martina Gerken
University of Goettingen, Albrecht-Thaer Weg 3, 37075 Goettingen, Germany; lmckenn@gwdg.de

According to EU Directive 2008/120/EC, pigs must have permanent access to a sufficient quantity of enrichment materials to enable proper investigation and manipulation activities. The present study was conducted to investigate whether male and female growing pigs show differences in their exploratory behaviour between various objects and materials. A total of 18 animals (9 females, 9 males) were tested over a period of four weeks when animals were between 5 and 9 weeks of age. The experimental animals were kept in unisexual groups in stables with litter and straw bedding (2 m²/animal). For behavioural testing, the animals were led into the test room in groups of three to avoid social isolation stress. There, three novel objects (the same for each animal) were laid out for the animals to explore. The objects consisted of: a rubber duck, one wooden and one rubber dog toy filled with grapes, an examination glove, ropes, a wallow, potting soil and dried leaves. Test objects were chosen based on their potential to trigger positive emotions in the animals. The animals were videotaped during the testing period of 7 minutes. The following traits were recorded by means of the time sampling method: standing, moving, exploring environment and exploring object. Every 20 seconds, the video was stopped and the behaviour of the animals was recorded. A total of 21 time sampling points were recorded for each animal and test. Each animal was tested 16 times (twice per object). Analyses of variance included the fixed effects of sex, object, age, the sex×object interaction and the random effect of the animal. Females and males showed no significant differences in the times spent exploring the presented objects and materials (females spent on average 8.33 out of 21 total time sampling points exploring the object while males spent on average 8.45 times exploring the object) (P=0.218). With regard to the attractiveness of the different objects, a significant difference was found (P=0.0001). Judging by the average exploring time, especially the natural materials (soil: 16.8, wallow: 13.8, leaves: 18.8) were explored significantly more than the artificial objects (i.e. rubber duck: 6.7, rubber dog toy: 8.9). The sex×object interaction was significant (P=0.003). It was found that the males explored the wooden and rubber dog toys more than the females (males 9.5 and 10.7 times, females 7.2 and 7.3 times, respectively) whereas the females explored the wallow to a larger extent (males 12.3 times, females 15.2 times). The findings generally underline the preference of pigs to engage with organic materials. Their shape and form can be manipulated to a large extent; in addition they offer a broad range of tactile and olfactory stimuli which may be inherently associated with natural foraging behaviour.

An evaluation of the hemp rope as environmental enrichment for mice

Karen Gjendal[1], Dorte Bratbo Sørensen[1], Maria Kristina Kiersgaard[2] and Jan Lund Ottesen[2]
[1]University of Copenhagen, Section of Experimental Animal Models, Faculty of Health and Medical Sciences, Thorvaldsensvej 57, 1871 Frederiksberg, Denmark, [2]Novo Nordisk A/S, Novo Nordisk Park, 2760 Måløv, Denmark; kgjendal@sund.ku.dk

Environmental enrichment is increasingly being used to enhance welfare among laboratory animals. However, when introducing environmental enrichment it is important to investigate whether positive and/or negative effects occur on behavioural and physiological parameters and thus on the welfare of the animals. This three-step randomized trial in male C57bl/6 mice investigated: (1) the effect of supplementing standard environment with a hemp rope on aggressive and social behaviour, stress and anxiety level; (2) the effect of supplementing standard environment with 1, 2 or 7 hemp ropes per cage on aggressive behaviour in mice only subjected to routine handling; and (3) the duration and frequency of climbing a hemp rope and material shredded from the hemp rope. The social interaction test measured agonistic and social behaviour, the elevated plus maze measured anxiety, and faecal corticosterone metabolites were used as a measure of stress in 224 male C57bl/6 mice (Part 1). The number of wounded companion animals and the number of wounds per animal were used as surrogate measures of aggression in 224 male C57bl/6 mice (Part 2), and climbing activity in a hemp rope was video recorded and scored manually in 56 male C57bl/6 mice (Part 3). An analysis of variance (Part 1) or a $\chi 2$ test (Part 2) was used to compare groups, and a repeated measures analysis was used to analyse climbing activity in the hemp rope (Part 3). Part 1 showed that male C57bl/6 mice housed with one hemp rope had higher levels of corticosterone (ng/g faeces) (4.56(0.07;9.05), $P<0.05$) and showed signs of more social behaviour than mice housed without a hemp rope (following (no): 2.00(0.52;3.48), $P<0.01$; following (sec): 3.83(1.12;6.55), $P<0.01$), while no difference was detected in anxiety levels. Part 2 showed no difference in number of wounded animals or wounds per animal when adding 1, 2 or 7 hemp ropes to the existing environment in male C57bl/6 mice undergoing minimal human handling ($P>0.05$), and part 3 showed that male C57bl/6 mice shredded (gram) the hemp rope with no decrease over time or when provided with a new and fresh rope (week 1 vs week 2: 0.75(0.18;1.32), $P<0.05$ and week 6 vs week 7: 0.62(0.055;1.20), $P<0.05$). The hemp rope was used for climbing, chewing, unravelling for nesting material and manipulation throughout the study, although the time spent climbing the rope decreased slightly over time. A hemp rope can be used as additional environmental enrichment amongst male C57bl/6 mice.

What ferrets want: studies into the enrichment priorities of ferrets (*Mustela putorius furo*)

Marsinah L Reijgwart[1,2], Claudia M Vinke[2], Coenraad FM Hendriksen[1], Miriam Van Der Meer[1], Nico J Schoemaker[2] and Yvonne RA Van Zeeland[2]

[1]*Intravacc, Animal Research Centre, Antonie van Leeuwenhoeklaan 9, 3721 MA Bilthoven, the Netherlands,* [2]*Utrecht University, Animals in Science and Society, Yalelaan 2, 3508 TD Utrecht, the Netherlands; m.l.reijgwart1@uu.nl*

Information on suitable enrichment options for ferrets is scarce. Therefore, the motivation of seven female ferrets for six different enrichment categories (sleeping enrichment, conspecifics, foraging enrichment, water bowls, balls and tunnels – in randomised order) was tested in a series of consumer demand studies. The ferrets had *ad libitum* access to food and water at all times in a freely accessible chamber with bedding. The ferrets' motivation was measured by daily increasing the weight on push doors until the maximum price paid (MPP) was reached. In addition, respondents of an international survey ranked the same enrichment categories from 1 (least important for my ferret) to 6 (most important for my ferret). In the consumer demand studies, ferrets showed a higher motivation for sleeping enrichment (MPP 1,450±120 g), water bowls (MPP 1,075±153 g), conspecifics (MPP 995±267 g), foraging enrichment (MPP 950±228 g) and tunnels (MPP 940±393 g) than for an empty chamber (MPP 539±187 g; LMM; $P<0.001$; $P=0.002$; $P=0.017$; $P=0.005$; $P=0.020$, respectively). The motivation for balls (MPP 754±215) was not higher than for the empty chamber (LMM; $P=0.175$). The 683 respondents of the survey ranked the categories differently: conspecifics (4.8±1.8), sleeping enrichment (4.6±1.5), tunnels (4.2±1.2), balls (3.2±1.5), foraging enrichment (3.1±1.4) and water bowls (2.4±1.4). These results show a discrepancy between the ferrets' motivation for the different enrichment categories in the consumer demand studies and what ferret owners think are the most important categories for their ferrets. This illustrates the need for more science-based recommendations on ferret enrichment.

Behavioral preference for different enrichment types in a commercial sow herd

Kristina Horback, Meghann Pierdon and Thomas Parsons
University of Pennsylvania School of Veterinary Medicine, 382 W Street Road, Kennett Square,
PA 19348, USA; thd@vet.upenn.edu

Increased public concern about farm animal welfare is driving legislative initiatives to change how sows are housed and managed. This study investigated the use and preference for enrichment items at a 5600 head commercial sow farm in eastern USA. Gestating sows were housed in static, pre-implantation groups of approximately 75 head per pen and fed via a single electronic sow feeding station per pen. Eighteen pens were monitored immediately after loading (Day 1), as well as 3, 5 and 14 days post-mixing. Each pen contained one enrichment item: a hanging toy with 4 rubber chew sticks, a hanging toy with four strands of cotton rope, or an 89×89 mm woodblock fixed to pen corner. Behavioral data on daytime activity was collected on-site in three 2-hour blocks between 08:00 – 10:00, 11:00 – 13:00 and 14:00 – 16:00 for each pen. Nighttime pen activity was collected on Day 1 in three 1-hour blocks between 22:00 – 23:00, 00:00 – 01:00 and 02:00 – 03:00. Behaviors recorded included proportion of observation time animals interacted with the object, proportion of animals in pen that interacted with the object, and posture (up/down) of each animal in the pen. Lesion scores were recorded prior to mixing and two weeks post-mixing as a proxy for social aggression. There were significant main effects of treatment (OBJ) ($P<0.001$), day of observation (DAY) ($P<0.01$), and time of observation (TIME) ($P<0.001$) on proportion of time sows interacted with enrichment items. Post-hoc analyses using Tukey's HSD ($P<0.001$) indicated that sows were recorded to make contact with the rope more frequently (M=61.2±3.4% of total observation time) than the rubber (M=30±3.5%) or woodblock pens (M=24.5±2.8%). Sows in all pens made contact with the items significantly ($P<0.001$) more often on Day 3 (M=53.9±4.6%) than Day 1 (M=28.2±2.9%). There was a significant main effect of OBJ ($P<0.001$) on the proportion of sows in pen interacting with the item, with a greater proportion of sows making contact with the rope (M=4.7±0.3%) than the rubber (M=2.6±0.2%) or woodblock enrichment items (M=2.3±2.1%). There was also a significant two-way interaction of OBJ and DAY ($P<0.01$) on frequency of object contact. These results indicate that sows can exhibit clear preferences for enrichment type, with the sows interacting with the rope toy significantly more often throughout the study. There was no difference in the lesion severity or body posture among the three enrichment types, suggesting that establishment of social hierarchy takes precedent over the pursuit of available enrichment. Additional studies are needed to understand how preferences for enrichment objects could be utilized to potentially impact sow productivity and welfare.

Piglet survival and causes of mortality in the SowComfort Pen: data from two commercial loose-housed sow herds

Inger Lise Andersen[1] and Elsbeth Morland[2]

[1]Norwegian University of Life Sciences, Animal and Aquacultural Engineering, P.O. Box 5003, 1432 Ås, Norway, [2]Fjøssystemer Sør AS, Olgar Dahls vei 3, 3175 Ramnes, Norway; inger-lise.andersen@nmbu.no

The SowComfort farrowing pen was developed with the aim to satisfy behavioural needs of the sow and to increase the resting comfort during farrowing and lactation. Due to promising results regarding sow-piglet communication and piglet survival in the first prototype pen, the objective of the present innovation project was to collect production data from this and two traditional similar sized pens with a creep area, no mattress, and without access to hayrack in two selected herds. Based on the preliminary experimental results of the prototype pen, we hypothesized that this pen would result in lower piglet mortality also in commercial farms. The SowComfort pen (7.6 m^2) consists of a nest area and an activity/dunging area. The nest area has solid side walls, sloped walls on three of the sides and a hay rack on the fourth wall allowing free access to hay or straw. The nest area had two zones with floor heating covered by a 30 mm thick rubber mattress. In the first commercial herd, we collected data from 162 healthy sows of different parities and their litters, of which 61 litters were from three different batches in an old pen system vs 101 litters were from four different batches in the SowComfort pen. In the second herd, we collected production data from 156 healthy sows and their litters distributed between three different batches kept in an old pen system. We collected production data from 343 healthy sows with different parities, distributed between 7 consecutive batches in the new pen. The data within each of the two herds were analysed separately as the feeders differed, and we used a generalized model in SAS. Production results from the first herd showed that % mortality of live born piglets was around 13% in both pen systems, but causes of mortality in the two pens differed. While the SowComfort pen resulted in lower mortality (1.7±0.5%) due to starvation (i.e. no milk in the stomach), more piglets were crushed in this pen comared to the old system (SowComfort pen: 11.8±1.4 vs old pen system: 9.6±1.2). Percentage of piglets per litter without knee lesions were significantly lower in the SowComfort pen (28.8±3.1) than in the old pen system (11.0±2.4; P<0.0001), indicating that the rubber mattress provides more protection than concrete floor with sawdust. In the second herd, piglet mortality declined significantly and steadily from 15.4±1.6% in batch 1 to as low as 11.7±1.6 in batch 7 (P<0.0001). Mortality of liveborn piglets was significantly lower in the SowComfort pen than in the old pen system (P<0.0001), and primiparous and second parity sows had the lowest mortality (P<0.0001).

Effects of stall improvement and regular exercise on body injuries of cows housed in tie-stall

Santiago Palacio[1], Steve Adam[2], Renée Bergeron[3], Doris Pellerin[4], Anne Marie De Passillé[5], Rushen Jeff[5], Haley Derek[6], Trevor Devries[3] and Elsa Vasseur[1]

[1]McGill University, Animal Science, 21111 Lakeshore, Sainte-Anne-de-Bellevue, H9X 3V9, Canada, [2]Valacta, 555 Boul des Anciens-Combattants, Sainte-Anne-de-Bellevue, H9X 3R4, Canada, [3]University of Guelph, Animal Bioscience, 50 Stone Rd E, N1G 2W1, Canada, [4]Université Laval, Départment des Sciences Animales, 2325 Rue de l'Université, G1V 0A6, Canada, [5]University of British Columbia, Dairy Research and Education Center, 6947 Lougheed Hwy, V0M 1A0, Canada, [6]University of Guelph, Ontario Veterinary College, 50 Stone Rd E, N1G 2W1, Canada; spalacio89@gmail.com

Ensuring proper comfort in the stall along with providing regular exercise may be a sustainable option to improve cow welfare. The objectives of this study were to evaluate how regular exercise (access to pasture during the season and/or winter exercise) and stall design improvements (e.g.: tie-rail adjustment, increase of chain length, more bedding) affected the welfare of lactating dairy cows housed in tie-stalls. Over 12 months 12 tie-stall farms (8 providing regular exercise, 4 that did not) were visited and the welfare of 20 cows/ farm was assessed 4 times. Visit 1 was conducted towards the end of the pasture season, visit 2, 9-30 days after stall improvements were applied, visit 3, towards the end of winter, and visit 4, 1 year after visit 1. Stall improvements were applied on half of the study cows within each farm with the most common change being re-adjusting the tie-rail. Cow welfare assessments were conducted at each visit and consisted of animal-based and housing measures, as well as a management questionnaire. Comparisons in body injuries between farms that provided exercise and farms that did not (Exc) as well as cows kept in improved stalls or unmodified stalls (Mods) were analysed in a mixed model. Farm was nested in Exc and included as a random effect and Exc, Mods and their interaction as fixed effects. On visit 1, results showed 31 and 13% less cows with neck and knee injuries, respectively, in herds that provided exercise access (P<0.01). On visit 2, we found 19% (P<0.05) fewer cows with neck injuries when provided with exercise and 13% (P<0.05) fewer when kept in improved stalls. Towards the end of winter (visit 3) access to exercise resulted in 16% fewer cows with hock injuries (P<0.05). In addition, exercise provision and improved stalls lowered the number of cows with neck injuries by 39% and knee injuries by 28% (P<0.05). On visit 4, access to exercise reduced cows with hock injuries by 39% (P<0.05). Our results showed that providing tie-stall cows with regular exercise can help improve cow welfare by reducing body injuries. Furthermore improving stall design might have an additional benefit in reducing neck and knee injuries during the winter season.

Effect of overstocking during the non-lactating (dry) period on the behaviour and welfare of dairy cows

Mayumi Fujiwara[1,2], Hannah M Ensor[3], Marie J Haskell[2], Alastair I Macrae[1] and Kenny M.D Rutherford[2]
[1]*University of Edinburgh, Royal (Dick) School of Veterinary Studies, Easter Bush, Midlothian, EH25 9RG, United Kingdom,* [2]*Scotland Rural College, West Mains Road, Edinburgh, EH9 3JG, United Kingdom,* [3]*Biomathematics and Statistics Scotland, Edinburgh, EH9 3FD, United Kingdom; mayumi.fujiwara@sruc.ac.uk*

The level of agonistic social interactions and daily lying times in dairy cows are known to be affected by space allowance and social regrouping frequency. The aim of this study was to investigate the effects of overstocking and frequent regrouping on the activity and stress levels of cows during the non-lactating (dry) period. Forty-eight Holstein cows (calving over a six month period) were dried off 8-9 weeks before the expected calving date and kept in cubicle housng for 5-6 weeks (Stage1: the first 21 days; Stage2: from day 22 to 3 weeks before calving), and moved to straw yards from approximately 3 weeks before cavng until calving (Stage3). Cows were allocated (balanced by parity) to either high (H) or low (L) stocking density groups after dry-off. During Stages1&2 H and L cows had 0.5 vs 1.0 yokes and 1.0 vs 1.5 cubicles/cow respectively. During Stage3 they had 0.3 vs 0.6 m feed face and 6 vs 12 m^2 lying space/cow respectively. Cows' activity (MI: MotionIndex (a measure of absolute acceleration over a given period), SC: step count) and lying behaviour (LB: daily frequency of lying bouts, LP: daily proportion of lying time) were recorded throughout the dry period using activity monitors. Faecal samples were collected at 5 time points to measure faecal glucocorticoid metabolites. Data were analysed by Residual Maximum Likelihood (REML) using Genstat on log transformed data (except LP). Means and standad errors (back-transformed where necessary) for MI, SC, LB, LP and cortisol are presented. There were no sgnificant differences between the two treatments in the variables measured. Primiparous cows were more active than multiparous cows (MI: 2,893.1±175.7 vs 2,281.1±114.6, W=10.12, P=0.003, SC: 1028.0±53.3 vs 846.5±36.8, W=9.00, P=0.004). Main effects of the experimental stage were found on LB (Stage1=7.9±0.3, Stage2=8.0±0.3, Stage3=9.4±0.4, W=74.22, P<0.001), LP (Stage1=0.53±0.01, Stage2=0.52±0.01, Stage3=0.62±0.01, W=265.47, P<0.001) and concentrations of faecal glucocorticoid metabolites (Stage1=262.5±21.9 ng/g, Stage2=348.9±28.4 ng/g, Stage3=358.5±35.0 ng/g, W=19.97, P<0.001). The results suggest that parity and stage of the dry period, rather than stocking density can affect cows' activity level, lying patterns and stress level.

Influence of reduced stocking density and straw on exploratory behaviour and lesions in fattening pigs

Lisa Wiesauer[1], Katharina Schodl[2], Christoph Winckler[1] and Christine Leeb[1]
[1]*Univ. of Natural Resources & Life Sciences, Gregor-Mendelstraße 33, 1180 Vienna, Austria,* [2]*Doctoral School of Sustainable Development (dokNE), Peter-Jordan-Str. 82, 1190 Vienna, Austria; lisawiesauer@gmx.at*

Tail and ear biting are common problems in current pig fattening systems with high stocking density and lack of environmental enrichment. The present study aimed to investigate the combined effect of increased space allowance and provision of straw in a rack on pig behaviour and lesions on a commercial farm with fully slatted floor. Even when experimental studies have shown the effect of tested measures before, a successful application on commercial farms is important to encourage farmers to provide organic enrichment material and to stop tail docking. The measures were implemented in half of the pens (improved pens (IP); 1.03 m^2/pig), while housing conditions were not changed in the others (control pens (CP); 0.76 m^2/pig). In total 575 animals in 44 pens (3 batches) were assessed at the beginning (I) and half-way through the fattening period (II). Continuous behavioural observation (10 minutes per pen) comprised the following behaviours: 'manipulation of pen', 'manipulation of wooden block', 'exploration of straw', 'tail biting' and 'ear biting'. Furthermore lesions of tails and ears were scored. Data were analysed using a linear model with observation (I, II), group (IP, CP) and their interaction as fixed effects. Pigs in IP expressed more exploratory behavior towards the pen (IP I: 66.1±45.0; IP II: 92.1±22.7 vs CP I: 30.7±33.5; CP II: 62.4±19.4; P>0.001; all incidences/100 animals/10 min) and the wooden block (IP I: 8.4±15.0; IP II: 29.0±29.2 vs CP I: 5.9±22.5; CP II: 15.1±15.7; P=0.033). Tail biting (IP I: 0.8±2.6; IP II: 4.7±5.8 vs CP I: 0.6±2.0; CP II: 10.8±11.4) and ear biting (IP I: 11.6±12.2; IP II: 10.8±11.6 vs CP I: 9.2±13.6; CP II: 40.8±22.9) occurred more frequently in CP than in IP (P>0.039 and P>0.001, respectively). Tail biting as well as ear biting increased over time with a significantly higher increase in CP (P=0.028 and P>0.001, respectively). However, no significant differences were found regarding tail and ear lesions. Behavioural observation was limited to 10 min/pen, but it was always carried out at the same time of day and order balanced across treatments. The results indicate that the combination of higher space allowance and straw as enrichment material reduced tail and ear biting behaviour significantly, even when this was mostly 'tail-in-mouth' behaviour not leading to injuries. The markedly higher space allowance in IP presumably allowed for easier avoidance of pen mates and access to important resources (e.g. water, wooden block).

Effect of xylazine for castration procedures on the behaviour of beef calves

L Carlos Pinheiro Machado F°, Bruna A Da Silva, Rolnei R Daros, Jose A Bran, Jessica M Rocha, Angelica Roslindo, Luciana A Honorato and Maria J Hötzel
Universidade Federal de Santa Catarina, LETA – Depto. de Zootecnia e Des. Rural, Rodovia Admar Gonzaga, 1346, Florianópolis, 88034-000, Brazil; pinheiro.machado@ufsc.br

Calf sedation suppresses animal's movement, which may prevent calves to follow their dams soon after castration under pasture conditions. The objective of this research was to evaluate the behavior of calves treated with sedative xylazine and postoperative behavior of castrated calves with or without xylazine. Two trials were conducted. In trial 1, the effect of xylazine without surgery on calves was tested in a cross-over design, three days apart. Group 1 (n=7) received xylazine (SX) and group 2 (n=8) was only manipulated (SM) in the corral; the procedure was inverted 3 days later. In trial 2, three groups were tested: castrated using xylazine (X), castrated without xylazine (C) and uncastrated (F, female). Surgical castration was performed after treatment with lidocaine for local anesthesia and ketoprofen for analgesia. In both trials calves were observed in the day before (P0) and for 3 h following procedure around the corral area (P1). They were moved with their dams back to pasture and observed for another 4 h (P2), and in the following day (P3) and in the next day (P4). Immediately after the procedure calves were observed for pain behaviors (posture, head and ear movements), and suckling length, urinating, defecating, grooming, self grooming, vocalization, play behavior, drinking, licking salt. In a ten-minutes interval scans it was recorded: idling, grazing, ruminating, exploring, suckling. UFSC's Ethical Committee approved the procedures. Data were analysed as repeated measures in a multivariate model with calves as random effect. In trial 1, SX calves idled longer than SM (90 vs 56%; $P<0.0001$), but ruminated (0 vs 23.5%; $P<0.0001$) and explored (0.7 vs 9.8%; $P<0.05$) less time, during P1. The number of events on abnormal posture was higher in SX than in SM calves (1.45 vs 0.16; $P<0.006$), as well as vocalization (23.5 vs 0.26; $P<0.03$), but self grooming was lower (4.87 vs 6.14; $P<0.02$). Xylazine affected the time elapsed from the end of procedure to first interaction with the mother in both trials (Trial 1: 74 min for SX and immediately for SM; Trial 2: X=140 min, $P<0.05$; C=58 min and F=14 min). In Trial 2 xylazine had effect on idling (59.9%), ruminating (6.8%) and exploring (0.9%) time in P1 ($P<0.05$), compared to C (30.3; 26.4 and 7.4%) and F (29; 20.2 and 14.4%). Frequency of abnormal posture was higher in X than in C and F calves ($P<0.02$), as well as ear movements ($P<0.05$). Other behaviours, in other periods, were NS for both trials. Lower activity after castration was related to xylazine, delaying recovery of calves under pasture conditions.

Tail biting and tail damage in pigs: a comparison between pigs with and without the tail docked

S. Q. Cui, A. Holten, J. Anderson and Y. Z. Li

University of Minnesota, 46352 State Hwy Morris, MN, 56267, USA; yuzhili@morris.umn.edu

Tail docking is a common practice to prevent tail biting in pigs in North America. Because tail docking is a painful procedure for pigs, the routine practice of tail docking is coming under scrutiny. This study compared pigs with tails docked and pigs with tails intact. Pigs (n=240, initial wt=24.9±2.9 kg) were housed in 8 pens of 30 pigs with a 3-space dry feeder and 2 nipple drinkers per pen on slatted floors with floor space allowance of 0.7 m^2/pig for 16 wk. Four pens housed pigs with their tails docked after birth, and the other 4 pens housed pigs with their tails intact. Pigs were weighed when entering the barn at 9 wk of age, every 4 wk thereafter, and at the conclusion of the study. All pigs were assessed for tail damage when entering the barn (d 0), d 3, wk 1, wk 2, wk 5, wk 8, wk 10, and wk 16 of the study, using a 0 to 4 scoring system (0=no damage; 1=healed lesions; 2=visible blood without swelling; 3=swelling or signs of infection; 4=partial or total loss of the tail). Pigs with tail scores equal to or greater than 3 were identified as victim pigs and were moved to hospital pens. All pigs were observed one day per week between wk 8 and wk 12. On every observation day, each pen was scanned for 4 h at 5-min intervals to record the number of pigs that performed tail biting, pig-directed behavior, eating, drinking, lying, or standing, and the location where tail biting occurred. Data were analyzed using the Mixed and Glimmix Procedures and the Frequency Procedure with χ^2 analysis. During the study period, all pens with undocked pigs and 2 pens with docked pigs had incidences of tail biting with more than one pig having a tail score of 2 or greater. These incidences occurred to undocked pigs as early as d 3 after they entered the barn, while these incidences did not occur to docked pigs until wk 8 in the barn. Undocked pigs spent more time tail biting (0.81 vs 0.39%; F=4.9, df=24, P=0.04) and eating (11.5 vs 10.6%, F=11.4, df=24, P=0.003) than docked pigs. On average, 40% of tail biting events occurred near the pen enclosure, 30% near the feeder, 20% in the open area, and 10% near the drinkers. There was no difference between tail docking treatments. More undocked pigs (18 vs 5%, χ^2=13.1; P=0.001) became victim pigs compared to docked pigs. Victim pigs that survived to market had lower ADG (0.82±0.02 vs 0.87±0.01; F=4.8, df=222, P=0.03) and market weight (120.9±2.66 vs 126.5±0.92; F=4.2, df=222, P=0.04) compared to non-victimized pigs. Results of this study suggest that tail biting occurs earlier in undocked pigs than in docked pigs. Since the majority of tail biting events occurred near the pen enclosure and the feeder, these locations should be considered when installing enrichment devices to reduce tail biting.

Increasing sows' comfort: do lying mats contribute positively to sow welfare?

Emilie-Julie Bos[1,2], Dominiek Maes[1], Miriam Van Riet[1,2], Sam Millet[1,2], Bart Ampe[2], Geert Janssens[1] and Frank Tuyttens[1,2]
[1]Ghent University, Faculty of Veterinary Medicine, Salisburylaan 133, 9820 Merelbeke, Belgium, [2]Institute for Agricultural and Fisheries Research (ILVO), Animal Sciences Unit, Scheldeweg 68, 9090 Melle, Belgium; emiliejulie.bos@ilvo.vlaanderen.be

Soft flooring may contribute to positive welfare, including comfortable lying behaviour in sows, especially considering the high prevalences of lameness in sows. Rubber mats may function as a practical and less expensive alternative to straw for providing a comfortable lying surface. We hypothesised that providing rubber mats to group-housed sows would improve their lying comfort, and that this would be reflected in their behaviour and lying posture. In total, three groups of 21±4 sows were monitored during three reproductive cycles. From day 28 until d108 of gestation, the sows were housed in group pens with an electronic feeding system (EFS). The floor of the pen consisted of 15.9 m^2 solid, rubber-covered lying area (A), 15.9 m^2 solid, concrete lying area (B), 8.0 m^2 slatted, rubber-covered walking area (C) and 10.6 m^2 solid floor around the EFS (D). The behaviour of the sows was recorded (scan sampling, with an interval of 30 minutes between 5.00-19.00 h and a scan at 23.00 and 3.00 h) during two periods in the reproductive cycle: just after moving them to the group housing (d28-d31, period (1) and around mid-gestation (d50-d54 (period 2). In addition, sows' gait was observed on d28 and d50. The effect of rubber lying mats on lying behaviour was analysed, regarding gait score. A preliminary data analysis was performed on the proportion of time in one of the areas and on the proportion of lying posture within an area. Preliminary results (based on 78% of the total recorded data) showed that the sows performed less inactive behaviour (e.g. standing immobile, lying and sitting) in period 1 than in period 2 (86.4 vs 93.3%; P=0.02), irrespective of being lame or not (P>0.1). Sows were inactive for 96% of the time they spent in area A or B. Inactivity was less prevalent in area C 38.0% (P<0.01) and area D 56.5% (P<0.01). Lying posture (sternal, semi-lateral and lateral lying) did not differ between lame vs non-lame sows (all P>0.36), nor for observation period (all P>0.08). Time spent lying sternal did not differ between the 4 areas (1.3%; P>0.3). Semi- lateral lying was less common in area C than in A, B and D (7.9 vs 28.8, 27.5, 25.4%; P<0.0001). Lateral lying was more common in areas A or B (64.6%) as compared to area C (13.1%, P<0.01) or D (19.4%, P<0.01). In conclusion, under the conditions of the present experiment, covering solid concrete floors with 20 mm rubber mats did not improve lying comfort as it did not affect lying preference or lying posture of lame and non-lame sows.

How are behaviours indicating positive and negative affective states in lambs affected by step-wise weaning?

Lena Lidfors and Joyce Broekman
Swedish University of Agricultural Sciences, Department of Animal Environment and Health, P.O. Box 234, 532 23 Kvänum, Sweden; lena.lidfors@slu.se

The aim of this study was to investigate in what way and for how long different behaviours of lambs were affected by weaning at 8 weeks of age. The study was carried out at Götala Beef and Lamb Research station SLU in Sweden. Twin lambs (n=15 pairs; 16 males, 14 females) from ewes of Finewool × Dorset and sire of Texel were housed in straw bedded pens with their mother (6 m^2) The animals were fed grass silage and lamb concentrate daily, and had ad lib. access to water and minerals. The lambs were weighed every week. Lambs could walk out of the home pen trough a small gate to enter a playground (8-16 m^2) where they could interact with 1-6 other lambs before weaning (BW). During weaning (DW, 50-65 days old, 28.8 kg) the mother was removed from the home pen and placed within hearing distance of the lambs, and the gate to the playground was closed. After weaning (AW, 55-70 days old) the mother was removed from the barn, and the gate to the playground remained closed. Behaviour was observed 3 sessions during 3 days BW, 5 sessions during 6 days DW and 5 sessions during 6 days AW. Recordings were done both instantaneously at 15 s intervals and continuously within each 15 s giving 24 min observation time per session. Statistical analysis was done with a Restricted Maximum Likelihood analysis giving Wald tests. There was a reduction during weaning in total play (P<0.01, X^2=10.17, BW=6.87, DW=2.16, AW=5.74 no. of recordings/h), social play (P<0.05, X^2=8.06, BW=4.19, DW=1.72, AW=4.10 no. of recordings/h) and locomotor play (P<0.05, X^2=7.97, BW=2.68, DW=0.44, AW=1.65 no. of recordings/h). During weaning there was an increase in vocalisation (P<0.001, X^2=283.74 BW=0.22, DW=58.23, AW=4.70 no. of recordings/h), and in social contact with the twin lamb (P<0.001). Percentage of lying declined during weaning (P<0.05, X^2=6.89, BW=52.78, DW=48.04, AW=51.96%). During and after weaning raised ear posture was shown more than before weaning (P<0.001, X$^2_2$=19.33, BW=17.7, DW=24.0, AW=22.6%), and asymmetrical ear posture was shown more during weaning than before weaning (P<0.001, X$^2_2$=42.41, BW=8.4, DW=10.9, AW=6.3%). Plane ear posture was shown less during weaning than before weaning (P<0.001, X$^2_2$=15.97, =27.1, DW=19.4, AW=21.0%). Backward ear posture was not affected by weaning (38.4%). In conclusion, removal of the mother from the home pen to another pen in the same barn, and concurrent reduction of space and separation of playmates by bars DW, evoked behavioural changes indicative of negative affect that lasted 2-4 days. The removal of the mother from the barn after another 5 days AW had minimal effects on the behavioural parameters.
hier

Raising undocked pigs: straw, tail biting and management

Torun Wallgren, Rebecka Westin and Stefan Gunnarsson
Swedish University of Agricultural Sciences, Department of Animal Environment and Health,
Box 234, 53223 Skara, Sweden; torun.wallgren@slu.se

Tail biting in pigs is common in pig production and has been suggested correlated to several behaviours. It is associated with reduced welfare and production losses. A common practice to reduce tail biting within EU is tail docking where part of the tail is removed; a painful procedure that does not eliminate the behaviour. According to the EU Directive 2008/120/EC routine tail docking is banned and other measures to reduce tail biting must replace docking. An alternative is to improve the pig environment by using straw and thus decrease development of tail biting. Straw usage has been difficult to implement since it is argued that straw provision is incompatible with fully slatted floors. In Sweden, tail docking and fully slatted floors are completely banned through national legislation. Furthermore, it is a legal requirement that pigs should have access to manipulable material. The implementation of straw usage in Swedish farms was investigated in a telephone survey to study straw usage and farmers' opinion on straw impact on tail biting and farm management. A total of 46 nursery and 43 finishing farmers were interviewed, all reporting providing pigs with enrichment material, most commonly straw (98%). Median straw rations provided in systems with partly slatted floor was 29 g/pig/day (8-85 g) in nursery and 50 g (9-225 g) in finishing farms. Straw was the only manipulable material in 50% of nursery and 65% of finishing farms while remaining farms used additional material, most commonly wood shavings (65%). 'Toys', e.g. balls and ropes, were used by 13% of nursery and 16% of finishing farmers as a supplement to other manipulable material. Of these, 62% only provided these 'toys' occasionally, e.g. at re-grouping or when tail biting had been observed. Problems in the manure handling systems caused by straw had occurred in 32% of the farms, of these 25% had problems at yearly and 7% monthly, or more seldom (58%). Tail biting had been observed in the production at least once by 50% of nursery and 88% of finishing farmers, an average of 1.6% finishing pigs were reported tail bitten per batch (0.1-6.5). Tail biting was observed ≤twice/year (78%) 3-6 times/yr (17%) and monthly (4%) by nursey and ≤2 times/yr (21%), 3-6 times/yr (37%), monthly (34%) and weekly (8%) by finishing farmers. The provided amounts of straw seem to be sufficient to keep tail biting at a low level in undocked pig herds (<2%/batch). The low incidence of straw obstruction in manure handling systems reported also implies that straw usage at this rate 30-50 g/pig/day) is manageable in pig production systems.

Effects of feeding management and lying area on the behaviour of group-housed horses

Joan-Bryce Burla[1], Anic Ostertag[1], Christina Rufener[1], Antonia Patt[1], Lorenz Gygax[2], Iris Bachmann[3] and Edna Hillmann[1]
[1]*ETH Zurich, Animal Welfare and Ethology Unit, Universitätstr. 2, 8092 Zurich, Switzerland,* [2]*Centre for Proper Housing of Ruminants and Pigs, FSVO, Tänikon, 8356 Ettenhausen, Switzerland,* [3]*Agroscope, Swiss National Stud Farm, Les Longs-Prés, 1580 Avenches, Switzerland; jburla@usys.ethz.ch*

Group housing is the most appropriate form of housing horses. However, increased agonistic behaviour, bearing an increased risk of injury as well as disturbed feeding and lying behaviour, is reported in practice. We investigated how hay provision and duration of forage availability affect agonistic behaviour in 50 groups with 4-21 horses (n=390). Each group was observed for 30 min before and for the first 30min after hay provision. Agonistic behaviour was affected by the feeding system and the duration of hay availability. Hay was provided mostly 2-3×/day but the duration of availability varied from 1.5-24 h/day, whereas straw was mostly available *ad libitum*. Agonistic behaviour of high intensity was lowest in the feeding system 'net' and highest in 'floor' (P=0.043). Threatening behaviour was lowest in 'feed stalls' and highest in 'floor', 'fodder rack', and 'feed fence' (P<0.001). Further, agonistic behaviour of high intensity decreased with increasing duration of hay availability (P=0.008, odds ratio=0.2). In conclusion, feeding places which are individually separated or distant from each other and thus enabling animals to keep their individual distance, reduced agonistic behaviour. We further recommend providing hay *ad libitum*, regardless of the feeding system. Recumbency is required for REM sleep and therefore essential. To assess the effect of space allowance of the littered area on lying behaviour, 38 horses in eight groups were exposed to four treatments for 11 days each in a systematically balanced order; T0: no litter, T0.5: 0.5× minimal area littered (MAL) according to Swiss legislation, T1: MAL, and T1.5: 1.5× MAL. Lying areas measured 1.5× MAL, non-littered areas in T0, T0.5 and T1 were covered with hard rubber mats. Lying behaviour was observed during the last 72 h of each treatment. Recumbency occurred only rarely in T0 (P=0.0002) and low-ranking horses were generally lying more often on rubber mats than high-ranking horses (P=0.044). The duration of recumbency per 24 h increased with increasing space allowance of the littered area (P<0.0001; median (range) min/24 h in T0: 18.4 (0-300.3); T0.5: 95.1 (0-232.7), T1: 123.9 (0-378.4), T1.5: 143.6 (0-403.2)). Furthermore, low-ranking horses were more often displaced from recumbency in T0.5 and T1 (P=0.005). In conclusion, undisturbed lying behaviour relied upon the provision of a soft and deformable surface and enlarged space allowance of the littered area was beneficial, especially for low-ranking horses.

Risk factors for Dead on Arrivals during broiler chicken transports

Leonie Jacobs[1,2], Evelyne Delezie[2], Luc Duchateau[1], Klara Goethals[1] and Frank Tuyttens[1,2]
[1]*Ghent University, Faculty of Veterinary Medicine, Salisburylaan 133, 9820 Merelbeke, Belgium,*
[2]*Institute for Agricultural and Fisheries Research (ILVO), Animal Sciences Unit, Scheldeweg 68,*
9090 Melle-Gontrode, Belgium; leonie.jacobs@ilvo.vlaanderen.be

The pre-slaughter phase is a critical phase of the production process, with severe implications for broiler welfare and profitability, such as death (Dead on Arrival %; DOA%). In order to effectively reduce DOA%, a better understanding of its risk factors is needed. Specifically thermal stress has been identified as an important underlying factor for DOA%. Thus, the aims of this study were to investigate the link between DOA% and pre-slaughter characteristics, and between DOA% and behavioural indicators of thermal stress. Transports (n=79) from 52 broiler farms to 5 slaughter plants in Belgium were assessed. Data were collected on DOA% (recorded by slaughter plant personnel), flock information (provided by the farmer), potential pre-slaughter risk factors (including weather conditions and pre-slaughter phase duration), body temperature, in-crate panting, and huddling prevalence (44 crates/transport). A multivariable linear regression model was built with an automatic stepwise selection procedure. Association between DOA% and behavioural data were tested with Spearman's rank correlation coefficients. Mean (±SD) size of transported flocks was 48,453±27,168 birds, with a mean age of 41.2±1.3d and weight of 2.58±0.4 kg. Mean pre-slaughter phase duration was 440±143 min. Median DOA% was 0.19% (range 0.04-3.34%). The model (R^2=29%) showed that more DOAs were found when farmers did not check chick quality upon arrival (0.51 vs 0.21% when checked; P=0.012), when birds were caught by acquaintances (0.46 vs 0.27% when professionally caught; P=0.01), when birds were heavier at slaughter (P=0.046), and when they were kept outside during lairage (0.38 vs 0.27% when laired inside; P=0.045). DOA% was positively correlated with body temperature (ρ=0.29; P=0.012), but did not correlate significantly with panting or huddling prevalence (both P>0.5). These findings suggest that broiler welfare may be improved by adjusting the farmers' management, including checking chicks upon arrival. Slaughtering broilers earlier in life (or at lower body weights), using a professional team for catching, and keeping the birds inside a warehouse during lairage, appear to be promising strategies for reducing DOAs as well. Additionally, a link between thermal stress (as indicated by body temperature) and DOA% was found, although this was not reflected in the behavioural indicators. In order to effectively reduce DOA% further research is needed to investigate whether the relations with the identified risk factors are causal.

Weaning and separation methods in suckling systems in dairy cattle

Cynthia Verwer
Louis Bolk Institute, Animals and environment, Hoofdstraat 24, 3972 LA Driebergen, the Netherlands; c.verwer@louisbolk.nl

In suckling systems, at the time calves need to be weaned and separated from their dam, both dam and calf display a pronounced behavioral response. By means of on-farm research we have identified characteristics signalling emotional distress resulting from weaning and separation and investigated practical management strategies to reduce this distress. In this study, weaning distress upon abrupt weaning, fence-line weaning and nose-flap weaning after ten weeks of damsuckling was evaluated by means of behavioural observations. At fence-line calves were gradually weaned between 10 and 12 weeks of age and gradually separated between 12 and 14 weeks of age. Nose-flaps were given at 10 weeks of age after which the calves were separated from the herd at 12 weeks of age. Cows were milked twice a day and calves moved freely in the herd or pasture and had *ad libitum* access to their dam, water and roughage. Calves were observed in the herd a day before treatment started (day -1), at the day treatment started (day 0 respective to initiation of treatment) and on the three consecutive days, as well as a week after initiation of treatment (days 1, 2, 3 and 7). In case of fence-line and nose-flap weaning, calves were also observed when a new stage of weaning and separation from the dam initiated, as well as the three consecutive days and week after this modification in handling. One-zero focal sampling was used to record the occurrence of the different behaviors in 48 intervals of 5 minutes. Behaviors comprised body posture, feeding-, social-, distress-, abnormal- and play behavior. Vocalizations were measured continuously. On the day that the treatment or a new stage of the treatment initiated, as well as on the observation days a week after such a change, the heart girth of the calves was measured as an indication for their weight. The response to abrupt weaning and separation resulted, compared to pre-weaning behavior, in increased restlessness (3.0 ± 2.8 vs 37 ± 3.7 intervals; $P<0.01$), excessive vocalizations (0.2 ± 0.2 vs 41 ± 1.4 intervals; $P<0.01$), progressively increasing abnormal behavior (0.7 ± 0.7 vs 5.2 ± 2.6 intervals; $P<0.05$) and weight loss (breast circumference -4.2 ± 1.4 cm; $P<0.04$). Fence-line weaning resulted, compared to abrupt weaning, in less excessive vocalizations (848 ± 203 vs 252 ± 99 vocalizations; $P<0.01$), less restlessness (70% of the intervals calm compared to <50%), almost no abnormal behavior (2.3 ± 1.0 vs 0.2 ± 0.1 intervals; $P<0.08$) and some weight loss (-0.6 ± 1.7 cm; $P<0.06$). Nose-flap weaning resulted, compared to abrupt weaning, in less vocalizations (848 ± 203 vs 67 ± 66 vocalizations; $P<0.01$) and weight gain (2.5 ± 0.5 cm; $P<0.06$). Both nose-flap and fence-line weaning reduced the behavioural impact of weaning off milk and separation from the dam.

The calf raising success rate for suckler beef cows in the presence of their yearling

Dorit Albertsen
University of Bristol, Clinical Veterinary Science, Langford, BS40 5DU, United Kingdom;
da13529@bristol.ac.uk

Separating calves from their mothers at five to nine months of age, frequently leads to stress induced immune suppression followed by an increase in disease in the calves. To address the problem, calves in this study were left with their mothers until maturity or at least 18 months of age, raising the question whether the next born calf would be compromised by the older sibling's presence. Sibling rivalry could have an impact on the survival of the young calf through disturbance of the bonding process between the mother and her newborn, by depriving the newborn of its colostrum or competing for the mother's milk. Data from 663 suckler beef cows, giving birth to 1820 live calves over seven years was analysed, comparing the successful raising of a calf up to its first birthday by cows who had retained their yearling to those who's yearling had been removed. Cows were measured between one and seven times. The variables calf raised/failed and yearling present/absent were explored per year and again by cow age to avoid repeated measures per cow. Due to the low number of failed raising attempts, only two years (2008 and 2012) and three age groups (three, four and five years) qualified for statistical analysis. In 2008 no difference was found between the raising success rates of cows in the presence or absence of their yearling (χ^2=0.087, df=1, P=0.493). In 2012 the raising success rate for cows in the presence of their yearling was significantly higher than for cows whose yearling was absent (χ^2=8.66, df=1, P=0.004). Three year old cows showed a trend towards having a lower raising success rate in the presence of their yearling (χ^2=2.68, df=1, P=0.074), while four year old cows showed a trend towards having a higher raising success rate in the presence of their yearling (χ^2=3.24. df=1, P=0.062). No difference in the raising success rate between individuals of either treatment was found in five year old cows (χ^2=1.104, df=1, P=0.225). Calf raising success rates did vary between years and between cow ages, but independent of yearling presence. It can be concluded that suckler beef cows are able to raise their calf in the presence of its older sibling, providing the option to reduce disease loss in youngstock by omitting premature separation of cow and calf.

Effect of different nest-building materials on sow behaviour in modern farrowing crates

Kirsi-Marja Swan, Anna Valros, Camilla Munsterhjelm and Olli Peltoniemi
University of Helsinki, Department of production animal medicine, P. O. Box 57, 00014, Finland;
kirsi.swan@helsinki.fi

Previous studies have shown benefits of facilitating nest-building in sows before farrowing. Yet, there is a lack of knowledge of the suitability of different materials for nest-building, especially in farrowing crates, where it is practically challenging to provide material. Our aim was to evaluate the suitability of different nest-building materials under practical conditions, in farrowing crates, by evaluating their impact on sow behaviour. We tested three different materials: two pages of shredded newspaper (NP), and two handfuls of sawdust (SD) or chopped straw (S). Material was given twice daily in the farrowing unit, from the day the animals were introduced to the crate until farrowing. Peripartal behavior of the sows was video recorded and analysed continuously starting 12 hours before the first piglet was born. Recorded behaviours included nest-building, bar-biting, eating and drinking and time spent in five different positions: sitting, standing, laying sternally, laying laterally and bottom up. We used Kruskal-Wallis test to compare the treatments using two different time periods: 12 and 2 hours before the first piglet was born. We found a difference in nest-building duration between the groups 12 h prepartum (P=0.019). The NP (n=6, Median 2:09:20) group performed more nest-building than the SD (n=8 Median 0:29:47) group (P=0.026). Furthermore a difference in duration of bar-biting 2 h prepartum was found (P=0.011). The S group (n=9, Median 0:13:08) performed more bar-biting compared to the NP (n=6 Median 0:01:03) group (P=0.009). No differences between the groups in other behavioural parameters were found. Based on the results we conclude that the choice of nest-building material is important. The sows appear to benefit more from newspaper as nest-building material than from sawdust and chopped straw. As the quantity of materials given was small, the positive effect of newspaper may be due to it being available to the sow for a longer period of time than other materials. Due to the small sample size, these results need to be considered preliminary, and we cannot conclude on practical implications of the materials.

Effect of a shelf-furnished screen on space utilisation and social behaviour on indoor group-housed cats

Emma Desforges[1], Alexandra Moesta[1] and Mark Farnworth[2]
[1]WALTHAM centre for Pet Nutrition, Freeby Lane, Waltham on the Wolds, Leics LE14 4RT, United Kingdom, [2]Plymouth University, A426 Portland Square, Drake Circus, Plymouth, Devon PL4 8AA, United Kingdom; emma.desforges@effem.com

The laboratory cats environment can be restrictive and may impact their welfare. Enrichment is often provided to alleviate welfare impacts but is seldom assessed for efficacy. This study investigated the effect of novel room furniture (a screen) on the expression of agonistic and affiliative behaviours and space utilisation amongst colony-housed laboratory cats. Video footage of cats (n=29) housed in social rooms (n=4) was collected for 2 days before (baseline phase), 4 days during (test phase) and 2 days following (removal phase) introduction of novel furniture. Space utilisation data were collected using 10 min scan sampling and analysed using a generalised linear mixed effects model for repeated proportional measures, with a fixed effect of experimental phase (baseline, test, removal) and a random effects structure of day within phase within room. Between phase comparisons were made using Tukey's HSD test at an overall level of 5%.Behavioural data were collected using continuous sampling for 3 h a day in 6×30 min episodes using a Poisson generalised mixed effects model with phase (baseline, test, removal), time point and their interaction as fixed effects, and with day nested in phase nested in room as random effects. Significantly more agonistic events occurred before the morning feed compared to after feeding within all phases (pre-feed mean = 0.227; post-feed mean = 0.026; P<0.0001). However no significant differences were observed before the morning feed compared to after feeding between phases indicating that the screen had no effect on reducing pre-feed aggression at the morning feed. Agonistic behaviours occurred significantly less following the morning feed during the test phase when compared to the baseline phase (test post-feed mean = 0.011; baseline post-feed mean = 0.029; P=0.0342). Significant differences were also observed on removal of the screen with agonistic behaviour increasing above baseline at the afternoon pre-feed time point, possibly indicative of aggression due to frustration or a rebound effect (removal pre-feed mean = 0.151; baseline pre-feed mean 0.048; P<0.0001). Affiliative interactions between phases were not significantly affected by screen presence. Given the ratio of the screen to existing shelving (0.58:0.42) a statistical significant proportion of cats (0.75) were found to be on the screen in the test phase of the study (P<0.0001). This study suggests that exploiting the unused vertical space by the addition of stand-alone shelving should be considered a valuable resource for the cat by increasing useable space and reducing agonistic interactions.

The challenge of meeting behavioural needs of pheasants and partridges in raised laying cages

Simon Turner[1], Sarah Brocklehurst[2], Victoria Sandilands[1], Jo Donbavand[1] and Stephanie Matheson[3]
[1]*SRUC, West Mains Road, Edinburgh EH9 3JG, United Kingdom,* [2]*Biomathematics & Statistics Scotland, Kings Buildings, Edinburgh, EH9 3FD, United Kingdom,* [3]*Newcastle University, AFRD, Agriculture Building, NE1 7RU, United Kingdom; simon.turner@sruc.ac.uk*

Captive reared pheasants and red-legged partridges are commercially important in Europe but rarely studied. Outdoor raised wire-bottom cages for breeding partridges are the norm and increasingly used for pheasants. Provision of cage enrichment is variable and not based on evidence of efficacy and mandatory minimum space allowances are not stipulated. Here we examined the effects of enriching raised cages on behaviour, welfare and egg production. A review of the behavioural ecology identified putative behavioural needs. Bird use of enrichments (perch, solid floor, hide, dust bath, claw shortener, grain pecking block) was studied (Experiment 1) using 72 pheasant (1 male:7 females) and 48 partridge cages (1 male:1 female). Usage of all enrichments apart from the claw shortener and pecking block over 28 days was >4% of scan samples and subsequently studied in Experiment 2. Pheasants (bitted to reduce feather pecking) were housed in enriched cages (80 cages, 1 male:6 females) with space allowances representing the UK maximum (0.43 m^2/bird), minimum (0.33 m^2/bird) and intermediate (0.375 m^2/bird), in barren cages (20) or in barren industry standard grass-based floor pens (42 pens, 1 male:9 females; 0.9 m^2/bird). Partridges were housed at 0.2 (conventional) or 0.4 m^2/bird in enriched or barren cages (96 cages, 1 male:1 female). All birds were housed for 12 weeks. Using linear or generalised linear mixed models, bird welfare was assessed at weeks 1 and 12 by measures of claw length, foot health, feather condition, presence of ectoparasitism and qualitative behavioural assessment (QBA; animal mood assessed by 6 observers from behavioural expression (observer effects included in LMM)). Cage size, enrichment and their interaction did not significantly affect pheasant welfare or productivity except QBA. Birds were interpreted as more content and relaxed in small and medium sized cages than the largest cages or floor pens (P<0.018). Significant but biologically small deteriorations in partridge foot and feather condition occurred in enriched cages (P=0.001), particularly at the small space allowance. Parasitism was more common in small enriched cages (P=0.023). QBA suggested that partridges were more relaxed and content in enriched cages (P=0.013). The enrichment and space allowances studied had little impact on the physical welfare of caged pheasants but QBA suggests that their welfare may be improved in cages compared to floor pens. For partridges, enrichment must be offered in combination with more space to benefit welfare.

Impact of playing classical music and scratching on avoidance distance in loose housed farrowing sows

Vivi A. Moustsen[1], Katrine P. Johansson[1,2], Björn Forkman[2], Mai Britt F. Nielsen[1] and Sine N. Andreasen[1]
[1]SEGES Pig Research Centre, Innovation, Axelborg, Axeltorv 3, 1609 Kbh V, Denmark, [2]University of Copenhagen, Dep. of Large Animal Sci., Grønnegårdsvej 8, 1870 Frederiksberg C, Denmark; vam@seges.dk

Neonatal piglet mortality, partly caused by crushing, causes economic loss and reduced welfare. Studies have shown that reactivity of the sow can influence the number of piglets being crushed. The effect of handling/scratching on sow reactivity has previously been found to result in more calm sows. Since handling takes time, an alternative and more economically feasible method is enrichment through sound. A number of studies support the calming effect of classical music both in humans and animals. The hypothesis is that classical music and/or scratching has a calming effect on sows and will result in a shorter avoidance distance. Data originates from two commercial herds with sows housed individually in farrowing pens, two sections from each farm, 111/110 sows in each of the four groups. A split-plot design was used, with section as hole plot (music +/− (+M-sows and-M-sows)) and farrowing pen as subplot (scratch +/− (+S-sows and −S-sows)). The +S-sows were scratched by the farm staff once daily for 15 seconds. The music was played continuously 06.00-18.00 from 5 days before to 5 days after expected farrowing. The playlist used was ''100 calm classics for study and concentration''. Three speakers were placed in the section, to allow even distribution of music throughout the section. To test the reactivity of the sows, a forced approach test was done by an unfamiliar person on the day of placement and before treatment was initiated, the day before expected farrowing and day 5 post farrowing. The test person was not blinded regarding treatment. The test person crouched in front of the sow and tried to touch her head; the sows were scored 0 if they could be touched and did not withdraw, 1 if they initially withdrew but could be touched within 15 s, 2 if they withdrew and could not be touched within 15 s. A Glimmix model (SAS) that included farm, music, scratching, batch and the day of avoidance was used to analyse the results. Scratching resulted in a highly significant decrease in avoidance behaviour in line with the hypothesis (+S=0.63 (SE: 0.03), −S=0.74 (SE: 0.03), P=0.02) whereas music had no significant effect (+M=0.68 (SE: 0.03), −M=0.68 (SE: 0.03)). However, it cannot be excluded that other noises reduced the possible impact of the music. Personel on farm, when asked about the effect of the two treatments, stated that they found sows in all treatment groups less reactive and easier to handle than sows in the non-treatment group. Also they did not consider the treatments as time-consuming or annoying. This research was funded by the EU FP7 Prohealth project (no. 613574).

Improvement of laying hens welfare by immunomodulator immunobeta

Nadya Bozakova[1], Lilian Sotirov[1], Tsvetoslav Koynarski[1], Ivan D. Ivanov[2], Stanimira Kalvacheva[1] and Plamena Boycheva[1]
[1]*Trakia University, Department of General Animal Breeding, Student Campus, Faculty of Veterinary Medicine, 600 Stara Zagora, Bulgaria*, [2]*Agricultural Institute, Section of Sheep Breeding, 600 Stara Zagora, Bulgaria; nadiab@abv.bg*

Assessing welfare of poultry reared under industrial conditions has gained increasing importance. An economically profitable approach for welfare improvement of laying hens under farm conditions is diet supplementation with immunomodulators like yeast additives. The purpose of this study was to evaluate the welfare of 1,056 Loman Brown laying hens whose feed was supplemented with Immunobeta at a dose of 4 g/kg using a mathematical assessment model. The yeast préparation Immunobeta contains three active ingredients: 30% beta-glucans, 25% mannan-oligosaccharides and 5% nucleotides. The laying hens' welfare was scored on the basis of hens' behaviour and plasma corticosterone levels using the modified method of Bozakova et al. The behaviour was observed on 4 consecutive days for 12 hours by a video camera counting the number of birds engaged in specific forms of activity: ingestive (ingestion of water or food), gregarious (moving, resting, egg-laying, dust bathing and feather cleaning), sexual and agonistic behaviour. The plasma corticosterone levels were assayed by means of commercial ELISA kit. Statistical processing of the results was performed by one-way ANOVA using the GraphPad InStat 3.06 software. After 2-months of Immunobeta supplementation, there were statistically fewer hens in the control group than Immunobeta-group for the following activities: egg-laying 57.27 ± 6.45 and 80.73 ± 7.03, $P<0.05$; feather cleaning 4.91 ± 1.67 and 11.46 ± 2.53, $P<0.05$; dust bathing 2.18 ± 1.05 and 7.64 ± 2.03, $P<0.01$. Concurrently, the amount of aggressive hens in control group was higher than in Immunobeta group: 14.73 ± 3.04 (2.79%) and 7.64 ± 1.71 (1.45%), $P<0.05$. Two types of behaviour are indicative for hens welfare: comfort behaviour (dust bathing, feather cleaning, improved egg-laying) and aggression/fear behaviour. The welfare score of Loman Brown laying hens from the control group was 66.67%, and that of the group supplemented with Immunobeta –80.00% due to the positive impact of the yeast préparation.

Does training reduce stress during transport in cats?

Lydia Pratsch[1,2], Jennifer Rost[2], Natalia Mohr[2], Josef Troxler[2], Johann G Thalhammer[1] and Christine Arhant[2]
[1]*Internal Medicine Small Animals, University of Veterinary Medicine, Department for Companion Animals and Horses, Veterinärplatz 1, 1210 Vienna, Austria,* [2]*Institute of Animal Husbandry and Animal Welfare, University of Veterinary Medicine, Department for Farm Animals and Veterinary Public Health, Veterinärplatz, 1210 Vienna, Austria; christine.arhant@vetmeduni.ac.at*

Transport can be a very stressful experience for cats, which might prevent regular visits to the veterinarian. Positive reinforcement training for handling procedures such as blood sampling reduces stress and facilitates handling in several species. The aim of this study was to investigate whether training cats with positive reinforcement reduces stress during car transport. We carried out a blinded randomized controlled trial with a paired sample design with 22 cats from a university cattery (training group (TG): 11; control group (CG): 11). Cats were randomly allocated to the TG or CG. The cats were transported twice with an interval of 6 weeks. The TG received 28 training sessions between the first and the second transport (mean duration per session: 8 min). The training consisted of seven phases (bottom half of carrier, complete carrier, stay in carrier, being carried, stationary car – engine off, stationary car – engine on, car ride). The transport consisted of a 10 minute car ride during which the cats were filmed. For transport, the cats were handled by a familiar person blinded to the group allocation. Stress levels of the cats were assessed with a modified version of the seven-level Cat Stress Score (CSS) every second minute. A cat was observed for one minute and each aspect (body, belly, legs, tail, head, eyes, ears, whiskers, activity) was rated separately. The ratings for this minute were then combined to a mean score. The person coding the videos was blinded to the treatment (intra-rater reliability: r_s=0.97, P<0.001). A mean score for each transport was calculated and DeltaTransport (DT) was obtained by subtracting the result of the first from the result of the second transport. The TG showed a significantly larger reduction of the CSS from the first to the second transport (DT TG: mean: -0.6, range: –1.1-0.18; CG: mean: –0.2, range: –0.6-0.2; Mann-Whitney U test: P=0.007). Transport training is a promising tool to reduce transport stress in cats.

Free-range use and behavior of slow-growing broiler chickens: effects of shelter type, enrichment, age and weather conditions

Lisanne M. Stadig[1,2], T. Bas Rodenburg[3], Bart Ampe[2], Bert Reubens[2] and Frank A.M. Tuyttens[1,2]
[1]Ghent University, Salisburylaan 133, 9820 Merelbeke, Belgium, [2]Institute for Agricultural and Fisheries Research, Burg. v. Gansberghelaan 92, 9820 Merelbeke, Belgium, [3]Wageningen University, P.O. Box 338, 6700 AH Wageningen, the Netherlands; lisanne.stadig@ilvo.vlaanderen.be

Increased free-range use by slow-growing broiler chickens could benefit their welfare. This study aimed to assess the effects of enrichment early in life, shelter type (artificial or natural), weather conditions and age on broilers' free-range use. Three production rounds (R) were performed, with 440 Sasso chickens each. Birds were housed indoors in 4 groups of 110 animals from d0-25, during which 2 groups per round received environmental enrichment (hay bales, strings, grain, mealworms). At d25, birds were moved to 4 mobile houses on a 1 ha field. From d28-70, birds had access to both grassland with artificial shelter (A-frames; AS) and short rotation coppice (SRC). SRC consisted of densely planted willows (15,000 trees/ha). Free-range use was observed 3 times daily on 15, 21 and 18 days in R1, R2, and R3, respectively. The number of animals outside, their location (AS; SRC) and distance from the house (0-2 m; 2-5 m; >5 m) were recorded. In R2 and R3, behavior of the outside chickens was also recorded. Weather conditions were recorded every 15 min. Data were analyzed using a GLM Poisson regression model, with a first-order autoregressive covariance-structure to correct for multiple observations over time within the same house. For multiple comparisons, p-values were corrected using the Tukey-Kramer method. On average, 26% of the birds were outside. Early-life enrichment tended to have a small positive effect on free-range use (0.4% more birds outside; P=0.052). At all distances, more birds were in SRC compared to AS (all P<0.001). In AS more birds were at 0-2 m than at 2-5 m or 5 m from the house (all P<0.005). In SRC more birds were at 0-2 m and 2-5 m than at >5 m from the house (all P<0.001). Free-range use increased with age, particularly in areas further from the house (P<0.001). In AS, rainfall and low solar radiation were related to more birds outside, while the opposite was true in SRC (P<0.001). Fewer birds were outside with increasing wind speed (both AS and SRC; P=0.049). A higher percentage of chickens was observed to forage in AS compared to SRC (50 vs 28%; P<0.001). The opposite was true for standing (6 vs 9%; P<0.001) and sitting, but for the latter the difference tended to decrease with distance from the house (P=0.065). We conclude that: (1) early-life enrichment had no profound effect on free-range use in broilers; (2) SRC was preferred over AS, suggesting this shelter was more suitable and attractive; and (3) behavior was related to shelter type.

Preference between recycled and clean sand bedding of freestall housed Holstein dairy cows

Jessie A. Kull, Randi A. Black, Nicole L. Eberhart, Heather D. Ingle and Peter D. Krawczel
The University of Tennessee, Department of Animal Science, 2506 River Drive, 353 Brehm Animal Science, Knoxville, TN 37996, USA; jkull@vols.utk.edu

Dairy cows spend 12-14 hours a day lying down, so understanding their preference for lying surface can improve welfare. The objective was to evaluate the preference of dairy cows for clean or recycled sand bedding. Data were collected from Holstein dairy cows (days in milk = 268.1±11.9 d; n=32) from August to September 2014. Cows were evenly distributed into 2 pens and housed at a stocking density of 50% in a 4-row, freestall barn at the University of Tennessee's research dairy (Walland, TN, USA). Stalls were bedded with control or recycled sand in an alternating manner. Clean, unused sand was the control. Recycled sand, same origin as the control, was reclaimed from the sand separator of the dairy's flushing system and stockpiled for the study. Sand was used once previously. Cows were previously exposed to recycled and control sand for one week during a no-choice phase. Groups were exposed to treatments in separate pens over 7 d. Sand was added daily to keep bedding level with rear curb, regardless of sand type. Cow behaviour within a stall was categorized as lying (any recumbent position), perching (front two hooves only), and standing (all hooves). Video data was reviewed at 10-min intervals to quantify behaviour during 5 24-h periods. Data were analyzed using the mixed procedures in SAS to assess the effect of bedding and behaviour on proportion time in pen with observations repeated by date. Total stall occupancy was greater for control (32.7±1.3%) compared to recycled sand (28.6±1.3%; P=0.03). When occupying a control stall, cows spent the majority of their time lying (27.8±0.9%) compared to perching (3.4±0.9%) or standing (1.5±0.9%; P<0.001). On recycled sand, cows spent 24.6±0.9% (P<0.001) of the time lying down, 2.7±0.9% perching (P=0.004) and 1.3±0.9% standing (P=0.15). Despite cows spending the majority of their time within a freestall lying, this behaviour was greater on the control sand (P=0.01). No preference was evident for perching (P=0.61) or standing (P=0.88) on either surface. There was a greater occupancy and proportion of time spent lying for control sand compared to recycled sand, but the magnitude of these differences may not be biologically relevant. Furthermore, the most common observed behaviour was lying in the stalls, which suggested either might be suitable. Caution should be used with this interpretation, as sand was recycled only once. This limited reclamation was still sufficient to potentially alter the composition of sand, driving the observed preference. If these changes in composition continue, then the strength of the preference may also change.

Influence of housing treatment on the activity of mice in a contrast and an exploratory test

Anette Wichman and Elin Spangenberg
Swedish University of Agricultural Sciences, Department of Animal Environment and Health, P.O. Box 7068, 750 07 Uppsala, Sweden; anette.wichman@slu.se

Living conditions can influence an individual's emotional state and thus also how it respond in different situations. In this study we investigated how the activity of 150 female C57BL/6N mice in two different tests corresponded to the housing environment they were kept in. The mice were housed in groups of three in five different housing treatments; barren (B; 800 cm^2 cage with bedding), barren stressed (BS; barren cage with exposure once per day to a mild stressor, e.g. food deprivation or wet litter for 30 min), standard (S; 800 cm^2 cage with bedding, cardboard house and nesting material), enriched (E; 2×1,800 cm^2 cages connected with a tunnel, with two cardboard houses, nesting material and two hammocks), extra enriched (EE; enriched cage with additional food treats once per day). The study was divided in two parts where the B, S and E treatments were used first and compared with one another and a second part where th BS and EE treatments were used and compared. Half of the mice went through an exploration test where the number of visits to different sections in the arena during five minutes was used as a measure of their exploratory activity. The arena measured 1×1 m with a central section where a start box was placed. Eight smaller sections surrounded the central area and each section contained a different type of object such as branches, hay, plastic tubes and paper rolls. The mice visited the arena once per day for a minimum of seven days. The second test used for the other half of the mice was a positive contrast test where the latency to run to food rewards of different values in a 1.4 m long runway were noted. The mice were trained on a baseline reward for 11 days (3 trials/day). Half received a high reward (control) and half a low reward (positive contrast). After the baseline training the reward was shifted from low to high for the contrast mice for the next 5 days, while the control mice remained on the high reward. In the exploratory test there was a significant difference in the number of visits to different sections on the first day in the arena (GLM; F=3.4, P=0.045) where E mice had 21±0.95 (mean±se) and B mice 17±0.88 visits. There was a tendency that EE mice had more visits (19.8±0.76) compared with BS (17.3±1.05) the first day in the arena (GLM; F=3.8, P=0.078). In the contrast test control E mice had the fastest running times on day 10 of baseline training (7.8±2.1 s) and were significantly faster than the B mice ((33.2±7.0 s) Mann-Whitney U test; P=0.001), while S mice where in between these with running times of 16.6±3.3 s. Thus, housing had some effect on the behaviour of the mice with enriched mice in general showing a more active response.

Perching behavior in broiler breeders

Sabine G. Gebhardt-Henrich[1], Michael J. Toscano[1] and Hanno Würbel[2]
[1]University of Bern, Center for Proper Housing: Poultry and Rabbits, Division of Animal Welfare, Burgerweg 22, 3052 Zollikofen, Switzerland, [2]University of Bern, Division of Animal Welfare, Länggassstr. 120, 3012 Bern, Switzerland; sabine.gebhardt@vetsuisse.unibe.ch

Broiler breeders are commonly kept without perches although perching has been shown to be a high priority behavior in laying hens. We tested whether fast (Ross 308) and relatively slow (Sasso) growing hybrids used aerial perches (P) and aviary tiers (A) during rearing and production and how it affected health and production. Two hybrids and 3 treatments (control, A, P) were employed in 18 pens. Pens contained 119 hens and 12 males according to legal stocking densities. Control pens consisted of litter, raised slats, group nests, 2 feeder and 1 drinker lines. In P pens 8 wooden perches at 25, 50, 75, 100 cm above the slats provided 14 cm of perching space per bird. A pens included 4 tiers with wooden bars 55, 68, 115, 138 cm above the slats. Birds were filmed every 5 weeks until 45 weeks of age and their locations on the structures were assessed 11 times during a 24 h period. The percentage of birds on the structures was analyzed in a generalized linear model using the beta distribution. Health parameters were assessed at 45 weeks and production was monitored continiously. Keel bones were palpated while blind to treatment. With increasing age, both hybrids spent the dark period on elevated structures instead on the litter ($t_{1,1069}$=867.4, P<0.0001) with more animals above the litter in A than in P pens (t_{14}=-4.1, P=0.001). Elevated structures present in all pens (grill over feeders, slats, tube above drinker) were used more in control pens than in A and P pens (t_{14}=-2.14, P=0.05). More birds perched in A than in P pens (t_9=5.23, P=0.0005). Production was not affected by the treatments except more floor eggs in A pens ($F_{2,12}$=534.3, P<0.0001). Keel bone fractures were less frequent than in laying hens but more frequent in A and P pens than in control pens (A:26.7%, P:31.7%, C:21.6%, $F_{2,14}$=4.97, P=0.023) and Sasso (39%) had more than Ross (15%) ($F_{1,160}$=10.2, P=0.002). Plumage was better in A and C than in P pens ($F_{2,14}$=7.3, P=0.007) and better in Sasso than in Ross ($F_{1,14}$=157.3, P<0.0001). In conclusion, our results suggest broiler breeders prefer perches for roosting at night over slats and drinkers although there were more keel bone fractures. Perches on aviary tiers were used more than aerial perches. Since production was not impaired, perches should be offered to broiler breeders.

Predation of free range laying hens

Monique Bestman and Judith Ouwejan
Louis Bolk Institute, Hoofdstraat 24, 3972 LA Driebergen, the Netherlands; m.bestman@louisbolk.nl

In NL 5.9 million of free range (FR) laying hens are being kept. Farmers see raptors killing chickens, find killed chickens and counting at the slaughterhouse show that up to hundreds of hens are 'missing' compared to the mortality records. Our purpose was to investigate the killing of chickens by raptors in qualitative and quantitative terms. On 11 farms (mean 11,900 hens) 79 observations × 1.5 hours were done (2-10 times/farm depending on slaughter date and presence of raptors) and 41 times the observer searched for killed chickens prior to the observation (1-5 times/farm). On 1 of the farms cameras were installed on a 'killing hot spot'. Presence of raptors was noted and interactions between raptors and chickens (incl. chicken and location info) and the result (chicken dead, escaped or euthanized) was described in qualitative terms. Chickens found dead were categorized as killed by raptor, mammal or other/unknown cause. Information about weather, farms, flocks (roosters, age, use of FR) and farm mortality records was collected. SAS (mixed model) was used to check for significance of mentioned characteristics related to the findings concerning the raptors and chickens. In July – November 141 raptors were seen: 109 buzzards, 5 hawks, 27 other. Number of attacks seen 'live' or on recordings was 16: 6 by buzzard and 10 by hawk. Number of dead chickens found was 44: 32 by raptor, 4 by fox and 8 by other cause. None of the flock, weather or farm factors was significantly related to the presence of raptors, number of chickens found dead in the FR or number of chickens caught during observations. Chickens attacked during observations were all healthy chickens that fought back or tried to escape. Bystander chickens fled or attacked the raptor. Chickens were not afraid of raptors sitting in trees, on poles or eating from prey. They scavenged on the remains after the raptor left. Within 1-3 days a killed chicken was reduced to a clean skeleton and the camera recordings learned that killed chickens were not always found back. The farmers reported 2-52 chickens killed by foxes, 70-160 chickens killed by raptors and 73-487 'missings' per flock. Raptors were causing more killings than foxes. Observed killings and number of chickens found dead in the FR were an underestimation of the actual killing. The presence of trees and bushes was not effective as shelter; raptors used them for their attacks and attacked chickens from the ground as well. Roosters did attack raptors, but were not always present on the right moment & place. A survey among all FR poultry farmers will be done to investigate the scale of the killings, economic damage and possible preventive measures. Although attacking chickens is a welfare issue for the victims, the presence as such of raptors did not seem to affect chicken behaviour.

Effects of keel bone fractures on individual laying hen productivity

Christina Rufener[1], Sarah Baur[2], Ariane Stratmann[1], Hanno Würbel[3], Urs Geissbühler[2] and Michael Toscano[1]
[1]Center for Proper Housing: Poultry and Rabbits, Animal Welfare Division, University of Bern, Burgerweg 22, 3052 Zollikofen, Switzerland, [2]Department of Clinical Veterinary Medicine, Clinical Radiology, University of Bern, Postfach 8466, 3001 Bern, Switzerland, [3]Animal Welfare Division, University of Bern, Länggassstrasse 120, 3012 Bern, Switzerland; christina.rufener@vetsuisse.unibe.ch

Although aviary systems enable laying hens to perform natural behaviour, falls and collisions likely increase the frequency of keel bone fractures (KBF) in hens. Since the direct impact of KBF on hen welfare is not fully understood, the aim of this project was to investigate the relationship between KBF and laying hen productivity as an indicator of welfare. Hens were housed in 10 identical pens in a commercial aviary system (15 focal hens + 210 non focal hens per pen; 5 pens with brown and 5 pens with white focal birds). Focal hens were orally administered a dye on three consecutive days to allow eggs to be linked to focal hens. At two time points (week of age (WoA) 28 & 32), eggs were collected and counted over a five day period to determine individual egg laying rates (n=1403). Egg mass, shell breaking strength and shell width were measured in all eggs laid on the first three days of egg collection (n=811). Radiographs were performed on the last day of data collection to detect fractures (F=fracture [WoA 28: n=66, WoA 32: n=102], nF=no fracture [WoA 28: n=84, WoA 32: n=47]). Linear mixed effects models were used for statistical analysis. Hens with fractures laid heavier eggs than hens without fractures (62.3 vs 60.7 g; $F_{1,144}$=41.66; P<0.0001). Fractures were not related to shell width in brown birds, whereas white birds laid eggs with thicker shells when a fracture was present (brown nF: 0.34 mm, brown F: 0.34 mm, white nF: 0.32 mm, white F: 0.33 mm; $F_{1,143}$=6.63, P=0.011). In WoA 28, egg shell breaking strength was higher (56.7 vs 55.7 N; $F_{1,145}$=4.476, P=0.036) and egg mass was lower (59.4 vs 63.9 g, $F_{1,144}$=252.31, P<0.0001) than in WoA 32. Brown birds laid heavier eggs than white birds (62.6 vs 60.7 g; $F_{1,7}$=12.92, P=0.009). Laying performance of birds with fractures tended to be reduced compared to birds with intact keels (93.9 vs 97.7%; $F_{1,147}$=3.855, P=0.052). Reduced laying performance in birds with fractures could explain higher egg mass and thicker shells since the same amount of calcium is available for fewer eggs. Ongoing work will provide data on individual hen mobility in order to link changes in behaviour and productivity due to fractures. Our current findings indicate the need for further research to investigate the relationship between keel damage and productivity as welfare and economic concerns.

Effect of different laying hen strains on daily egg laying patterns and egg damage in an aviary system

Silvia Villanueva[1], Ahmed Ali[1], Dana Campbell[2] and Janice Siegford[1]
[1]*Michigan State University, East Lansing, MI, USA,* [2]*University of New England Australia, Armidale, NSW, Australia; villan36@msu.edu*

Proper space management to stimulate natural species-specific behavior in poultry production systems is a widely explored issue. One consideration for laying hens is a nesting site, usually provided in the form of an artificial nest box. However, single nest boxes are typically provided at less than one per hen; therefore competition for these egg laying sites can result. There has been limited research examining nest box use and laying patterns of various strains of laying hens in commercial-style aviaries. The objectives of the current study were to compare daily patterns of nest box use among 4 laying hen strains (Hy-Line Brown, Bovan Brown, DeKalb White and Hy-Line W36) and to understand if strain influenced egg laying outside the nest box (i.e. system-laid) or damaged eggs. Observations of hens (4 aviary units/strain, 144 hens/unit) were conducted at 36 wk of age by the same observer over 3 consecutive days (total of 6 observation sets classified into morning, midday, and evening). Nest boxes were located in the top tier of the aviary units and divided into left and right halves by a partition. During each observation, the number of hens in each side of the nest box and the number, location, and condition of eggs were recorded. Hens of all strains laid most eggs in the nest box (90.53-95.57%); however brown hens consistently laid more eggs outside the nest box than white hens ($P<0.05$). More total damaged eggs were found in units with brown hens (5.33 ± 2.7 vs 1.26 ± 0.76, $P<0.05$). Higher nest box occupancy in both strains of brown hens was correlated with more system-laid (r_s: 0.39-0.62, $P<0.03$) and damaged eggs (r_s: 0.43-0.53, $P<0.02$). Brown hens both occupied more nest box space in the morning compared to white hens (82.97 vs 34.66% of space, $P<0.05$) and also laid more eggs in the nest boxes at this time (91.35 vs 68.73% of nest box laid eggs, $P<0.05$). White hens continued to occupy nest box space at midday (26.27% of space), and laid more eggs than brown hens at this time (29.17 vs 7.68% of nest box laid eggs, $P<0.05$). Brown hens preferred the right side of the nest box, and laid more eggs on this side, whereas white hens preferred the left side and W36 laid more eggs on the left than the right side ($P<0.05$ in all cases). These findings indicate that different strains of poultry have different patterns of nest box use and laying behavior. In brown hens, heavy morning nest box use was related to egg laying outside the nest box and more damaged eggs, suggesting insufficient space for oviposition in the nest box. Modifications of facilities tailored to each strain are imperative in order to properly accommodate production and welfare needs of each individual strain.

Do domestic birds avoid the fecal odours of conspecifics?

Bishwo Pokharel[1], Vinicius Santos[1], David Wood[2], Bill Van Heyst[2] and Alexandra Harlander[1]
[1]University of Guelph, Animal Biosciences, 50 Stone Road East, Guelph, N1G 2W1, Canada,
[2]University of Guelph, School of Engineering, 50 Stone Road East, Guelph, N1G 2W1, Canada;
pokhareb@uoguelph.ca

Although feces avoidance behaviour has been demonstrated in a variety of animals, it is poorly understood if this behaviour occurs in domestic birds and how this behaviour operates. Aversion testing to ammonia (a normal gas by-product of feces) in laying hens demonstrates that hens display a preference for fresh air over an air/ammonia gas (from cylinders) mix. However, this may not reflect the air quality conditions on a farm, where birds are chronically exposed to feces which release a variety of gaseous chemicals along with ammonia. Our goal was to determine how latency, time spent and interruptions of foraging (searching for raisins and mealworms in a feed mix) of 20 adult laying hens are affected when exposed to selected air/ammonia gas mixtures. Gas mixtures included ammonia from certified cylinders (artificial source [A]) and natural gas mixtures from laying hen feces [F] at different concentrations (25, 45 ppm), as well as fresh air [FA]. Testing occurred in a two-compartment chamber, where behaviour was video recorded and compared for each gas mixture treatment. The chamber consisted of an entrance (EC) and feeder (FC) compartment separated by a sliding door. A and F gaseous stimuli were introduced with a constant air flow through a set of holes in the FC floor. After habituation, each bird underwent 3 testing sessions – after 5 sec in EC, the door was lifted to allow entry into the FC for 5 min – for each A/ F concentration on separate days (15 sessions/ bird) in a systematic test order. To minimize carryover effects, birds were tested on alternate days. Planned comparisons using orthogonal contrasts were used to test for foraging differences between treatments in the GLIMMIX (SAS) procedure. Estimated means± standard errors are presented. Birds showed increased latency (min) (A=0.38±0.02; F=0.37±0.02; FA=0.22±0.02; P<0.01) and reduced total time (percentage of total time) spent foraging (A=42.13±3.39; F=44.37±3.39; FA=55.60±4.00; P<0.01) in both A and F compared to FA treatments. Foraging bout lengths (min) did not differ between FA and 25 ppm F stimuli [FA=0.79±0.02; F(25 ppm)=0.64±0.02]. Birds were more likely to forage longer (fewer interruptions of bouts) in 25 and 45 ppm F treatments than in 25 and 45 ppm A treatments (min) (A=0.23±0.02; F=0.42±0.02; P<0.01).The preference for domestic birds foraging/feeding in FA is consistent with other studies. Differences in interruptions of ongoing foraging behaviour indicated that birds discriminate between artificial and natural sources of ammonia. In contrast to our assumption, the presence of conspecific feces may act as a positive stimulus for good foraging locations.

Effect of switching sows positions during lactation on performance and suckling behaviour of early socialized piglets

Jaime Figueroa, Sergio Guzmán-Pino, Tamara Tadich and Paulina Poblete
Universidad de Chile, Fomento de la Producción Animal, Avenida Santa Rosa 11735, 8820000,
La Pintana, Santiago, Chile; jaime.figueroa@u.uchile.cl

Under natural conditions, piglets have the opportunity to share with conspecifics from different sows while their mothers share a common grazing space. Nevertheless, in intensive pig production systems, social interactions are restricted during the suckling period and piglets only have contact with animals from the same parity crate. Social interactions can be encouraged by removing the barrier between litters while keeping sows in their own farrowing cages, and with piglets suckling from their own mother over 90% of the time. The aim of this study was to analyse the effect of switching position of sows when their piglets share a common space over piglet's performance and suckling behaviour. A total of 16 sows were selected and distributed in pairs from similar parity number. The barriers that separate litters in each sows pair were removed at d2 after farrowing. Half of sows were switched with their pair at d14, and the other half remained in their original position. All piglets (n192) were weighted at the beginning (d2), middle (d14) and end (d23) of the trial to calculate their body weight and average daily gain (ADG). Creep feed intake was also registered from d15 to d23. Piglets were video recorded (Video-cameras, SENKO S.A.) from d2 to d23 in order to analyse suckling frequency and duration. Performance and behaviour parameters were analysed taking into account sows treatment (switched or not at d14) and the period (pre o post switching) by using SAS statistical software. Piglets' body weight between treatments differed at d2 (1,749 vs 1,542 g; P=0.013) but not at d14 (4,857 vs 4,594 g; P=0.107) or d23 (after switching, 7,274 vs 7,144 g; P=0.566), for piglets coming from control or switched sows, respectively. ADG did not differ during the first (2-14d) or second (14-23d) periods (256 vs 249 g/d; P=0.532 and 267 vs 276 g/d; P=0.716) respectively. Creep feed intake neither differed between piglets coming from control nor switched sows (66.92 vs 62.94 g/d; P=0.559). In terms of suckling behaviour, animals coming from control or switched sows did not differ in their suckling frequency (26.5 vs 25.9 events/day; P=0.986) or duration (772 vs 749 seconds; P=0.884) after d14. Group housing may be a positive husbandry management in terms of welfare facilitating early social interactions between conspecific animals before weaning, not altering piglets' performance or suckling behaviour during lactation even if sows which share litters are switched in their initial positions.

Behavioural responses of commercially housed pigs to an arena and novel object test

Amy Haigh, Chou Jen Yung and Keelin O'Driscoll
Teagasc, Pig Development Department, Pig Development Department, Teagasc Animal and Grassland Research and Innovation Centre, Moorepark, Fermoy, Co. Cork, Ireland; amy.haigh@teagasc.ie

Individual variation in personality traits has been recognised in a number of species and is believed to have important consequences in terms of an animal's coping strategy and ability to handle stress. This study investigated the reaction of 80 pigs to novel arena (NA) and novel object (NO) tests at 14 weeks of age. Pigs from eight pens (n=4 pens per experimental room) that were part of an experiment comparing straw and plastic toys as enrichment were used. Within each pen of 50 pigs, the following 10 pigs types (TYPE) were selected for testing: 3 pigs that had severe tail/ear bites, 1 confirmed tail/ear biter, 1 pig that successfully gave a saliva sample each week of the experiment (good focal, GF), 1 pig that was unable to be trained to give a saliva sample (bad focal, BF), 2 pigs that consistently approached humans (Approach+), and 2 pigs that consistently avoided humans (Approach-). No pig fell into more than one TYPE. Pigs were individually placed in a novel arena, then behaviour continuously recorded for 5 min. Immediately following this, a novel object (brush head) was placed in the arena, and behaviour continuously recorded for a further 3 min. The floor of the arena was divided into a nine square grid, to facilitate recording of pig movement. Data were analysed using mixed models (fixed effects: enrichment type, day, sex and TYPE; random effect: room), fishers exact, and Mann-Whitney tests. The NA and NO tests were analysed seperately. During the NA test TYPE had an effect on duration of exploratory behaviour (P<0.05), which was lower in bitten pigs (02:10±00:13; mm:ss) than non-bitten (03:02±00:15; P<0.01). More bitten pigs tended to perform high pitched vocalisations (HPV; 50 vs 14%; P=0.06), and these performed more HPV per pig (10.5 vs 6.6; P=0.07) than non-bitten. In the NO test, there was a tendency for bitten pigs to direct attention without touching the NO for longer than all other types (10:48±01:59 vs 04:38±02:46; P=0.06). However, focal pigs (GF and BF combined) spent more time physically interacting with it than all others (01:42±00:21 vs 01:05±00:14; P<0.05). Type of pig did not affect latency to interact with the NO. Overall, the behaviour of bitten pigs was more indicative of fearfulness than non-bitten pigs. These data imply a link between being bitten and fearful behaviour. While fearfulness may be a result of being bitten, an alternative interpretation could be that pigs may be more vulnerable to being victims as a result of fearfulness. Further work is required to elucidate the relationship.

Nursing behaviour and teat order in pigs reared by their dam or nurse sow

O Schmitt[1,2], E Baxter[1], L. Boyle[2] and K O'driscoll[2]

[1]SRUC, Animal Behaviour and Welfare, Animal and Veterinary Science Research Group, West Mains Road, Edinburgh EH9 3JG, United Kingdom, [2]Teagasc, Pig Development Department, Teagasc Moorepark Research Centre, Fermoy, Co. Cork, Ireland; oceane.schmitt@teagasc.ie

Using nurse sows to rear surplus piglets from large litters (>14 born alive) is a fostering strategy intended to optimise piglet pre-weaning survival. However, moving piglets between litters may negatively affect their behaviour during subsequent nursing bouts. This study compared nursing behaviour of piglets subject to different nurse sow stategies. At one day-old, the heaviest 3-4 piglets born into large litters (R: Removal, n=9) were fostered either onto a nurse sow 21 days into lactaion (N1: one-step srategy, n=10); or onto a nurse sow 7 days nto lactation (N2: two-step strategy, n=9). At that time, N2 offspring were fostered onto a nurse sow 21 days into lactation (N3: two-step strategy, n=9). In addition to removing the heaviest piglets, cross-fostering was allowed on R litters when up to 25% of remaining piglets were <1 kg (RC: Removal Cross-fostered, n=10). Two full nursing bouts were observed by a simple observer for each sow at D-1 (birth day), D0 (fostering day), D1, D2, D6, D9, D16 and D23. Teat order stability was estimated using a validated formula to calculate the probability of a piglet sucking the same teat across observations (Psuck). Nursing bouts were not observed on D0 for the nurse sows due to delayed acceptance of fostered piglets. Data were analysed using a mixed model approach, accounting for repeated measures. In all treatments the number of fights per piglets and the percentage of piglets involved in fights during nursing were higher on D1 compared to any other day ($P<0.001$) and decreased over time ($P<0.001$). Piglets fostered onto N1 had more fights than piglets who remained with their dam (N1=0.32±0.02, R=0.24±0.02, RC=0.24±0.02; $P<0.05$). For all the treatments, the percentage of piglets missing a nursing bout was higher on D1 than any other day ($P<0.01$) and was positively correlated with the number of fights per piglet ($P<0.01$). Overall Psuck was higher for RC and lowest for N1 and N3 (RC=0.60±0.05, N1=N3=0.46±0.05; $P<0.05$). Psuck was numerically lowest during the initial observation period (D1-D2) and highest at D9, exept for N3 which did not change over time ($P<0.05$). Thus, since fostering piglets onto a nurse sow resulted in reduced nursing bouts for 24 h and high levels of fighting associated with risk of missing nursing bouts, repeated fostering of piglets should be minimised. However, conflicts for teat ownership were quickly reduced, even if teat order did not remain stable over lactation.

Does split marketing affect animal welfare in organic fattening boars?

Jeannette C. Lange, Nikolaus E. Fuchs and Ute Knierim
University of Kassel, Farm Animal Behaviour and Husbandry Section, Nordbahnhofstr. 1a, 37213 Witzenhausen, Germany; jlange@uni-kassel.de

The intention to improve animal welfare by refraining from the painful castration of boars may be compromised by more agonistic and mounting behaviour among entire boars. While it has been reported that split marketing may increase this problem, it is possible that effects under organic husbandry conditions differ. Therefore, we compared agonistic and mounting behaviour of boars and barrows on 5 commercial organic farms before ('pre') and 2-3 days after the first marketing ('post') and scored skin lesions and lameness in order to assess possible effects of castration and split marketing on animal welfare. Video based continuous behaviour sampling in 17 boar and 14 barrow groups, with group sizes varying from 10 to 25, kept over a total period of 2.5 years, was carried out on group level during light hours within 48 hours. Following the Welfare Quality Protocol®, the altogether 252 boars and 213 barrows before and 140 boars and 119 barrows after first marketing were scored. Linear mixed models in R were used with the fixed factors castration and marketing status, as well as age for agonistic interactions and age and season for mounting behaviour. Random factors were farm, batch, group, and for lesions additionally the individual. Castration significantly affected the frequency of biting and head knocking (t=-4.57, P<0.001): pre: 2.0/boar×h (SD: ±1.1) vs 1.4/barrow×h (±0.7); post: 2.1/boar×h (±1.1) vs 1.2/barrow×h (±0.8), while marketing status did not (t=1.12, P=0.27). Fighting and mounting showed similar patterns and statistic results. The weighted lesion sums of one body side were not significantly different (t=-0.76, P=0.46 and t=0.48, P=0.63): pre: 5.4 (±4.3) in boars vs 5.3 (±4.0) in barrows; post: 6.3 (±4.9) in boars vs 5.5 (±4.5) in barrows. Severe lameness was observed in 1 boar and 1 barrow pre and 1 boar and 1 barrow post, shortened stride or other undefined difficulties while walking in 7 boars and 7 barrows pre and in 4 boars and 8 barrows post marketing. It can be concluded that also under organic husbandry conditions increased agonistic and mounting behaviour occurred in boars which, however, did not lead to an increased bodily impairment and might therefore be judged as tolerable. Furthermore, split marketing did not negatively affect pig welfare as regards the investigated measures.

Application of Social Network Analysis in the study of post-mixing aggression in pigs

Simone Foister[1], Andrea Doeschl-Wilson[2], Rainer Roehe[1], Laura Boyle[3] and Simon Turner[1]
[1]Scotland's Rural College, Roslin Institute Building, Midlothian, EH25 9RG, United Kingdom, [2]The Roslin Institute, R(D)SVS, University of Edinburgh, Roslin Institute Building, Midlothian, EH25 9RG, United Kingdom, [3]Teagasc, Pig Development Unit, Teagasc Research Centre, Moorepark, Fermoy, Co. Cork, Ireland; simone.foister@sruc.ac.uk

Post-mixing aggression is a substantial welfare issue in the commercial pig industry. Various aggressive behavioural traits (e.g. reciprocal fighting) are correlated with lesion scores (LS) at individual and group levels. However total durations or frequencies of involvement in aggressive behaviour are gross measures that may not well describe an individual's position within an interacting network of animals. Social network analysis (SNA) provides a novel approach to studying aggression by exploring the effect an individual's position within its network and the overall group structure has on the accumulation of LS. This study aimed to assess whether more complex network measures could provide new insight into accumulation of LS at individual and pen levels. Network graphs were created by plotting reciprocal aggressive interactions that occurred between 15 pigs in each of the 78 pens during the 24 hours following mixing into new social groups. LS were obtained 24 h post-mixing for anterior, central and posterior regions of the body. In the results 'LS' refers to the combined total of these lesion scores unless otherwise stated. All graphs and network measures were calculated using the R package 'igraph'. At individual level, the following measures were calculated; degree, strength, closeness, betweenness, eigenvector and cluster coefficient. Pen level measures included; number of strongly connected components (SCC), density, and centralized values of closeness, betweenness and eigenvector. Results revealed that basic network measures such as degree (number of 'ties' to another animal) and strength (sum of the 'weight' of that 'tie' e.g. duration), were highly significantly correlated to lesion scores (0.26 and 0.33, P<0.001 respectively), confirming results of previous studies. More complex network measures also provided significant correlations with LS. Of the pen level centrality measures calculated, only SCC was correlated with LS (-0.27, P<0.01), indicating that the presence of small highly connected subgroups corresponds to low LS within the pen. Furthermore, pen level density was significantly correlated with anterior LS (0.29, P<0.01). All individual level centrality measures were significantly correlated with individual LS (0.33 to -0.06, P<0.001 to <0.05). The results suggest that the structure of an animal's network and the position within that network are related to LS and that SNA can provide novel insights into the structure underlying post-mixing aggression.

Weaner piglet activity measured with commercial motion sensors in groups with and without tail biting

Sabine Dippel[1], Moritz Leithäuser[1,2], Lars Schrader[1], Christine Leeb[2] and Christoph Winckler[2]
[1]Friedrich-Loeffler-Institut, Dörnbergstr. 25/27, 29223 Celle, Germany, [2]BOKU, University of Natural Resources and Life Sciences, Vienna, Gregor-Mendel-Strasse 33, 1180 Vienna, Austria; sabine.dippel@fli.bund.de

Tail biting is a major welfare problem in pig production and early detection and intervention is crucial for preventing tail injuries. Recent findings suggest that it often occurs a few weeks after weaning. Therefore the aim of this study was to validate a low-tech method for automated assessment of weaner piglet activity and to test if changes in activity are related to tail biting (TB). A total of 20 groups of weaners (mean 26 animals) were scored weekly for tail lesions and behaviour was recorded for 21 DAYS by video and commercial infrared motion sensors (IMS). The proportion of active (walking, standing, sitting) pigs was assessed from video using instantaneous sampling (15 min intervals). IMS output was recorded in 10 s intervals and summarised as mean activity (%) per 24 h-day. For validating IMS with video data, logistic regression models, sensitivity and specificity were calculated (8 groups, mean 358 observations/group). Mean activity was tested for differences between groups with (TB; n=12; ≥1 assessment with ≥5 tail lesions) and without (noTB; n=8) tail biting using linear mixed models with DAY and TB × DAY as additional fixed factors, and pen, room and batch as random factors. IMS validation showed a significant relationship between the proportion of active weaners as obtained from video analysis and activity detected by IMS (OR 1.53; $P<0.001$). When a cut-off point of '>10% of animals active' (from video) was used, sensitivity and specificity of IMS was 67 and 84%, respectively. Sensitivity increased to 78% at >90% as cut-off, while specificity decreased to 54%. Activity recorded with IMS showed significant differences between noTB and TB groups with a higher activity in TB than in noTB groups (estimates: noTB=-6.16, DAY=0.54; both $P<0.01$; TB×DAY n.s.). Across the entire recording period, mean (standard deviation) activity was 51.6 (10.3)% in noTB groups and 53.7 (10.6)% in TB groups. In conclusion, commercial IMS are feasible tools for assessing activity levels of weaners with satisfactory sensitivity and specificity at rather low proportions of active animals. It could support farmers in early detection of tail biting. However, further validation using time line analysis is recommended before implementation in commercial farms. In order to prevent tail biting, farmers should pay extra attention including early intervention measures to weaner groups which are more active than others.

Effect of experience on the use of outdoor areas in a commercial free-range laying flock

Leonard Ikenna Chielo, Tom Pike and Jonathan Cooper
University of Lincoln, School of Life Sciences, Brayford Pool, LN6 7TS, United Kingdom;
lchielo@lincoln.ac.uk

Free-range hens do not normally have access to outdoor areas until laying has been established, and this delay can be a factor in subsequent range use. The aim of this study was to investigate the changing patterns in the use of outdoor areas in free range hens from release. This was achieved by a study of a single commercial flock of 8,000 Hyline laying hens from first release at 20 weeks of age for 15 weeks. This study was carried out from July to October 2013 in Heckington Fen, Lincolnshire, UK. To minimise observer influence, data was collected using digital camera, and twenty 10×10 m sampling quadrats were established in each of three outdoor areas; the apron (0-10 m from shed, without enrichments); the enriched area (10-50 m from the shed with enrichment resources including trees, dust-baths, perches, and shelters); and the outer range (50 m and beyond from the shed with short grasses and no tree cover). Data was collected at hourly intervals and the total number of hens and the behaviour of up to five selected hens were recorded from each photographic frame. The environmental temperature, relative humidity and wind speed were recorded every ten minutes using a weather station. Data was analysed using GLM ANOVA to determine the effects of time of day, week and outdoor area on the number and behaviour of hens. The results showed that outdoor use increased significantly ($P<0.001$) as the weeks progressed especially in the earlier weeks. The hens used only the apron in the first two weeks but progressed to the enriched and outer range areas in the third and fourth weeks of access respectively. Use of the outdoor area became stable from 3 weeks of access. Hen density was significantly higher ($P<0.001$) in the apron area (21.3 hens/frame) than the enriched (12.3hens/frame) or outer range (2.8 hens/frame) throughout this period. More hens were found outside in the early morning observations ($P<0.001$) with 17.8 hens/frame at 09:00 h, then 16.2 at 10:00, 11.7 at 11:00, 7.6 at 13:00 and 7.3 at 14:00 h respectively. Temperature ($P<0.05$) and relative humidity ($P<0.05$) had significant positive effects whereas wind velocity ($P<0.05$) had a significant negative effect on the use of range. Walking (36.3%), pecking (12.6%), standing (23.2%) and foraging (20.1%) were the most commonly recorded activities with standing and pecking occurring most in the apron area and walking and foraging recorded most in the outer range. In this study, the free range hens take several weeks to exploit the outdoor environment, and (as has been reported in other studies) the outer range was poorly utilised. Age on first access to range may be a factor in this poor utilisation and would warrant further investigation of the robustness of our findings.

Use of space patterns and its impact on the welfare of laying hens in free-range

Ane Rodriguez-Aurrekoetxea[1] and Inma Estevez[1,2]
[1]Neiker-Tecnalia, Animal Production, 01080 Vitoria-Gasteiz, Spain, [2]IKERBASQUE, Basque Foundation for Science, Maria Diaz de Haro 3, 48013 Bilbao, Spain; arodriguez@neiker.net

Access to an outdoor area is considered of fundamental relevance for the welfare of laying hens. Existing studies on the use of the outdoor area have focus in determining the frequency of use. However, there is insufficient information regarding the use of space patterns of laying hens in fee-range systems. The aim of this study was to explore the factors influencing use of space patterns of commercial free-range laying and their relation with welfare indicators. Three free-range laying hen flocks were studied during a production cycle by collecting spatial locations of 150 hens/flock individually tagged. Production cycle was divided in three age periods (AP1: 20-36, AP2: 37-53 and AP3: 54-69 wks of age) for statistical analysis. Net, total, maximum and minimum distance travelled was calculated indoors. Mean, maximum and minimum distance to the house was calculated for the outdoor area, while home range and activity centers were calculated for the indoor and outdoor areas. At the end of production welfare and morphometric measures were collected. The results indicate that use of the outdoor area was lower during midday ($P<0.05$), but remained stable across age periods ($P>0.05$, mean use $32.60\pm15.3\%$). Tagged hens were classified according to their use of the outdoor area (high, medium, low or never) per age period, and showed that 49.5% were never observed using the outdoor area, percentage that was superior to all other categories ($P<0.05$). In addition, experience during the AP1 determined the level of use of the outdoor area at later age periods ($P<0.05$). Most use of space parameters considered did not varied according to age period ($P>0.05$), only the activity center indoors increased ($P<0.05$), while mean distance from the hen house tended to increase ($P=0.053$). However, birds with higher frequency of use of the outdoor area had larger home ranges and activity centers ($r=0.956$, $P<0.0001$; $r=0.964$ $P<0.0001$, respectively) and showed lower plumage damage ($r=-0.337$, $P<0.001$). Birds with higher mean distance to the hen house appeared to have a lower incidence of footpad dermatitis ($r=-0.307$, $P<0.001$). On the contrary, birds showing higher total walked distance indoors showed a higher incidence of footpad dermatitis ($r=0.329$, $P<0.01$). These results suggest that birds using the outdoor area during AP1 were more likely to use the outdoor area throughout production and that those visiting the outdoor area more frequently also used larger areas. In addition, individual spatial patterns were shown to have some impact on the incidence on foot pad dermatitis and plumage condition.

Carcass lesions as indicators of injurious behaviour and welfare lesions in pigs

Nienke Van Staaveren[1,2], Bernadette Doyle[2], Alison Hanlon[1] and Laura Boyle[2]
[1]*University College Dublin, School of Veterinary Medicine, Belfield, Dublin 4, Ireland,* [2]*Teagasc Animal and Grassland Research and Innovation Centre, Pig Development Department, Moorepark, Fermoy, Co. Cork, Ireland; nienke.vanstaaveren@teagasc.ie*

Tail and skin lesions are suggested as potential 'iceberg' indicators, capable of providing information on the overall health and welfare of pigs. The aim of this study was to investigate the potential of carcass tail and skin lesions to reflect not only tail biting and aggression but other injurious behaviour (e.g. ear or flank biting, mounting) on farms and thereby act as an 'iceberg indicators' for pig welfare. Welfare assessments were conducted on 31 integrated pig farms by observing 18 randomly selected pens of 1st and 2nd stage weaner and finisher pigs (6 pens per stage). Pens were observed for 10 min and the number of pigs with tail, ear, flank and skin lesions was recorded. All occurrence behaviour sampling was used to record frequency of tail, ear, and flank biting, fighting and mounting (5 min). The average percentage of pigs in a pen affected by welfare lesions was calculated, as was the average frequency of behaviours per pig, in each production stage. One batch of pigs from each farm was observed at the slaughterline (204±25.8 pigs) and carcass tail lesions (0-4) and skin lesions (0-3) were scored according to severity. An average carcass tail (TLS) and skin lesion (SLS) score for each batch was calculated. Spearman rank correlations were calculated between TLS/SLS and the measured welfare conditions and behaviours on farm. Average TLS was 0.83±0.02 (range: 0.53-1.26) while average SLS was 0.87±0.04 (range: 0.37-1.28). TLS was correlated with the prevalence of healed skin lesions in the finisher (r=0.4, P<0.05) and a tendency was found in 2nd stage (r=0.3, P=0.1). In addition, TLS was correlated with the frequency of mounting behaviour in the 2nd stage (r=0.04, P<0.05) and tended to be correlated with fighting in 1st stage (r=0.4, P=0.06). SLS tended to be correlated to the prevalence of tail lesions in the finisher stage (r=0.3, P=0.07). A positive correlation was found between SLS and frequency of mounting in the 1st stage (r=0.6, P<0.01). SLS tended to be correlated with flank biting in the 1st stage (r=0.3, P=0.07), ear biting in the 2nd stage (r=-0.3, P=0.08) and fighting in the finisher stage (r=-0.3, P=0.1). Our results indicate that carcass tail and skin lesion scores can reflect injurious behaviours in pigs on the farm, however further work is needed to elucidate this relationship. Further work is also needed to validate the use of these welfare lesions on the carcass as 'iceberg' indicators for pig welfare on farm.

Evaluation of an animal welfare labelling scheme for broilers in the Netherlands using animal based measures

Sabine Hartmann[1], Paul Schwediauer[2] and Manuel Flätgen[3]
[1]FOUR PAWS International, Linke Wienzeile 236, 1150 Wien, Austria, [2]University of Natural Resources and Applied Life Sciences (BOKU), Division of Livestock Sciences, Department of Sustainable Agricultural Systems, Gregor-Mendel-Straße 33, 1180 Wien, Austria, [3]Tierarztpraxis Flätgen, Kraxenberg 8, 2851 Krumbach, Austria; sabine.hartmann@vier-pfoten.org

The international animal welfare organization FOUR PAWS has implemented a new international animal welfare labelling scheme for broilers in the Netherlands. Farmers are amongst others required to use slower growing strains (in this project: Hubbard JA 757), stock at a max. density of 15 birds/m^2 (day 1-21) and 11 birds/m^2 (from day 21) and to spread part of the feed by hand. Birds over 21 days of age must have access to a roofed outdoor area. To be certified, farms currently must comply with resource-based criteria only. Since 2012 in total 62 producers have been evaluated by one trained observer of an independent certification body who additionally used an inspection protocol adapted from the Welfare Quality® Protocol. The aim of this study was to evaluate the results of on-farm (n=136) and abattoir audits (n=111) and derive thresholds for future monitoring of animal based measures (ABM). Behavioural measures were assessed on flock level, individual birds were assessed on-farm and on abattoir for clinical parameters. Animals showed signs of fear in 37%, of feather pecking in 1.2% and of cannibalism in 2.5% of the flocks (81 audits). Huddling, panting and crowding were observed in none of the audits. Lameness was assessed on 45 birds per flock (score 0-4, n=88). The average prevalence of sound birds (score 0) was 91%. All other ABM (a=none, b=mild, c=severe) were assessed on a different set of 30 animals per flock (n=52). On average 0.4% of birds had lesions (b), 1.5% incomplete plumage (b), 1.2% dirty plumage (b) and 0.2% breast skin alterations (b). Prevalence of score c was 0% in all four measures. Mean prevalence on abattoir for breast blisters was 0.1%, hock burn 1%, severe lesions 1%, bruises on wings 3% and animals rejected 0.4%. Footpad dermatitis (FPD) was calculated as one score for 100 birds per flock (100 × [(0.5 × number of feet with score 1) + (2 × number of feet with score 2)] / total number of scored feet), with a mean score of 13. Only 1.8% of flocks scored >80 (0 = no feet with lesions; 200 = all feet with severe lesions). In discussion with experts intervention-levels (IL) were suggested. The following 3 ABM were identified above IL: >6% of birds with lesions score b (found in 14% of audits), >1 bird with lameness score 4 (in 7% of audits) and >10% of birds with plumage score b (in 8% of audits). The results provide a sound basis for the future inclusion of ABM in the certification process.

Human attitudes and their relevance to animal welfare

Grahame Coleman

University of Melbourne, Animal Welfare Science Centre, Parkville, 3010, Australia; grahame.coleman@unimelb.edu.au

Social science contributes to the study of animal welfare in two broad ways: elucidating firstly the ways in which human populations think and behave and, secondly, the ways in which individual people think and behave. The key aspect of both of these is that way in which human behaviour affects the welfare of domesticated animals. Often, human attitudes to farm animal housing and husbandry, companion animal behavioural problems and zoo animal welfare are treated as behaviours in their own right without reference to the actual human behavioural outcomes and their relevance to animal welfare. The aim in this paper is to present a framework that explicitly links human attitudes to welfare outcomes and to illustrate it using published and unpublished research by the Animal Welfare Science Centre on farmer attitudes to their livestock, public attitudes to livestock practices, companion animals and visitor attitudes to zoo animals. The framework draws specifically on the Fishbein and Ajzen theory of planned behaviour that identifies behavioural attitudes as the proximal driver of behaviour and explicitly addresses the secondary role of demographics and human values. There are several reasons for taking this approach. First, public attitudes may be regarded as indicators of people's willingness to buy an animal product or visit a zoo for example. Many stakeholders such as food retailers, legislators, regulators and producers respond to public attitudes by changing practices either to capitalise the opportunity to obtain some advantage or to correct a perceived problem in the absence of any real understanding of the true relevance of these attitudes to animal welfare. Our data show that these attitudes predict a range of non-purchasing behaviours that may impact on animal welfare. Second, there are many behaviours that directly impact on animal welfare because they involve the way in which people engage in human-animal interactions. In the case of livestock, we have found that people's attitudes directly influence the way in which they handle their animals and that changing these attitudes directly improves animal welfare and productivity. The method by which human behaviour can be changed is to use a cognitive-behaviour approach that explicitly targets the beliefs that underlie behaviour and provides explicit information of the appropriate behaviours. Little has been done to adopt this approach with companion animals, but we are beginning to assemble data on human attitudes to companion animals and recreational horses that have the potential to lead to similar outcomes. The key message from all of this is that social science can identify the critical human factors that impact on animal welfare and provide the specific information that is needed to change human behaviour.

Student perceptions of animal welfare in intensive and extensive animal production environments

Janice M. Siegford, Bethany Larsen, Paul B. Thompson, Laurie Thorp and Dale Rozeboom
Michigan State University, 480 Wilson Rd, East Lansing, MI 48824, USA; thom@anr.msu.edu

Students in four undergraduate classes at Michigan State University (MSU; n=238) were given educational tours of animal production facilities and completed pre-and post-tour open-ended questionnaires eliciting their views on production methods and animal welfare in the food system. Students toured both intensive and extensive facilities at MSU. Intensive tours were done at either the MSU Swine Farm (swine at all stages from farrow to finish, including breeding boars and sows in gestation stalls) or the MSU Poultry Farm (laying hens in aviaries and enriched cages). The extensive facility toured was the MSU Student Organic Farm, where both pigs and hens had relatively unrestricted outdoor access. In both cases, tours were conducted as moderated experiential exercises led by MSU faculty and were intended to give students increased understanding of methods and issues in contemporary food systems. We present preliminary analysis of baseline patterns in student responses as well as the effectiveness of a tour experience in shaping perceptions of animal welfare. Overall, before the tours, student perceptions of intensive systems were overwhelmingly negative (as evidenced through phrases including 'necessary evil', 'lack of movement', 'poor living conditions', and 'inhumane') while perceptions of extensive systems were overwhelmingly positive (as evidenced through phrases including 'natural', 'freedom to move', ' care', 'humane', and 'happy'). Prior to the tours, about 50% of students stated they had no experience with intensive animal agriculture and about 40% had no experience with extensive agriculture. Many students with no prior experience of these systems admitted developing their perceptions from media sources. Mirroring the lack of experience before the field trip, 46% of students had no substantive definition for animal welfare, with 6% stating they did not know what animal welfare meant. However, non-substantive definitions halved to 23% after the tour, indicating a doubling in knowledge of animal welfare. Before the tour, roughly equal proportions of students (~18%) referenced each of the three dimensions of Fraser et al.'s animal welfare model: animal natures, bodies, and minds. After the field trip, discussion of animal natures doubled to 35%, while discussion of animal bodies grew to 25%. The discussion rate of animal minds remained steady after the field trip (17%). Across the categories of animal natures, bodies, and minds, students defined animal welfare based both on how we treat animals as well as how animals feel. This tour-based experiential learning opportunity in the context of an undergraduate course appears to have improved student understanding of animal welfare.

Perceptions of current and future farmers and veterinarians regarding lameness and pain in sheep

Carol S Thompson[1], Joanne Conington[1], Joanne Williams[2], Adroaldo J Zanella[3] and Kenny M D Rutherford[1]

[1]SRUC, Animal and Veterinary Sciences, West Mains Road, EH9 3JG, Edinburgh, United Kingdom, [2]School of Health in Social Science, The University of Edinburgh, EH8 9AG, United Kingdom, [3]Universidade de São Paulo, Faculdade de Medicina Veterinária e Zootecnia, São Paulo, 87 CEP 05508 270, Brazil; carol.thompson@sruc.ac.uk

Accurate identification of disease and pain is paramount to good animal welfare. This study assessed how farmers (F), veterinarians (V) and students (agriculture (AS) and veterinary (VS)) perceived lameness and pain in sheep. Movie clips (20 seconds each) of four sheep with varying levels of lameness (one 'sound', one 'mildly lame' and two 'moderate/severely lame') were shown to F (n=68), V (n=46), AS (n=89) and VS (n=56) in order to investigate how they perceived lameness and its associated pain. After each clip, participants completed a short questionnaire, which asked them to rate, using a 100 mm visual analogue scale, the level of: (1) lameness (L); and (2) pain (P) they felt the sheep was experiencing; and (3) their own emotional response (ER). Participants were also asked whether they would catch the sheep to inspect its feet. Data were analysed (Genstat 15) using REML and Spearman Rank Correlations. Strong positive correlations were found between ratings of L, P and ER for all groups (L vs P: r≥0.93, L vs E: r≥0.81, P vs E: r≥0.80; all P<0.001). Differences were found between groups for L, P, and ER ratings. For example vets provided higher ratings than did farmers for the two moderate/severely lame sheep, and veterinary students provided higher ratings than did agriculture students for the sound, mildly lame and one of the moderate/severely lame sheep (e.g. for one of the moderate/severely lame sheep (mean±SE): L: V=78.5[b]±3.1, VS=79.8[b]±3.1, F=70.6[a]±2.9, AS=69.4[a]±2.8, W=62.8, P<0.001; P: V=74.0[c]±3.2, VS=70.9[b]±3.3, F=67.2[ab]±3.0, AS=62.3[a]±2.9, W=52.6, P<0.001; ER: V=64.1[c]±4.0, VS=59.4[bc]±4.0, F=51.0[a]±3.7, AS=32.4[d]±3.5, W=76.6, superscripts denote a significant difference at P<0.05, calculated using a pairwise comparison). Participants who rated L, P and ER higher were more likely say they would catch the sheep for a foot inspection (e.g. for one of the moderate/severely lame sheep (mean±SE): L: Yes=80.8±2.4, No=26.9±5.4; P: Yes=74.1±2.4 No=18.7±5.6; ER: Yes=58.5±2.7, No=14.9±6.4, P<0.01). These results indicate that participants viewed lameness as a painful condition, and that the decision to catch is based, at least in part, upon the perceived severity of the condition. Individual perceptions about lameness and pain severity may dictate treatment outcomes for lame sheep.

A prospective exploration of farm, farmer and animal characteristics in human-cow relationships: an epidemiological survey

Alice De Boyer Des Roches[1,2], Luc Mounier[1,2], Xavier Boivin[1,2], Emmanuelle Gilot-Fromont[3,4] and Isabelle Veissier[1,2]
[1]*INRA, UMR1213 Herbivores, Centre de Theix, 63122 Saint-Genès-Champanelle, France,* [2]*Université de Lyon, VetAgro Sup, UMR1213 Herbivores, 1 av Bourgelat, 69280, Marcy l'Etoile, France,* [3]*Université de Lyon, Université Lyon 1, UMR CNRS 5558 – LBBE, Campus de la Doua, 69622 Villeurbanne, France,* [4]*Université de Lyon, VetAgro Sup, 1av Bourgelat, 69280 Marcy l'Etoile, France; alice.deboyerdesroches@vetagro-sup.fr*

Human-animal relationships are essential for dairy farming, impacting on work comfort and efficiency, and on milk production. Many studies have demonstrated the multifactoriality of these relationships. Prospectively, we aimed at assessing the relative importance of various factors expected to be associated with poor human-animal relationships. On 118 dairy farms, we applied the Welfare Quality® standardized avoidance distance test to 4.418 cows at the feeding rack. The sample of farms covered a wide range of situations (location, size, breed, housing and milking systems). We used Poisson regression to analyze the links between the number of cows that accepted being touched, and farm characteristics, animals, management and farmers' attitudes (measured by a questionnaire using psychometric evaluation scales). The median was 9.10% of the cows touched by the observer (min-max, 0.0-35.0%). The best fitting multivariable Poisson regression model explained 32.7% of the variability between farms ($P<0.001$). The proportion of cows accepting being touched increased by 1.86 and 2.31 respectively on farms where the main calving location was at pasture or in a calving pen ($P<0.001$), by 1.57 on farms where farmers did not clean or add litter after calving ($P<0.01$), by 1.67 ($P<0.05$) when the proportion of lean cows in the herd was above 45%, by 1.37 ($P<0.01$) when worker/cow ratio was above 0.05, by 1.34 and 1.87 ($P<0.05$) when farmers considered 'health' or 'human-cow relationships' as most important issues for farm success, by 1.48 ($P<0.05$) when farmers' had more than 10 years of experience, and was reduced by 0.86 ($P<0.001$) when farmers showed more negative behavioral attitudes toward cows (e.g. agreed more about aversive contact). In conclusion, the human-animal relationship was not found to be associated with farm characteristics but varied with farmers' attitudes and management. We confirm that cows' fear of people is linked to negative attitudes displayed by caretakers towards cows, and is reduced in farms where there was more human contact. Our study also suggests further exploring the key role of factors linked to calving conditions, as cows are more likely to be afraid of people when disturbed at calving.

The burden of domestication: a representative study of behaviour and health problems of privately owned cats in Denmark

Peter Sandøe[1,2], Annika Patursson Nørspang[2], Thomas Bøker Lund[1], Charlotte Reinhard Bjørnvad[3] and Björn Forkman[2]
[1]*University of Copenhagen, Department of Food and Resource Economics, Rolighedsvej 25, 1958 Frederiksberg C, Denmark,* [2]*University of Copenhagen, Department of Large Animal Sciences, Grønnegårdsvej 8, 1870 Frederiksberg C, Denmark,* [3]*University of Copenhagen, Department of Veterinary Clinical and Animal Sciences, Dyrlægevej 16, 1870 Frederiksberg C, Denmark; pes@sund.ku.dk*

The ways in which domestic cats are kept and bred have changed dramatically. Many cats are kept indoors, most are neutered and a significant number are selectively bred which has consequences for their behaviour and health – and ultimately for their welfare. This representative study of the living conditions of privately owned cats in Denmark is the first to quantify the risks and document the prevalence of risk factors. Thus the focus of the paper is to investigate how indoor confinement, neuter status and selective breeding affect the prevalence of behaviour and health problems in domestic cats. 415 cat owners were recruited and answered questions regarding their oldest cat through a questionnaire study aimed at a representative sample of the Danish population (conducted in accordance with Danish data protection legislation). The effects of confinement, neuter status and selective breeding on levels of reported behaviour and health problems were examined with multivariate logistic regression (significance level: 0.05) where all three factors along with cat-age were inserted as explanatory variables to account for potential confounding. Confined cats had significantly more behavioural problems than free-roaming cats (e.g. 35.5% of confined cats were found to destroy furniture compared to 16.8% of free-roaming cats, and 19.1% of confined cats displayed signs of boredom compared to 9.0% of free-roaming cats), and more free-roaming cats (54.7% compared to 37.1% confined cats) had none of the behavioural problems asked about. Entire cats had significantly more behavioural problems compared to neutered cats (e.g. only 28.6% of intact cats had none of the behavioural problems asked about compared to 52.5% of neutered cats). On the other hand few differences in behavioural problems were seen between purebreds, domestic shorthair and mixed breed cats. The level of disease reported by the owners was similar between confined cats and free-roaming cats, whereas significantly more purebred cats (36.1%) had one or more of a number of diseases commonly found in cats compared to domestic shorthair cats (7.2%) and mixed breed cats (14.2%). Confinement and, being intact, leads to an increase in behavioural problems. Purebred cats are more likely to have health issues compared to domestic shorthair and mixed breed cats.

Predicting human directed aggressive behaviour in dogs

Rachel Orritt, Todd Hogue and Daniel Mills
University of Lincoln, Brayford Pool, Lincoln LN67TS, United Kingdom; rorritt@lincoln.ac.uk

Human directed aggressive behaviour (HDAB) in dogs is widely regarded as an international public health issue and welfare concern, yet there is little consensus on the proper management of this behaviour. Identifying risk factors for HDAB has been the primary response of the academic community to this issue, in the hopes of informing dog bite prevention strategies. The current authors identify three major shortcomings of research in this area. Firstly, many studies use biased clinical samples and fail to incorporate an appropriate control population. Secondly, the majority of studies prioritise the investigation of biting behaviour and it is uncommon to investigate less severe forms of HDAB. Finally, studies are often limited to dog-specific risk factors, neglecting consideration of the owner variables, the wider environment and context, and potential protective factors. Using methodologies from the psychological study of violent human behaviour, this study sought to investigate a broader range of factors using a more rigorous methodology, in order to better understand the circumstances that predict HDAB. In 2015, a large scale online survey was used to collect data from 1941 dog owners describing self-defined 'problem' (n=834) and 'non problem' (n=1761) dogs (in terms of HDAB). Following exploratory univariate analysis, a model was constructed using binary logistic regression to predict 'problem' dog ownership. Additionally, within-subjects analysis (McNemar's tests) was undertaken, as a proportion of participants had owned both types of dog (n=654). The model constructed using binary logistic regression correctly predicted 77.9% of cases (Omnibus $X^2(32)=773.017$, P<0.001). Between 32.9 and 44.1% of variability was explained, representing a substantial improvement upon previous attempts to predict HDAB. Nineteen factors were incorporated into the final model, following exclusion of factors that did not improve the predictive utility of the model. Two factors included in the model were not correctly predictive on their own, but added to the variability explained and were therefore retained. Some examples of predictive factors include male sex (dog), pastoral breed, experience of neglect/cruelty, owner response to unwanted behaviour, types of fearful behaviour and purpose of acquisition. Many of the predictive factors were also indicated in within-subjects McNemar's analysis. Results are comparable with a previous survey by the same authors (n=865), indicating further reliability. In summary, the results from this study represent a significant step forward in the identification of reliable risk and protective factors for HDAB in dogs, and support the development of an evidence based risk assessment for HDAB, to be used on a case-by-case basis.

Tickling pet store rats: impacts on human-animal interactions

Megan R. Lafollette[1], Sylvie Cloutier[2], Marguerite E. O'haire[1] and Brianna N. Gaskill[1]
[1]Purdue University, Comparative Pathobiology, 625 Harrison St, West Lafayette, IN 47907, USA,
[2]Canadian Council on Animal Care, Assessment and Certification Program, 190 O'Connor St,
STE 800, Ottawa, ON, K2P 2R3, Canada; lafollet@purdue.edu

Rats find initial interactions with humans frightening, leading to negative affect, poor welfare, and difficult handling. Playful handling, or 'tickling', a technique that mimics rat rough-and-tumble play, can be used to reduce fear and improve welfare. When tickled, rats produce 50-kHz ultrasonic vocalizations (USV) indicative of positive affect. Some rats consistently vocalize more (high-callers) and react differently to novelty than rats that vocalize less (low-callers). Pet store rats experience high-levels of novelty and potentially infrequent, inconsistent, and aversive human interaction reinforcing a fearful human-rat relationship. We hypothesized that tickling pet store rats would improve unfamiliar human-rat interactions, particularly for high-callers, as compared to controls or low-callers. We sampled 36 female rats across 2 replicates to accommodate caging restrictions. In each replicate, we randomly allocated 6 rats as minimally handled controls, and 12 rats as tickled. Tickled rats were split into high-callers (most vocalizations) and low-callers (least vocalizations) groups based on USVs produced during tickling for 3 days (5 min/rat/day). We applied handling treatments for 4 additional days (15 s/rat/day). Finally, we assessed all rats with an unfamiliar human approach (60 s), manual restraint (30 s) and approach test (60 s). We expected restraint to be moderately stressful. We analyzed data, presented as back-transformed, where applicable, LSM ± SE, using GLM and post-hoc Tukey tests and contrasts. High- (1.0 ± 0.28) and low-callers (1.1 ± 0.23) required fewer attempts until a successful restraint was made than controls (1.9 ± 0.23; $P<0.05$). Similarly, less time was required to restrain high-callers (7.2 ± 1.2 s) and low-callers (8.8 ± 1.1) than controls (12.5 ± 1.1; $P<0.05$). During the approach test after restraint, low-callers took longer to contact the unfamiliar hand (53 ± 1.2 s; $P<0.05$) and reared for less total time (7.2 ± 1.1 s; $P<0.05$) than high-callers (20 ± 1.2 s; 11.7 ± 1.1 s), indicating increased fear, but neither low- or high-callers were different from controls (25 ± 1.2 s; 11.2 ± 1.1 s). During both approach tests, controls (3.5 ± 0.36) and high-callers (3.5 ± 0.36) had more line crossings than low-callers (2.2 ± 0.35; $P<0.05$). Short-term tickling improves ease of restraint with an unfamiliar handler. High-calling rate is associated with reduced fear in an unfamiliar approach test, in comparison to a low-calling rate. Overall, our results suggest that applying tickling to pet store rats, especially high-callers, may improve some human-rat interactions.

Positive human-bird interactions in farmed ostriches: a way forward for improved welfare and production?

Maud Bonato[1], Anel Engelbrecht[2], Irek Malecki[1,3] and Schalk Cloete[1,2]
[1]*University of Stellenbosch, Department of Animal Sciences, Private Bag XI, 7600 Matieland, South Africa,* [2]*Directorate Animal Sciences: Elsenburg, Private Bag X1, Elsenburg 7607, South Africa,* [3]*School of Animal Biology M085, The University of Western Australia, WA 6009 Crawley, Australia; mbonato@sun.ac.za*

Ostrich farming originally commenced in the mid-19th century in South Africa but poor production results are still being observed worldwide (i.e. high embryo mortality, low chick survival and suboptimal growth rates) implying that aspects of management and husbandry may not be optimal to birds welfare. Limited research has been undertaken to determine best practices in ostrich farming to improve production and welfare. We investigated the effect of three different husbandry practices, namely standard husbandry (S; n=67); additional human presence (as compared to S) involving regular physical contact, audio and visual stimuli (I1; n=66), and additional human presence, audio and visual stimuli but no physical contact (I2; n=71). These treatments were carried out from day-old to 12 weeks old. Chick survival, live weight, immune responses, and behavioural responses towards human at 10 months old were compared between treatments. At 6 weeks old, chicks exposed to I1 and I2 were heavier than chicks exposed to S (I1: 7.47±0.18 kg; I2: 7.06±0.15 kg and S: 6.21±0.13 kg, P<0.001) but no difference was observed at 12 weeks of age. By 6 weeks old, I1 and I2 chicks survived better than S chicks (S: 79%, I1: 89%, and I2: 88%, and S: 79%, P=0.011), but no difference was observed at 12 weeks. Furthermore, following injection with Newcastle disease vaccine at 5 months old, a higher dilution rate was needed to cause complete inhibition of antigens in I2 and S chicks as compared to I1 chicks, suggesting a stronger immune response in I1 chicks. Behavioural observations conducted on the juvenile birds showed no difference in terms of willingness to approach or be touches by a human observer, but S chicks were more inclined to display aggressive behaviour (i.e. hissing at the human observer). Finally, a pilot docility test originally developed for sheep and cattle performed on a group of 30 chicks showed no difference between the three groups, in terms of how long it took the operator to get the bird into a marked area within an arena test, or the time the bird was contained in the marked area. From these results it was evident that exposure to additional human contact resulted in higher initial live weight, survival and immune responses. Further investigations are underway to better understand the full implication of extensive human care and standard husbandry practices performed at an early age on short- and long-term stress responses, as well as reproduction when the birds reach sexual maturity.

Qualitative analysis of live-feeding practices in YouTube videos reveals complex animal welfare issues

Tiarnan Garrity[1], Mark J Farnworth[2] and Jill R D Mackay[3]
[1]R(D)VS, University of Edinburgh, EH25 9RG, United Kingdom, [2]University of Plymouth, Drake Circus, PL4 8AA, Plymouth, United Kingdom, [3]SRUC, West Mains Road, EH9 3JG, United Kingdom; tiarnan.garrity@gmail.com

Live feeding is often practiced in Herpetoculture but raises serious animal welfare concerns on the part of the prey species. Despite this, it thrives in countries where it is still legal and video material is often shared by enthusiasts as examples of good practice. In this study we utilised freely available clips of live-feeding practices that were on YouTube in order to explore the reasons behind conducting and sharing live feeding practices; attitudes to the welfare of prey and predator species; and to further explore the human-animal bond. 20 YouTube videos were chosen following a strict selection criteria. Basic data from the videos were recorded and the first fifty comments (all unsolicited) on each video was categorised based on whether they felt positively or negatively about the three agents in the video; owner, snake and prey animal. 10 videos featured rodent prey and 10 videos featured lagomorph prey species. Videos with a high view count had more comments (r_s=0.764, P<0.001), more likes (r_s=0.821, P<0.001) and dislikes (r_s=0.749, P<0.001). Prey species had the greatest effect on comment content. In a two-sample t-test there was a significant difference in the proportion of comments which were classified as 'anti-human' between species, with lagomorph videos receiving more 'anti-human' comments (0.28±0.107) than rodent videos (0.10±0.108), $t_{(17)}$=-3.77, P=0.002. Cohen's d, indicating the standardised difference between means, was 1.68 with 95% CI, generally considered to be a large effect size. However there were no significant correlations between the number of views a video had and the proportion of comments classed as anti-human, pro-snake and neutral. There were some correlations between the attitudes of comments, with a strong negative correlation between comments expressing anti-human and pro-snake sentiments (r_s=-0.834, P<0.001) and a strong positive correlation between pro-human and neutral-prey sentiments (r_s=0.652, P=0.002). In summary, the popularity of a video on YouTube did not have much bearing on the attitude displayed by the commenters. Video content was the best predictor of the comment attitude, with lagomorph videos receiving more comments which treat the human in a negative manner. Curiously, the attitudes of the commenters to the animals didn't significantly change with species or video type, just the attitudes to the humans in the video, which may imply public recognition of the owner's role in protecting prey-animal welfare. Overall, qualitative analyses of spontaneous human-animal interactions on YouTube can be informative of human-animal relations.

Brazilian citizens' levels of awareness and support for contentious dairy production husbandry practices

Maria J. Hötzel[1], Angélica Roslindo[1], Clarissa S. Cardoso[1] and Marina A.G. Von Keyserlingk[2]
[1]*Laboratório de Etologia Aplicada e Bem-Estar Animal – LETA, Universidade Federal de Santa Catarina, Rod. Admar Gonzaga 1346, 88034-001, Florianópolis, SC, Brazil,* [2]*Animal Welfare Program, The University of British Columbia, 2357 Main Mall, Vancouver, BC, BC V6T 1Z4, Vancouver, Canada; maria.j.hotzel@ufsc.br*

We surveyed a convenience sample of 400 Brazilian citizens to explore their awareness and opinions about two dairy practices: zero-grazing and cow-calf separation; a secondary aim was to assess the influence of provision of information on respondents' opinions regarding these practices. Men (n=192) and women (n=208) were recruited in public places (e.g. airports and bus stations) with the majority living in urban areas (97%) and had no association with animal production (92%). Participants were presented a balanced science based view of the main positive and negative welfare and production outcomes of zero grazing (n=200) or cow calf separation (n=200) and asked for their position (Reject/Indifferent/Support), and to provide the reason(s) justifying their position. Participants were then presented a short statement describing either zero-grazing or cow calf separation, depending on what question they responded to in the first part. Two closed questions (Q) followed each of these statements: Q1 'Are you aware of this practice?' and Q2 'What is your position regarding this practice?', with choices Yes/Somewhat/No and Reject/Indifferent/Support, respectively. Most people were unaware and rejected zero-grazing (69 and 76%) ad separation (67 and 62%). We noted no association between awareness and the degree of support for the practice (Pearson's X^2=2.8, d.f.=2; P=0.25). Provision of information resulted in more people rejecting the practice: zero-grazing 86% (from 76%; Pearson's X^2=6.6, d.f=2; P=0.04) and separation 69% (from 62%; Pearson's X^2=15.6, d.f.=2; P=0.001). Reasons for rejecting zero-grazing fell into 4 themes (in descending order): naturalness, ethical imperatives (e.g. sustainability), animal welfare, and product quality. Reasons for rejecting Separation also fell into 4 themes: animal welfare, naturalness, the need for contact between the pair, and ethical imperatives. The inability for animals to express natural behaviours was an important concern for participants in this survey (cited by 27% of respondents). This survey indicates that Brazilians living in urban environments, with no association with dairy production, are generally unaware that many cows do not have access to pasture and that cows are separated from their calf at birth – two practices that are common on Brazilian dairy farms. However, when asked – with and without provision of additional information – participants did not support these practices.

Contexts and consequences of dog bites: clarifying myth from evidence

James Oxley[1], Rob Christley[2] and Carri Westgarth[2]
[1]*Independent Researcher, 35 Farnes Drive, Gidea Park, Romford, Essex, United Kingdom,*
[2]*Institute of Infection and Global Health and School of Veterinary Science, University of Liverpool,
Liverpool, United Kingdom; james_oxley1@hotmail.com*

Dog bites are the companion animal behaviour of most considerable public health concern. The majority of research gathers information about reported dog bites through hospital accident and emergency departments or newspapers. However, this data is limited to severe bites and little is recorded about the actual context of bite events and the behaviour of the dog pre, during and post incident. Therefore, the aim of the study was to gather preliminary information on the variety of contexts of dog bite incidents and consequences for both victim and the dog. A convenience-based online questionnaire, distributed through social media, targeted individuals who self-identified as been previously bitten by a dog, were over the 18 and lived in the UK. A total of 484 dog bite victims completed the online survey about their most recent bite incident (ranging from 1957-2016). 72.5% of respondents were female and were aged between 1-77 years when bitten, with 52.6% being under 30 years of age. Nearly all victims described an incident where they were bitten by one dog (96.7%) and only one bite occurred (86.0%). In the majority of incidents (59.5%) the victim was with another person at the time of the bite. Immediately prior to the incident, it was slightly more common for the victim to approach the dog (55.3%) than the dog to approach the victim (44.7%).Whether the dog approached the victim was associated with a history of aggression (P<0.001); 58.9% of dogs with a history of aggression approached the victim compared to 25.7% of those without a history of aggression. Prior to the incident the behaviour of the dog most frequently noted was described as either being 'excited/active' (23.6%) or 'aggressive' (17.2%). The body regions most bitten were the hands and/or lower arms (54.4%). Over half (52.3%) of victims did not require medical treatment and even if treatment was sought only 10.2% of these were admitted to hospital. The dogs involved were mainly noted as adult known breeds and 69.8% were either medium or large in size. In the majority of cases the dog involved was reportedly male (52.4 vs 24.4% female) and known to the victim (66.1%). In 53.3% of incidents nothing happened to the dog involved and only 8.7% of dogs received training/behavioural treatment as a consequence. This study highlights that the majority of dog bites are not captured by common research data methods. It also suggests that the contexts of dog bite incidents are varied and often do not involve the victim initiating contact with the dog.

Using REF 2014 as a tool to assess the impact of applied animal behaviour science on animal welfare

A Lawrence

SRUC, West Mains Road, EH9 3JG, United Kingdom; alistair.lawrence@sruc.ac.uk

Given the close relationship between applied animal behaviour science (AABS) and animal welfare (AW) it is not surprising that the field has had an interest in the impact of our science in improving animal welfare with a number of authors debating the extent to which AABS has made a difference to AW. The topic of research impact is also currently of interest to governments; for example, funding to universities in the UK is partly distributed on the basis of an assessment of research quality. In the current round of research assessment (REF 2014) an additional requirement was made to provide 'impact case studies' (ICSs) that demonstrate how a piece of science has achieved impact. The format was for ICSs to demonstrate how a peer reviewed piece of work led to a specific and measurable impact. These ICSs are publicly available on the REF web-site, and they provide an opportunity to explore the impact of our science in improving AW. Specific scores for ICSs are not available but it can be assumed that those that are listed have been carefully vetted by the submitting organization given the competitive nature of the process. This poster will provide a summary of a survey of the REF database when searched for ICSs that fall within the scope of AABS and AW. In brief: (a) There were 27 (out of over 6,000) ICSs that related to AABS and AW; this small number most likely reflects the modest size of the AABS/AW research community in the UK; (b) more than 50% of the studies came from farm animals (where historically funding has been allocated); there is also work on laboratory, companion and captive animals; (c) 8 (>30%) were classified as being on pain related issues, 5 (~20%) on aspects of housing, and smaller numbers on handling (3), transport and slaughter (3), human-animal interactions (2), well-being (1) and applications of behaviour recording (including automation of behaviour recording (4); (d) There was a broad and impressive range of impacts claimed including: influencing UK, European Union (EU), and other (e.g. commercial) policy making, including forming legislation; changing husbandry practice (e.g. handling of rodents, management of lameness in cows); changes to housing (e.g. aerial perches for hens, loose farrowing for sows); policy and practice relating to transport and slaughter; use of behavioural recording (e.g. oestrus detection in cows) and public engagement. In summary the REF 2014 provides a useful tool to understand better both the impact of our science and the process of achieving that impact (e.g. the need to collect data on impact during the research process). ISAE could build on this by developing international impact case studies that describe to relevant stakeholders (including the public) how our science has impacted on specific animal welfare issues.

Attitudes to free-roaming cats and to trap-neuter-return procedure in Sweden

Maria Andersson, Elin Hirsch, Moa Nygren, Margaux Perchet, Jenny Loberg and Christina Lindqvist
Swedish University of Agricultural Sciences, Department of Animal Environment and Health, Box 234, 53223 Skara, Sweden; maria.andersson@slu.se

This study aimed to investigate the public's perception and knowledge about free-roaming cats (FRC) in Sweden. An anonymous internet survey was posted in the university network and on cat-forums. The survey was open during 5 consequtive days in February 2015. 3,627 persons responded to the questionaire, out of which 78% were cat-owners, and 88% women. 82% of respondents think that Sweden has a problem with free-roaming cats and generally rate that homeless cats have a bad welfare. Cat owners were to a larger extent aware of the legislation concerning husbandry (84%) than non-owners (15%) (χ^2=312.25, P<0.05). Moreover, a majority of respondents (69%) had no knowledge about trap-neuter-return (TNR) procedures. 95% of people who were aware of the presence of FRC in their area also thought that FRC are a problem for cats (χ^2=56.29, P<0.05). 75% responded that owners are responsible for the prevention of homeless cats and that a transfer into a new home would be the ideal management choice (79%). Regarding alternatives for prevention of FRC problems, respondents suggested that establishment of a legal requirement of chip marking (29%), followed by neutering requirment of out-door cats (25%), stricter legislation which would directly forbid animals' abandonment (17%) and more severe punishment for people abandoning their animas (16%) were the four most proposed options. We would like to conclude that people are willing to solve the issue of free-roaming cats, and have suggestions to proceed. This study shows the need to further study the population of FRC and the effeciency of management alternatives in Sweden.

Relationships between temperament in dairy cows, somatic cell counts and farmers's attitude in two types of mechanic milking

Freddy Enrique García, Diana Katherine Flórez, Aldemar Zuñiga and Jaime Andrés Cubides
Colombian corporation of agricultural research, Health and animal welfare, Km 14 vía Mosquera,
Colombia; fgarciac@corpoica.org.co

The temperaments of cows (TC) during milking time have direct effects on mastitis rates and performance of production. The aim of this research was to evaluate the TC during milking time at two types of mechanic milking and its relationship with farmers's attitude and somatic cell counts. This study was conducted during April and November 2014 in 10 farms (5 fixed and 5 mobile milking machine) and the most of breed of cows was Holstein Friesian. All farms had similar production systems and milking procedures. The TC was evaluated with the methodology used by together with the perception of each farm milker. Temperament indicator data were used to classify individual animals as Calm, Intermediate or Excitable. The variable somatic cells counts (SCCs) were evaluated with the fossomatic cell counter and the results were screened for normality using Kolmogorov-Smirnov test and it was necessary were transformed by the natural logarithm. Dependence between variables was evaluated with X^2. Finally with all factors (TC, attitude, type of miliking and subclinical mastitis) were included in factorial analysis of variance. The results showed relationships of dependence between the presence of subclinical mastitis and type of milking and the attitude of farmer during milking. The temperament of cows on SCCs had significant statistical differences ($P<0.05$), likewise its interactions with farmers's attitude. The excitable temperament cows showed higher SCCs (Mean 773,555.6; SD ±106,068.98) than calm cows (552,405.84±94,696.3). In the presence of aversive milker the excitable temperament cows had higher SCCs (332,745±12,765) than intermediate cows (189,305±15,184.12). Others studies also associated that aversive management alters the behaviour of dairy cows in the milking parlour and its productivity. In conclusion, this study confirm the importance of evaluation TC and farmer's attitude in the management of mastitis control linked to animal welfare. This first approach in Colombia indicate the importance of develop more studies related with applied ethology.

Farmers' opinions about sheep management and welfare

Carolina A. Munoz[1], Grahame J. Coleman[1], Angus J. D. Campbell[2], Paul H. Hemsworth[1] and Rebecca E. Doyle[1]
[1]*The University of Melbourne, Animal Welfare Science Centre, Alice Hoy Building, 3010, Australia,* [2]*The University of Melbourne, Faculty of Veterinary and Agricultural Sciences, Werribee, 3030, Australia; rebecca.doyle@unimelb.edu.au*

Two focus groups were conducted to collect qualitative information on the opinions and concerns of extensive Merino sheep farmers. These discussions were conducted with the purpose of identifying the underlying beliefs that influence farmers' behaviour, and the consequences for the productivity and welfare of their sheep. Both discussions were approximately 60 min, followed semi-structured agenda and comprised of eight and seven farmers respectively (35 to 65 years). Farmers reported that monitoring of the flock was usually performed visually, without approaching the mob; all agreed that they visually assessed behaviour and body condition as indicators of individual sheep welfare, but did not assess either in close proximity. Understanding whether these distant observations are indicative of the welfare of sheep is important to determine. Farmers only visually assessed pasture quality and availability, and perceived barriers to more frequent/detailed management included time and labour availabilities. There was strong acknowledgement that behaviour and attitudes towards moving sheep dictated a farmer's success, with direct quotes including 'if you say sheep are stupid and you are getting angry at them, it is probably because you are doing something wrong'. As farmer attitudes influence the human-animal relationship, farmer attitudes will likely affect sheep behaviour and fear. It was consistently acknowledged that the treatment of sheep by shearing and mulesing contractors was something farmers had little control over. Farmers also acknowledged that some husbandry procedures may be painful, and believed that the use of pain relief (Trisolfen) makes a 'huge difference' in how the sheep recover. Diseases like foot abscess and eye injury were identified as likely to cause pain, but veterinary intervention was rare and reserved only for particular diseases or to assess rams before mating. Understanding which sheep behaviours farmers use as indicators of pain may help to understand these opinions. Farmers participated in workshops and training courses, indicating a desire for further education. In summary, farmer attitudes to sheep were positive and welfare was expressed as an important consideration; however, the nature of extensive farming conditions prevent close, frequent interactions with sheep and a perceived lack of control in some aspects sheep management are likely barriers to welfare monitoring/improvement. This qualitative study informs a current longitudinal study measuring the welfare and behaviour of sheep in relation to farmer attitudes.

The welfare of extensively managed sheep: a survey identifying issues and indicators

Amanda K Doughty[1], Grahame J Coleman[2], Geoff N Hinch[1] and Rebecca E Doyle[2]
[1]University of New England, Armidale, 2351 NSW, Australia, [2]The University of Melbourne, Animal Welfare Science Centre, Parkville, 3010 VIC, Australia; amanda.doughty@une.edu.au

A survey was designed as a first step in developing a framework for the welfare assessment and monitoring of extensively managed sheep in Australia with the aim of identifying potential causes of welfare compromise, useful welfare indicators and current monitoring practices. The survey was developed and conducted online using Survey Monkey. It was distributed to contacts in industry, research institutions/universities, government agricultural organisations, the RSPCA and social media outlets. Participants were asked between 14 and 26 questions that varied from scaled (1-5) to open ended. A range of 20 welfare issues were provided for consideration including thermal comfort and painful procedures. A total of 952 people completed the survey in its entirety, representing four stakeholder groups: Public (53.6%), Producer (27.4%), Scientist (9.9%), and Service provider (9.1%). Descriptive statistics, principle components analysis, restricted maximum likelihood analysis and Pearson's correlations were conducted. Sheep welfare was considered to be important by all participating groups in this survey (average score of 3.78/4), and respondents felt the welfare of grazing sheep was generally adequate but improvement was desired (2.98/5). Flystrike (4.25/5), nutrition (4.07) and predation (3.95) were considered to pose the greatest risk to sheep welfare, whereas yarding (3.01) and the use of sheep dogs (2.87) were perceived to have the lowest risk. Stakeholder category significantly influenced respondents' perceptions of grazing sheep welfare (P<0.001) and their self-reported issues, with the public believing off-farm issues posed the most significant risk (n=189, 19.9%). Women were more concerned about welfare than men and consistently rated welfare issues to be of greater significance (all P<0.05). Key indicators recognised by all respondents were those associated with pain and fear (3.98/5), nutrition (4.23), food on offer (4.40), mortality/management (4.27) and number of illness/injures in a flock (4.33). Again, there were stakeholder and gender differences with women and the public rating most indicators of greater importance than other respondents (all P<0.05 and P<0.01, respectively). Within the Producer group the majority (>90%) indicated that they monitored sheep for all listed issues. Overall these results suggest that there is general agreement on the ranking of importance of key welfare issues and indicators of grazing sheep across a wide range of stakeholders. Therefore, broad agreement may be possible on factors suitable for inclusion in a welfare assessment scheme and monitoring framework developed in the future.

Perceptions of the causes of poor welfare in a pasture based dairy industry

J. Marchewka[1,2], J. F. Mee[2] and L. Boyle[2]
[1]Institute of Genetics and Animal Breeding, PAS, Department of Animal Behaviour, ul. Postepu 36A, Jastrzębiec, 05-552 Magdalenka, Poland, [2]Teagasc, Animal and Grassland Research and Innovation Centre, Moorepark, Fermoy, Co. Cork, Ireland; joanna.marchewka@teagasc.ie

Abolition of the EU milk quota and subsequent expansion in the EU dairy industry may pose threats to cow welfare particularly in low cost pasture-based systems. ProWelCow (DAFM RSF – A 14/S/890) is a desk based project which aims to investigate risks to dairy cow welfare and to develop strategies to protect it. As a first step this study aimed to investigate the perceptions of key stakeholder groups on the main causes of poor welfare in Irish pasure-based dairy cows. The survey was conducted with randomly selected dairy farmers (F; n=115) at two national farming events (National Ploughing Championships and Moorepark Open Day 2015) and cattle veterinarians (V; n=60) at the Cattle Association of Veterinary Ireland conference by interview. Teagasc dairy advisors were asked to complete the survey themselves (A; n=48) at an 'in-service' training day. The 223 respondents were asked to identify the main causes of poor welfare in cows from the following list: lameness (L), social stress due to overcrowding (SS), mastitis, metabolic disorders, infectious diseases, poor body condition score (pBCS), cold stress and calving difficulties. A Chi-Square Fisher test (PROC FREQ, SAS), was used to investigate whether distributions of response frequencies differed for the 3 main causes of poor welfare between all respondents and between stakholders groups. Three main causes of poor welfare of cows in pasture-based dairy systems differed in importance (SS: 25.9%, pBCS: 16.7%, L: 15.3%; X^2=7.5; (df=2); P=0.02). SS, as a primary cause of poor welfare was equally improtant for F, V and A (F: 7.9%, V: 8.3%, A: 9.7%, X^2=0.5; (df=2), P=0.8). pBCS was rated as a primarey cause by the majority of F (F:12%, V: 2.3%, A: 2.3%; X^2=24.5; (df=2); P<0.0001), while L was rated as a primary cause by the majority of V (F: 6.9%, V: 7.9%, A: 0.5%; X^2=13.8; (df=2); P=0.001). Stakeholders agreed about the importance of SS as the main cause of poor welfare, which is perhaps surprising in dairy cows in pasture-based systems. There was a lack of consensus regarding pBCS nd L. This probably reflects the differing focus and areas of expertise beween the three stakeholder groups. However, all listed issues are important causes of poor cow welfare in expanding, low-cost, pasture-based systems. Given these results, further research is warranted focused on these identified primary causes of poor welfare in Irish pasure-bas dairy cows.

Visitor and keeper beliefs about the effects of visitors on zoo animals

Samantha Chiew[1], Sally Sherwen[2], Paul Hemsworth[1] and Grahame Coleman[1]
[1]The University of Melbourne, Animal Welfare Science Centre, Monash Road, 3010, Parkville, VIC, Australia, [2]Zoos Victoira, Department of Wildlife Conservation and Science, Elliott Ave, 3052, Parkville, VIC, Australia; schiew@student.unimelb.edu.au

Research has shown that zoo visitors can influence the behaviour and welfare of a range of zoo animals. However, the literature indicates that most research has been conducted on mammals, mainly non-human primates which has limited our understanding of visitor effects across taxa. The aim of this study was to gather information from zoo visitors and keepers on their beliefs about the effects of visitors on zoo animal behaviour and welfare, to identify species that may be susceptible to visitor effects. Two 90-min focus groups with visitors to Melbourne Zoo (six adults in each group) and one-on-one interviews with keepers (45 keepers across the Zoos Victoria properties) were conducted with the aim of encouraging all participants to express their views on visitor effects. The study was approved by the university's Veterinary and Agricultural Sciences Human Ethics Advisory Group. Data were analysed as qualitative data. Results revealed a wide range of opinions about visitor effects on zoo animals. Focus groups participants and most zoo keepers (96%) believed visitors can affect zoo animal behaviour and welfare, ranging from positive, negative and neutral effects based on animal behaviours such as avoidance, approach and unresponsiveness respectively. However, visitor proximity and behaviour were reported to affect the behavioural response in the animals. A total of 14 species were identified by focus groups to be affected by visitors (71% mammals, 14% aves, 7% reptiles and 7% invertebrates). Of these species, 36% mammals and 21% aves, reptiles and invertebrates were perceived to be susceptible to visitors based on animal behaviour while 43% mammals and 7% aves were perceived to be positively affected by visitors. Species perceived to be unaffected by visitors were all mammals including meerkats, kangaroos, giraffes and lemurs. Keepers identified a much larger range of species to be affected by visitors, with a total of 39 species identified (67% mammals, 20% aves, 10% reptiles and 3% invertebrates), and as expected keepers were able to elaborate in greater detail about visitor effects on each species compared to focus groups. 64% of these species were perceived to be both positively and negatively affected by visitors (72% were mammals, 21% aves and 8% reptiles). These findings suggest a greater concern or interest in mammals. Further research on visitor effects on mammals is clearly required but this research should examine both positive and negative effects. These results also provide hypotheses for future research on the manner in which visitors may affect zoo animal behaviour and welfare.

Dignity of the animal: a new concept in animal welfare legislation

Beat Wechsler, Katharina Friedli and Heinrich Binder
Federal Food Safety and Veterinary Office, Centre for proper housing of ruminants and pigs,
Tänikon 1, 8356 Ettenhausen, Switzerland; beat.wechsler@agroscope.admin.ch

In 2008, a revised version of the Swiss Animal Welfare Act entered into force. Unlike its predecessor, it protects not only the welfare but also the dignity of the animal. The amendment was necessary, as the term 'dignity of living beings' had been added to the Swiss Federal Constitution in 1992, mandating the Confederation to legislate on this issue. Consequently, the dignity of the animal had to be taken into account in the Swiss animal welfare legislation. In Article 3 of the Animal Welfare Act the term 'dignity of the animal' is defined as follows: 'Inherent worth of the animal that has to be respected when dealing with it. If any strain imposed on the animal cannot be justified by overriding interests, this constitutes a disregard for the animal's dignity. Strain is deemed to be present in particular if pain, suffering or harm is inflicted on the animal, if it is exposed to anxiety or humiliation, if there is major interference with its appearance or its abilities or if it is excessively instrumentalised.' In contrast to the inviolability of human dignity, this definition does not guarantee that the dignity of the animal takes priority in all circumstances. To make an assessment in the individual case, the interests of the animal and the severity of the strain have to be weighed against possibly overriding interests. The weighing of interests is not an empirical method but a normative procedure. To address the need for integration of social science and natural science in debates on the dignity of the animal, the Federal Food Safety and Veterinary Office has set up a study group composed of ethicists, representatives of the penal system, and experts in animal welfare science. Over the last years, this study group not only gave recommendations on how to apply the concept of the dignity of the animal in individual cases but also elaborated a model procedure to ensure that weighing of interests is carried out appropriately and uniformly. In my presentation, I will explain the background of the concept of the dignity of the animal as defined to the Swiss animal welfare legislation. Moreover, I will use the example of sows kept in farrowing crates and thus prevented from performing normal nest-building behaviour to explain the cornerstones of the model procedure for the weighing of interests elaborated by the study group.

Different consumer segments' interest in information about dairy cattle welfare measures

Sophie De Graaf[1], Filiep Vanhonacker[2], Jo Bijttebier[1], Ludwig Lauwers[1], Frank Tuyttens[1] and Wim Verbeke[2]
[1]Institute for Agricultural and Fisheries Research (ILVO), Burg. v. Gansberghelaan 92, 9820 Merelbeke, Belgium, [2]Ghent University, Coupure links 653, 9000 Ghent, Belgium; sophie.degraaf@ilvo.vlaanderen.be

Concerns about farm animal welfare are growing, but interest in animal welfare and related information, is not necessarily homogenous across consumers. Different consumer segments may exist, requiring different communication strategies for animal-friendly products. To unravel opportunities for marketing animal-friendly milk, a survey among 786 consumers was conducted in Flanders, Belgium. Segments were identified based on respondents' intention to purchase animal-friendly milk (IP) and on their evaluation of the current dairy cattle welfare in Flanders (EV). Next, between- and within-segment differences for interest in information about welfare (measures and criteria of the Welfare Quality® protocol (WQ) for dairy cattle) were identified by asking to what degree respondents wished to be informed when buying milk. Six segments were identified, accounting for 24% (S1), 8% (S2), 24% (S3), 4% (S4), 16% (S5), 24% (S6) of the sample. The smallest segments (S2 and S4) were considered as 'extreme'. S2 indicated very high IP and negative EV whereas S4 indicated very low IP and positive EV. S1 was more similar to S2, with high IP, and positive EV. S6 leaned towards S4, with low IP and neutral EV. S5 and S3 both indicated neutral IP, where S5 reported positive, and S3 rather negative EV. Interest in information differed (ANOVA; $P<0.001$): S4 (means ranging from 27-36), S5 (27-59) and S6 (33-45), indicated least interest, followed by S1 (44-78) and S3 (44-73) and S2 (66-94) (post-hoc; all $P<0.05$). Regarding interest in information for the WQ measures, most segments gave the lowest score to 'Number of difficult calvings' (mean±SD: 40±30) and highest to 'Showing negative behaviour' (61±34), 'Access to pasture'(60±32), and 'Showing positive behaviour' (60±33). Most segments scored the WQ criteria 'Expression of social behaviour' (55±33) and 'Expression of other normal behaviour' (55±33) lowest, and 'Freedom of movement' (65±32) and 'Absence of diseases' (63±34) highest (post-hoc; all $P<0.05$). We conclude that consumer interest in information increased with an increasing IP, with S1 and S2 as potential loyal consumers of animal-friendly products, and S4 as least promising. Generally, the measures 'Access to pasture' and 'Showing positive/negative behaviour' and the criteria 'Freedom of movement' and 'Absence of diseases' were promising selling propositions in future communication about animal-friendly milk and are therefore important to measure on-farm in dairy cattle welfare assurance schemes.

Swedish trotters – problems for horses and improvements to the sport according to trainers

Anna Lundberg[1], Lisa Lönnman[1] and Agneta Sandberg[2]
[1]Swedish University of Agricultural Sciences, Animal Environment and Health, Box 234, 532 23 Skara, Sweden, [2]Swedish Trotting Association, Svensk Travsport, 161 89 Stockholm, Sweden; anna.lundberg@slu.se

There are about 100,000 trotters in Sweden, of these 18% are in training. Until now, the reasons for Swedish racehorses leaving the industry, what happen to them after their career and why many of them never compete, has not been fully known. Injuries and poor performance are found reasons but lack of horse motivation and behavioural problems are other explanations. The purpose of this study was to get information to the low number of active horses and to obtain suggestions for improvements of the sport. The study was conducted through a questionnaire aimed at racehorse trainers. Answers were given by 524 trainers responsible for in total 1246 horses that have ended their career and 456 horses that never came to start in a race. Data were compiled and are described as percentage or as most common answer given. Most horses were seven years old when they finished their career. The most common reason why the racing career ended was injuries sustained during training or competition (26.5%) followed by a lack of motivation of the horse (13.3%). Poor performance was the most common reason why the horse never competed (25.7%). The next most common reasons were injuries sustained outside of training (18.6%) and unsuitable temperament or handling problems (14.3%). An alternative for horses when ending their career was becoming a riding horse (25.4% of the horses) but the most common outcome was slaughter or killing (29.2%). For horses that never reached competition slaughter or killing was the most common end (40.9%). A majority of trainers (76.4%) thought there are improvements to be made to the sport to increase animal welfare. Actions includes increased second-hand value of trotters, creation of trotter retraining courses, increased status of the equine sport, increased animal welfare work within the sport, increased slaughter value of the horses, an increased number of educational races for young horses, increased age limit and increased number of possible starts for older horses were all ranked highly. The trainers commented that there is a 'wear and tear mentality' and that they have welfare concerns regarding the horses after their careers due to the low status and value of the trotters. The study concludes that the trainers find several risks for reduced welfare due to the low values of trotter horses when they no longer can fulfill their purpose as racehorses. It is shown that factors such as injuries, low motivation, unsuitable temperament and handling problems influence whether and when the horses leave the sport. These factors are important to eliminate to decrease welfare problems and wastage of trotter horses.

An assessment of handling techniques used with dogs and cats during veterinary appointments

Lauren Dawson, Cate Dewey, Elizabeth Stone, Michele Guerin and Lee Niel
Ontario Veterinary College, University of Guelph, 50 Stone Rd. E., Guelph, ON, N1G 2W1, Canada; dawsonl@uoguelph.ca

Dogs and cats often exhibit signs of stress and fear during veterinary visits. As such, veterinary behaviourists recommend the use of low-stress handling techniques to mitigate fear and improve patient welfare. Using a welfare assessment scheme developed for use in veterinary clinics, our objectives were to assess the feasibility and reliability of the assessment measures, and to collect baseline data on which handling techniques are used by veterinarians and technicians during dog and cat appointments. Testing was completed in 30 companion and mixed animal veterinary clinics in Ontario, Canada. In each clinic, two video cameras were unobtrusively installed in each examination room to capture veterinary appointments. Up to six randomly selected appointments per clinic (n=152: 90 dogs, 62 cats) were analyzed for the use of recommended and discouraged handling techniques. During a single day visit, it was only possible to record sufficient appointments to meet our original criteria (3 canine, 3 feline) in 10 veterinary clinics. Based on weighted kappa statistics for tool performance, handling scoring showed substantial inter-observer agreement (K_w=0.74 and 0.70; 1 experienced vs 2 naïve, trained observers) and almost perfect intra-observer agreement (n=52 videos, K_w=0.84; 1 experienced observer only). Fear-reducing techniques, such as providing rewards and positive reinforcement (53% of dogs, 10% of cats), increasing surface traction (22% dogs, 34% cats), and using pheromones (0% dogs, 5% cats), were infrequently employed in veterinary appointments. Techniques for handling aroused patients, such as using towel or blanket wraps (1% dogs, 18% cats), were also infrequently observed; however, it is possible that animals were calm and these methods were unnecessary. Despite recommendations to allow cats to independently exit their carriers, cats were tipped, lifted, or pulled out of their carriers in 74% of cat appointments. Methods that are discouraged and potentially aversive, such as the use of heavy manual restraint (4% dogs, 11% cats) and scruffing of cats (7%) were also rarely observed. Overall, tool feasibility was poor because few appointments were recorded in some clinics; however, inter- and intra-observer reliability were reasonable. Although aversive techniques were rarely observed, limited use of recommended handling techniques suggests there is room for improvement with regard to reducing animal stress. Further research is necessary to assess whether feedback on performance leads to improvement during future assessments.

Pig farmers attitudes towards sick/injured pigs in relation to care and rehabilitation, the use of medication and euthanasia

Alessia Diana[1,2], Sylvia Snijders[3], Nola Leonard[1] and Laura Boyle[2]
[1]*UCD, School of Veterinary Medicine, Dublin, Ireland,* [2]*Teagasc, Moorepark, Fermoy, Co. Cork, Ireland,* [3]*University of Westminster, WBS, London, United Kingdom; alessia.diana@teagasc.ie*

Antibiotic (AB) resistance is a major concern worldwide. Knowledge about potential drivers of AB use and prescribing practices in the pig industry is required to address problems of mis and over use. We hypothesised that pig farmers attitudes towards the treatment and care of injured or sick pigs may influence their use of AB. Semi-structured face-to-face interviews were conducted with 30 producers who managed integrated pig units in Ireland and which supplied records to the Teagasc PigSys database. Ethical approval was obtained and pig producers were invited for interview by their Teagasc pig advisor. Consent was obtained at the time of the interview. Interviews with participants were recorded and transcribed using random numbers from 1 to 30, one per participant to ensure anonymity. Themes were then identified from the data. The importance of comfort and care in the rehabilitation of animals following injury was only expressed in relation to sows. Here the importance of straw bedding and a solid floor in ensuring recuperation from lameness/injury was mentioned by several producers (n=7). The relevance of analgesics to recovery from illness or injury was only mentioned by 2 producers who used pain relief for sows that were 'sick after farrowing'. Injectable ABs were the most common method of treating illness or injury (n=12), followed by euthanasia (n=7). The practice of euthanising pigs was driven by economic (n=6) or humane grounds (n=9). The latter also expressed the belief that it was more humane to euthanise than to hospitalise sick or injured pigs. However, there was a general dislike for the practice of euthanasia. Many producers mentioned that it is too expensive to call a vet to treat or to euthanise sick or injured pigs. Only one producer used a vet to euthanise pigs. Several producers did not agree with the use of hospital pens (n=7) either because they believed that pigs are usually abandoned/forgotten in such pens (n=2) or because hospital pens are places where diseases are harbored and re-circulated through the herd. Almost half of the producers expressed the view that medications (in-feed and/or injections) were the only solution to deal with pig health problems and to ensure good welfare. A 3rd of producers believed that the only solution to serious diseases challenges was to de-stock the herd. Low profit margins and lack of awareness of the role which husbandry and the environment have to play in the prevention of diseases as well as the role which comfort and care of pigs play in the recovery process are major barriers to reducing the reliance on medication in the pig industry.

Adolescents' perceptions of animals: a study of spontaneous categorisation of animal species

M Connor[1], J Williams[2], C Currie[3] and A Lawrence[1]
[1]SRUC, Animal and Veterinary Science Group, Roslin Institute Building, EH25 9RG, United Kingdom, [2]University of Edinburgh, Clinical and Health Psychology, Teviot Place, EH8 9AG, United Kingdom, [3]University of St Andrews, School of Medicine, St Andrews, LY16 9TF, United Kingdom; melanie.connor@sruc.ac.uk

Spontaneous perceptions and implicit associations (also often referred to as affect) are part of attitude formation. Most studies focus on explicit attitude evaluations, which may be biased due to people constructing preferences while responding to explicit questions. It is known that early experiences of animals can positively influence later attitudes to animals and their welfare, however, there have been no experimental studies of young peoples' spontaneous perceptions of animals. The present study investigates adolescents' spontaneous categorisation of 34 species including, farm, wild and pet animals by means of a card-sorting technique. During a face-to-face interview participants were firstly asked to sort the cards without providing any sorting criteria. Afterwards we assessed explicit evaluations of 16 selected animals in terms of their utility, likeability and dis-likeability. 105 British adolescents, 54% female and 46% male, mean age 14.5 (SD=1.6) participated in the study. Card-sorting data were analysed using multidimensional scaling (MDS) techniques of proximity data (to obtain a visual representation of adolescent's categorisation) and also hierarchical cluster analysis. The explicit evaluations were analysed using property fitting by means of regression analysis. Results indicate that the categorisation data can be best presented in a 3-dimensional space, indicated by a stress-I value of 0.1. Property fittings show that adolescents used likeability and perceived utility of animals as criteria to categorise animal species with 40.6% of variance of MDS Dimension 1 being explained by likeability and 80.9% of the variance of MDS Dimension 2 explained by perceived utility. MDS Dimension 3 could not be explained by property fitting, but results of hierarchical cluster analysis found two main clusters: birds; and all other species. The latter further divides into sub-clusters, which could be labelled as wild, pet, and farm animals suggesting that adolescents may also use a classification strategy to group the animals. These findings suggest that adolescents already perceive animals differently according to specific properties including the animals' perceived utility, likeability and its 'family'. In addition the data provide support for Serpell's suggestion that affect and utility are important dimensions underlying attitudes to animals.

Animal welfare and its relationship with worker welfare in dairy farms

Guilherme Amorim Franchi[1], Fernanda Victor Rodrigues Vieira Vicentini[2] and Iran José Oliveira Da Silva[2]
[1]Institute of Agricultural Sciences (490h), Universität Hohenheim, Department of Animal Husbandry and Breeding in the Tropics and Subtropics, Garbenstrasse 17, 70593 Stuttgart, Germany, [2]Escola Superior de Agricultura 'Luiz de Queiroz'/Universidade de São Paulo, Núcleo de Pesquisa em Ambiência, Avenida Pádua Dias 11, 13418-900 Piracicaba, Brazil; amorimfranchi@gmail.com

Despite prominent developments in dairy cattle systems in Brazil, it is observed that some aspects related to animal welfare (AW) and the human-animal relationship did not achieve the same breakthrough. Thus, actions regarding generation of positive changes in human-animal relationship, which are closely related to the AW and productivity in dairy cows, are required. Therefore, this research aimed to: examine the interrelationship AW and worker welfare (WW) and evaluate the influence of technical training on AW and WW levels in dairy farms. The research was divided into 4 stages, performed by the authors: selection of 6 grazing dairy farms in Sao Paulo/Brazil; first AW and WW assessment using adapted Welfare Quality protocol and questionnaire containing closed questions concerning positive and negative affects possibly expressed by workers in their daily routines, respectively; organization of technical cattle handling and welfare training with stockpersons and farmers from selected farms and, finally, a second AW and WW assessment. Farm workers answered each questionnaire's question (e.g. 'Does my work make me feel happy?') using a five-point scale, from 'Never' (1 point) to 'Always' (5 points). Each farm's WW score was represented by the arithmetic mean of respective workers' scores. Besides, authors also recorded workers' behaviour towards cows during milking using Hemsworth's behaviour assessment protocol on second and fourth stages. After the fourth stage, it was noticed that 5 farms enhanced their animal welfare statuses from 'Acceptable' to 'Good'. All farms presented an increase in the mean positive affects score from 59.7±9.2 to 64.6±4.3 points. Whereas, there was a decrease in the mean negative affects score from 34.2±9.4 to 26.8±4.7 points. Besides, negative human-animal interactions (e.g. pulling, hitting with hand and/or object) during milking reduced around 60% from first to second assessment. The technical training played an important role on these outcomes, once it mainly focused on the flaws of the workers and gave opportunity to the participants to share their perceptions and questions. Finally, it is suggested that satisfied and skilled stockpersons feel more pleasure and are more willing to carry out their daily tasks within a livestock farm, which may positively influence conditions offered to the animals, such as good handling, nutrition and health.

Milk yield effects from stockperson behavior toward cows in holding area

Haruka Saito, Tetsuya Seo and Fumiro Kashiwamura
Obihiro University of Agriculture and Veterinary Medicine, Obihiro Hokkaido, 0800834, Japan;
sukisukiushian@gmail.com

Cows and stockpersons per farm are increasingly numerous in Japan. This study was conducted to examine the influences of stockperson behavior toward cows at holding areas on milk yields. For this study conducted at the experimental university dairy farm, cows were milked twice daily at a milking parlor by stockpersons. Each milking was conducted by three stockpersons. The behaviors of 15 stockpersons toward cows at the holding area were observed when moving cows to the parlor. Observations were conducted 34 times during morning milking for 2 months. The herd consisted of 67 Holstein cows (26 first-calving, 21 second-calving, and 20 third-calving andover). Individual milk yields per milking of were recorded on the morning of the observation day. Behaviors were classified as negative (N) or positive (P). The N behaviors were attempts to move cows, categorized as either touch (NT), vocalization (NV), or gesture (NG). The P behaviors were communications with cows without attempting to move them, categorized as either touching (PT) or vocalization (PV). The frequencies of these behaviors were recorded without distinguishing individual cows. All stockpersons used N on all 34 observations, but none used P for over half of the observations. Spearman rank correlation was used between the number of N and milk yields. The difference between the average individual milk yields (kg/milking) when P were observed and when P were not observed was analyzed using Wilcoxon tests. When the test showed significance, interaction between the number of P (times/cow/milking) and the milk yields (cow/milking with observed P) was analyzed using Spearman correlation. Regarding N behaviors, although the numbers of NT, NV, and NG were not correlated significantly with milk yields of all cows, the number of NV (1.12 times/cow/milking) was correlated only with first-calving cow milk yields (n=34, rs=-0.43, P<0.05). As P behaviors, average individual milk yields in all cows were significantly higher (P<0.001) when PT was observed (16.3 kg) than not (15.9 kg). The average individual milk yields in first-calving cow were significantly higher (P<0.01) when PT was observed (14.5 kg) than not (14.3 kg). The number of PT (0.01 times/cow/milking on average of milkings with observed PT) was correlated with the milk yields (n=14, rs=0.67, P<0.05). Furthermore, average individual milk yields in second-calving cow were significantly higher (P<0.05) when PV was observed (16.9 kg) than not (16.5 kg). No significant difference was found between the number of P and third-calving cow (or more) milk yields. Results show that less negative vocalization and more positive behaviors such as touching and vocalization increase milk yields, especially in younger cows.

Shared and contrasting concerns and perceptions regarding gestation stall housing for sows among Brazilian stakeholders

Maria Cristina Yunes[1], Marina A G Von Keyserlingk[2] and Maria J Hötzel[1]
[1]*LETA, Universidade Federal de Santa Catarina, 88034-001 Florianópolis, Brazil,* [2]*AWP, University of British Columbia, V6T 1Z4 Vancouver, Canada; mcyunes@hotmail.com*

Gestation stall housing for pregnant sows is being phased out in many parts of the world through legislation or industry initiatives. In Brazil, one of the largest global pork producers, gestation stall housing is still common. The objective of this study was to investigate the views of Brazilians, affiliated (Aff) or not (N-Aff) with livestock production, on the use of individual gestation stall housing for sows. A convenience sample recruited by e-mail and social media was invited to answer an online survey. After answering some demographic questions, participants were provided the option of accessing a short text (231 words) describing the housing system and a 90 s video showing 12 intercalated segments of pregnant sows housed in individual or group housing. Respondents were then asked to state their position on housing pregnant sows in individual stalls (Reject/Support/Indifferent) and to provide the reason(s) justifying their position. Among N-Aff (n=173) respondents, 87% rejected, 8% supported and 5% were indifferent to housing gestating sows in individual stalls; among Aff (n=176) respondents, 70% rejected, 23% supported and 7% were indifferent (Aff vs N-Aff: X^2=16.9, df=2, P=0.001). More respondents that accessed the information (87%) rejected the stalls than those that did not (69%) (X^2=17.3, df=2, P=0.001). Preference for the systems was, among N-Aff and Aff respondents, respectively: gestation stalls 8 vs 17%; group system 35 vs 39%; other systems 49 vs 41%; don't know: 8 vs 2% (Aff vs N-Aff: X^2=13.1, df=3, P=0.005). Qualitative analysis revealed that concern regarding animal welfare, most often in reference to animal sentience, freedom of movement and ethics, was the main justification presented by 85% of all respondents that rejected gestation stalls. Many associated prevention of expression of natural behaviours with negative emotions. Justification for support of individual stalls were improved production, handling and animal health, and reduced aggression. N-Aff respondents used animal welfare related justifications for their positions more often than Aff respondents; in contrast Aff respondents used production related reasons more often than N-Aff respondents. Our results highlight a disconnect between stakeholders that are associated with livestock production systems and those that are not. These findings highlight that the public's opposition to gestation stalls for sows reflects an ethical position regarding the treatment of livestock, and should not be interpreted as support for group housing in confined systems.

More than Refinement – improving the validity and reproducibility of animal research
Hanno Würbel, Thomas S. Reichlin, Bernhard Voelkl and Lucile Vogt
University of Bern, Division of Animal Welfare, Vetsuisse Faculty, Länggassstrasse 120, 3012 Bern,
Switzerland; hanno.wuerbel@vetsuisse.unibe.ch

Refinement is one of the principles of the 3Rs (Replacement, Reduction and Refinement) concept and refers to measures which minimise suffering and improve the welfare of research animals. Whether any suffering imposed on animals is ethically justifiable, however, depends on the expected benefit of the research. Unless a study produces results that are scientifically valid and reproducible, animals may be wasted for inconclusive research, and any suffering imposed on them may be ethically unjustifiable. Recent evidence from our own research and that of colleagues indicates considerable risks of bias across all areas of animal research, questioning both scientific validity and reproducibility. Risks of bias are caused by flaws at all levels of research including the design, conduct, analysis and reporting of experiments. Thus, studies may be based on samples that are too small or idiosyncratic; they may lack independent replicates, violate good research practice (e.g. randomisation, blinding, sample size calculation), and use inappropriate statistics; or they may report results selectively or not at all (publication bias). All of this may compromise the scientific validity of results reported in the literature, thereby exacerbating the current 'reproducibility crisis'. We recently conducted a survey amongst all Swiss animal researchers revealing that most of them were uncertain as to how to avoid risks of bias. They also tended to overestimate the quality of their own research, and they were unaware of guidelines aimed at improving the conduct and reporting of animal research (e.g. ARRIVE guidelines). Perhaps an even greater problem, however, is the narrow scope of single-laboratory studies. Due to the specific and standardised conditions within laboratories, results of single-laboratory studies may often have very little external validity. A survey of data from 50 independent studies on the effect of hypothermia on lesion volume in animal models of stroke revealed that treatment effects of single studies varied widely (reduction of lesion volume between 0% and 100%), and only few of them (16%) included the overall effect size (reduction of lesion volume by 48%) in their confidence interval. By contrast, simulations showed that multi-laboratory studies with as few as three laboratories, using the same total number of animals, would result in a twofold increase of the predictive validity of treatment effects. Improving experimental design and conduct to enhance the validity and reproducibility of animal research, therefore, matters for scientific, as well as ethical reasons; it avoids wasting animals for inconclusive research and imposing unnecessary suffering on laboratory animals.

Validating play behavior of cattle as a positive welfare indicator: a review of research
Jeff Rushen and Anne Marie De Passille
University of British Columbia, Box 220, 6947 Highway 7, Agassiz, BC V0M 1A0, Canada;
rushenj@mail.ubc.ca

Play behaviour has been suggested as a positive welfare indicator. We review studies that validate this suggestion by examining whether locomotor play by dairy calves increases when their welfare is improved. Locomotor play (principally running, jumping and bucking) by calves is most clearly related to improved feeding methods. Increasing the amount of milk fed to unweaned calves increases energy intake and locomotor play. Calf growth during the pre-weaning period is positively correlated with the amount of locomotor play they show. When calves are weaned off milk their energy intake decreases during the subsequent week and locomotor play is almost absent. Delaying weaning to a later age improves energy intake and maintains locomotor play. When calves are reared with their mothers and then separated (a situation that has been shown to result in a negative judgment bias), there is a drop in energy intake and an increase in distress vocalizations. The amount of locomotor play is negatively correlated with vocalizations, suggesting that the decrease in locomotor play reflects the calf's emotional response to the separation. Providing calves with an alternative milk source before separation (to allow greater nutritional independence from the mother) improves energy intake after separation, reduces distress vocalization and maintains the level of locomotor play. A variety of non-nutritional threats to welfare also reduce locomotor play. Calves housed in smaller pens show less locomotor play than those in larger pens. Dehorning calves without adequate pain control reduces locomotor play but this effect is overcome when dehorning is done with adequate pain control. Locomotor play is higher in calves that respond to novelty with a greater balance of exploratory behavior compared to fearful behavior, suggesting a link with their emotional responses. Locomotor play appears to be linked to exploratory behavior. The treatments did not affect the duration of walking in the same way suggesting that these effects were specific to locomotor play rather than resulting from changes in general activity. Together, the results show that locomotor play is increased when welfare is improved in unrelated ways suggesting that this positive welfare indicator may be a more general response to changes in welfare than are signs of poor welfare. We argue that this is an important advantage with positive welfare indicators. Since locomotor play of calves can be measured automatically using accelerometers, this behavior could feasibly be included in on-farm welfare assessments.

Behind those eyes: associating eye wrinkle expression and emotional states in horses

Sara Hintze[1,2], Samantha Smith[3], Antonia Patt[4], Iris Bachmann[1] and Hanno Würbel[2]
[1]Agroscope, Swiss National Stud Farm, Les Longs-Prés 2, 1580 Avenches, Switzerland, [2]University of Bern, Division of Animal Welfare, Länggasstrasse 120, 3012 Bern, Switzerland, [3]University of Edinburgh, Easter Bush, Midlothian EH25 9RG, United Kingdom, [4]University of Maryland, College Park, MD 20740, USA; sara.hintze@vetsuisse.unibe.ch

Establishing valid indicators of emotional states is one of the biggest challenges in animal welfare science. Here we investigated how distinctively valenced situations affect eye wrinkle expression in horses. Wrinkles above the eyeball are common in horses but may differ between and within individuals. They are caused by contraction of the inner eyebrow raiser, which is known to be contracted when people are frightened or sad, and are more strongly expressed when horses are in pain. The aim of the present study was to induce positive and negative emotional states and to assess whether positive states would reduce the expression of eye wrinkles while negative ones would increase it. Sixteen horses were confronted with two presumably positive situations (food anticipation (FA) and grooming (G)) and two negative situations each (food competition (FC) and waving a plastic bag (PB)) in a balanced order with each horse being exposed to only one situation per day. Each situation lasted for 60 s (TRT) and was preceded by a 60 s control phase (CON). Throughout CON and TRT pictures of the eyes were taken. For each horse four pictures per situation (FA, G, FC, PB) and phase (CON, TRT), and balanced for eye side were randomly selected (n=512) and scored in random order and blind to treatment for six outcome variables: qualitative impression, number, angle and markedness of eye wrinkles, presence of eye white, and eyelid shape. Data were analysed using mixed-effects models and ordered logistic regression with 'situation', 'phase', and 'eye side' and their three-way interaction as fixed effects. Intra-observer reliability was very high for all outcome measures. We found that the angle between a horizontal line through the eyeball and the highest wrinkle caused by contraction of the underlying inner eyebrow raiser was consistently affected by the different situations: positive situations reduced the angle through muscle relaxation while negative situations increased it through muscle contraction ($F_{3,22}$=5.922, P=0.004). Moreover, eye white was visible less often in positive situations and more often in negative situations (χ^2_3=8.33, P=0.04). The other four outcome measures were not affected by the different situations. We conclude that emotional states may affect characteristics of eye wrinkle expression which may therefore be a promising indicator of horse welfare but further research is needed to validate the outcome variables in the same and in different situations.

Playful pigs: evidence of consistency and change in play depending on litter and developmental stage

SM Brown[1], R Peters[2] and AB Lawrence[3]

[1]University of Edinburgh, Roslin Institute, Penicuik EH25 9RG, United Kingdom, [2]Aberdeen University, Kings College, Aberdeen AB24 3FX, United Kingdom, [3]SRUC, West mains Road, Edinburgh EH9 3JG, United Kingdom; sarah.brown@ed.ac.uk

Play behaviour has applied relevance as a potential indicator of high levels of animal welfare as it tends to be expressed only under good or 'optimal' environmental conditions. This study aimed to confirm our previous findings of litter differences (LD) in play behaviour pre-weaning and to determine if these LD persisted into the early post-weaning period. Seven litters of commercially bred piglets were farrowed and raised in a free farrowing system (PigSAFE). Observations were taken on 6 days pre- and 5 days post-weaning. Piglets were weighed at weekly intervals from birth to weaning and on day 28 post-weaning. A fixed effects model was used to compare between and within litter variation in play in the pre and post weaning period with one value per individual (being the average of the transformed values from each of the observation days pre or post weaning) to remove any confounds of repeated measures. LD in counts of total play were observed in the post-wean period ($F_{6,71}=2.47$, P=0.03) with a tendency towards LD in the pre-wean period ($F_{6,71}=2.11$, P=0.06). Further analysis found this to be mainly due to differences in counts of locomotor play (Pre $F_{6,71}=11.91$, P<0.001; Post $F_{6,71}=7.69$, P<0.001), particularly run (Pre: $F_{6,71}=10.62$, P<0.001; Post: $F_{6,71}=7.48$,P<0.001). There was no evidence of a correlation between play behaviour in the pre- and post- weaning stages. We compared change in play counts pre- and post-weaning to determine the 'weaning effect' on play behaviour. LD were observed in the change in total play counts over weaning ($F_{6,71}=4.02$, P=0.002), particularly in the categories of locomotor play ($F_{6,71}=12.33$, P<0.001) and object play ($F_{6,71}=2.23$,P=0.05) but not in social play. A trend towards a negative relationship between change in locomotor play and average daily gain in the pre weaning period was observed (r=-0.709, P=0.074) suggestive that piglets growing well in the pre-weaning period may be more negatively affected by weaning in terms of their play behaviour. The results generally confirm our previous work showing LD in play behaviour in pre-weaning piglets. No evidence of an association between LD in play behaviour pre- and post-weaning was observed. Of note is what we have classed as the 'weaning effect', where litters play behaviour responds differently to weaning. This cannot be explained as an age-related effect as all litters were weaned and observed at the same age points. The indication of a drop in play post weaning associated with a high ADG pre-weaning could indicate that piglets with a higher 'dependency' on the sow may be more negatively affected by the weaning process.

Inducing positive emotions: cardiac reactivity in sheep regularly brushed by a human

Priscilla Regina Tamioso[1], Guilherme Parreira Silva[2], César Augusto Taconelli[2], Hervé Chandèze[3], Stéphane Andanson[3], Carla Forte Maiolino Molento[1] and Alain Boissy[3]
[1]Animal Welfare Laboratory – LABEA, Federal University of Paraná, UFPR, Department of Animal Science, Rua dos Funcionários, 1540, Juvevê, 80035-050, Curitiba, Paraná, Brazil, [2]Federal University of Paraná, UFPR, Department of Statistics, Avenida Coronel Francisco H. dos Santos, 210, Jardim das Américas, 81531-970, Curitiba, Paraná, Brazil, [3]UMR1213 Herbivores, Institut National de la Recherche Agronomique – INRA, INRA Theix, Saint-Genès Champanelle, 63122, Saint-Genès Champanelle, France; priscillatamioso@gmail.com

Welfare concerns the absence of negative and presence of positive experiences. We assessed cardiac indicators of sheep through a heart rate monitor, as well as ear postures and tail wagging. Thus, 38 female Romane ewes were trained to be brushed by a familiar human (B) on the neck, withers, chest and belly, or exposed to human presence (H). The ewes belonged to two lines: more (R+) or less (R-) reactive to social isolation. Heart rate (HR) and heart rate variability (RMSSD, RMSSD/SDNN and LF/HF ratios) were analysed using linear models. The models considered treatment, genetic line and phase (pre- (2.5 min), during (3.0 min) and post-exposure (2.5 min)) as fixed effects, including their interactions. The HR during and after brushing was lower than before brushing (P<0.01). No differences in RMSSD were found, but the RMSSD/SDNN ratio during the exposure was higher than before or after (P<0.05). The RMSSD/SDNN ratio in R- ewes was higher than in R+ ewes (P<0.01), revealing a stronger activation of the parasympathetic system in R- sheep. In R+ line, the B ewes had a higher HR than the H ewes (P<0.01) whereas in R- line the difference was reversed (P<0.01). In R+ line, the LF/HF ratio of the B ewes was lower than in the H ewes (P<0.01). Preliminary results on ear postures also indicate a positive perception of brushing, as sheep showed a higher duration of horizontal ears, equal to 105.82 (0/178.63)s, and wagged their tails for 28.83 (0/151.60)s when brushed. The H ewes performed raised up ears for longer, equal to 60.22 (11.12/171.32)s, and wagged their tails for 1.14 (0/8.68)s. Such behavioural variables will be further analysed. There is a need to better investigate the differences between R+ and R- before concluding that the emotional reactivity can modulate the autonomic responses to positive events.

Spreading happiness: induced social contagion of positive affective states and behaviours in monkeys via audio/video playback

Claire F. I. Watson[1,2] *and Christine A. Caldwell*[1]
[1]*University of Stirling, Psychology, Stirling, FK9 4LA, United Kingdom,* [2]*Kyoto University Primate Research Institute, CICASP, 41-2 Kanrin, Inuyama, Aichi, 484-8506, Japan; cfi.watson@gmail.com*

Research on the social contagion of yawning and negative affective states, e.g. anxiety, has proven productive. Contagion of positive affective states has been relatively neglected, especially social affiliation. The effect of social contagion, both negative and positive, from neighbouring conspecifics on captive animal welfare has been studied, e.g. pigs anticipating reward had positive effects on neighbour's welfare; affiliative calls by neighbour groups are associated with increased affiliation in marmosets. In human psychology a trend directing focus towards positive states led to findings that inducing emotional contagion of positive affect can improve group mood. In addition to elucidating cognitive mechanisms, precursors to empathy, investigation of inducement of social diffusion of positive affective states in animals offers potential for welfare application through enrichment. Social contagion of positive affective states through playback of conspecific social affiliative behaviour patterns has yet to be examined. We investigated, whether social contagion of affiliation could be induced, via manipulation, by playback of affiliative conspecifics: positive conspecific vocalisations or video of social affiliative behaviour in marmosets (*Callithrix jacchus*). Playback of positive affiliative vocalisations led to relatively long-lasting, non-contingent, increases in affiliation), compared to control condition individuals, exposed to silent playback. Playback was without humans present, and mutually exclusive to observation sessions, hours outwith playback. We found video playback of affiliative behaviour, allogrooming, to be associated with contingent increases in social affiliative, prosocial behaviours in marmosets relative to playback of control video [n=16; affiliation composite; P=0.005, food-share, P=0.003, allogroom, P=0.007, approximate-exact permutation tests: resampling method]. Each focal individual received 2 trials/day: control and allogroom, for 13 days. Observation, continuous focal sampling, lasted 5 min from start of stimulus presentation [95 sec: 9 silent video clips interspersed with blank screen (two marmosets: allogrooming, allogroom condition; in close proximity, control condition), screen in front of homecage, onscreen marmosets smaller than life-sized]. Marmosets are the most frequently used New World primate in UK laboratories. Auditory and visual playback of social affiliative behaviours represents ecologically valid potential sensory and non-contact social enrichment.

Investigating positive emotional contagion in rats

Jessica Frances Lampe[1], Oliver Burman[2], Hanno Würbel[1] and Luca Melotti[1]
[1]University of Bern, Division of Animal Welfare, Länggassstr. 120, 3012 Bern, Switzerland,
[2]University of Lincoln, School of Life Sciences, Riseholme Hall, LN2 2LG Lincoln, United Kingdom;
jessica.lampe@vetsuisse.unibe.ch

Emotional contagion is the tendency of individuals in a group to emotionally converge. Using play behaviour and positive (frequency-modulated 50 kHz) ultrasonic vocalisations (USVs) as proxy measures of positive emotional state and good welfare, we investigated whether positive emotions can be transferred between rats. 48 adolescent male Lister Hooded rats were housed in groups of 3 non-littermates, and one rat per group received positive or control treatments to assess if this affected play and USVs in the home cage. After habituation to the experimental procedure, 3 tests were performed on separate days. The 'treated' rat was taken to a separate room for 2 min, where it was either tickled by the experimenter, given chocolate rewards, or placed in the treatment arena without further action (control condition). Testing was randomized and counterbalanced across days to avoid order effects. USVs and play events in the home cage – attack to nape initiating play (AN) and solitary scampering (SC) – were counted blind to treatment during 10 min before and after the treatment. At cage level, there was an interaction between treatment and time (USVs: $F_{2,30}$=6.4, P<0.01; AN: $F_{2,30}$=17.3, P<0.001; SC: $F_{2,30}$=5.6, P<0.01). USVs decreased after all treatments, however after positive treatments this decrease was smaller than after the control treatment (η_p^2: 0.28 vs 0.76). AN and SC decreased after the control treatment only. We also investigated social direction of AN by comparing treated rats initiating play with untreated ones (TU), untreated rats initiating play with treated ones (UT) and untreated rats initiating play with untreated ones (UU). Treatment, social direction and time interacted ($F_{12,272}$=4.58, P<0.001). While all 3 groups showed a decrease in play after the control treatment, after tickling UT play increased, and TU and UU play remained unchanged. After the chocolate treatment, TU and UT play remained the same and only UU play decreased. With regard to SC, treatment, rat type (Treated or Untreated) and time interacted ($F_{7,192}$=2.68, P=0.02), showing that both T and U rats scampered less after the control treatment, yet not after both positive treatments. These results provide some evidence of short-term positive emotional contagion from one individual receiving a positive treatment to its social group. Positively treated rats, particularly after tickling, appeared to promote emotional contagion by becoming the target of more play initiations by untreated rats. This effect is unlikely to be due solely to odour stimuli brought with the treated rat as both control and tickled rats were handled by the experimenter.

Does the anticipatory behaviour of chickens communicate reward quality?

Nicky Mcgrath[1], Oliver Burman[2], Cathy Dwyer[3] and Clive Phillips[1]
[1]University of Queensland, School of Veterinary Sciences, Gatton, QLD, 4343, Australia,
[2]University of Lincoln, School of Life Sciences, Brayford Pool, Lincoln, LN6 7TS, United Kingdom,
[3]Scotland's Rural College (SRUC), Animal & Veterinary Sciences,Easter Bush, EH25 9RG, United Kingdom; nicmcgrath@hotmail.com

The anticipatory behaviour of animals may facilitate an understanding of what an animal wants. Our goal was to determine if behaviour in anticipation of different rewards was differentially expressed. We investigated whether certain behaviours were characteristic of anticipation of food and non-food rewards, and whether associated cues led to increased activity levels. Twelve chickens experienced a Pavlovian conditioning paradigm using sound cues to signal the availability of two food rewards (mealworms, normal food), one non-food reward (dustbath), and a sound-neutral event (signalled by a sound, but no reward given). A muted-neutral treatment (no reward / sound cue) controlled for any behaviour specific to sound cues. Behavioural responses and the number of transitions between behaviours were measured during a 15 second anticipatory period, before birds accessed rewards in an adjoining compartment by pushing through a door. These responses and latency to access the rewards were analysed using linear and generalised linear mixed models. Differences in pushing/pecking at the door (frequency: Dustbath 4.87[a], Mealworm 3.18[b], Normal Food 2.23[b], Sound Neutral 0.30[c], Muted Neutral 0.03[d], $\chi^2(4)=228.99$, P<0.001), looking towards the reward chamber (duration (s) Dustbath 17.49[a], Normal Food 13.54[b], Mealworm 11.57[b], Sound Neutral 7.05[c], Muted Neutral 5.10[c] $\chi^2(4)=62.08$, P≤0.001) and standing (not walking) (duration (s): Sound Neutral 9.92[c], Muted Neutral 7.49[bc], Normal Food 7.39[bc], Mealworm 7.05[b], Dustbath 3.06[a], $\chi^2(4)=36.28$, P<0.001) reflected the perceived value of the rewards, with birds appearing to be more motivated to access the dustbath compared with food rewards. Rewarded sound cues elicited increased transitions between behaviours, compared with neutral events (Dustbath 10.16[a], Mealworm 10.13[a], Normal Food 9.22[ab], Sound Neutral 7.89[bc], Muted Neutral 6.43[c], $\chi^2(4)=72.05$, P<0.001). The sound-neutral treatment induced increased head movements, previously associated with anticipation of rewards (duration (s): Sound Neutral 1.58[b], Muted Neutral 0.58[ab], Normal Food 0.48[a], Mealworm 0.27[a], Dustbath 0.00[a], $\chi^2(4)=25.56$, P<0.001). Latency to access rewards conveyed the relative value of rewards (Dustbath 7.30[a], Mealworm 10.06[ab], Normal Food 16.53[ab]). Our experiment indicates that chickens increase their activity (behavioural responses and transitions) in anticipation of rewards and that this response is not food specific. This outcome extends our knowledge of reward-related anticipatory behaviour.

Pharmacological manipulation to validate indicators of positive emotional states in dogs

Linda J Keeling, Yezica Norling, Claudia Von Brömssen, Carina Ingvast-Larsson and Lena Olsén
Swedish University of Agricultural Sciences, Box 7068, 750 07 Uppsala, Sweden; linda.keeling@slu.se

Building upon approaches used previously to validate indicators of pain and fear, we used psychopharmacological drugs in a situation designed to elicit positive anticipation in 8 female beagle dogs. The hypotheses were that drugs considered to promote a more positive mood (methylphenidate, clomipramine) would increase the behaviours previously observed in a presumed positive situation whereas a drug considered to dampen mood (diazepam) would result in a decrease of these candidate behavioural indicators of a positive emotional state compare to a placebo. Each dog, acting as its own control, freely entered a cubicle facing a theatre where one of three familiar rewards (a meatball, a familiar person, both presumed positive, and a neutral wooden block) was made visible in a balanced order. Each dog was tested twice, 1.5 and 5 h after receiving the drug with 6 days washout between drugs. Body posture and behaviour in the test situation was recorded for 5 sec before and 7 sec after the stimulus became visible to the dog. Behaviour was also recorded for 60 min in their undisturbed group at midday. Statistical analysis used linear mixed models. The behaviour of each dog pre-exposure was used as the baseline for its response to the stimulus. Irrespective of the drug, dogs showed more tail wagging, lip licking, lowered head and flickering glaze to the familiar person (P<0.05 for all) compared to the wooden block, confirming the results of a previous study. There were significant drug-stimulus interactions for the behaviours nose stretching (more with methylphenidate) and turning the head (to left side with diazepam) to the familiar person compared to with the placebo drug (P<0.05 for both). There were clear effects of the drugs on behaviour in the group situation e.g. decreased lying (29 vs 48%) and increased whining (7 vs 1) were observed with diazepam compared to placebo (P<0.05 for both). We could not support our hypothesis. The few drug-stimulus interaction effects during testing, compared to the pronouced effects of the drugs on behaviour when undisturbed could be because the training enabled the dogs to focus on their tasks at these drug doses. The drugs may also be acting on the ability of the dog to express its emotional state rather than on what it actually feels. The agitated behaviour after diazepam administration was unexpected. This drug is usually characterized as a sedative in dogs, but the behavioural changes indicate the opposite, showing that it is difficult to extrapolate effects of drugs described in humans to what dogs may feel or how they express emotions under drug influence.

Promoting positive welfare in chimpanzees through videos of chimps

Dean O'driscoll, Kirk Dennison, Deborah Wells and Peter Hepper
Queen's University Belfast, Animal Behaviour Centre, Psychology, Belfast, BT7 1NN, United Kingdom; p.hepper@qub.ac.uk

Sensory stimulation has been extensively used as a means of environmental enrichment to improve animal welfare. Visual stimuli may promote positive welfare in captive apes as they are are particularly attuned to the visual world. Here we examined the effect of video playback of chimps on the behaviour of a chimp group at Belfast Zoo. The group, 2 adult males, 2 adult females, 2 juvenile females and 2 female youngsters were played a video, presented by TV monitor, for 60 minutes and the behaviour of each chimpanzee recorded every 150 s using a standardised chimp ethogram. Four videos were used: Control, blank screen(C); Real-time playback of the chimp group (RT); Pre-recorded playback of the chimp group(PRO); and, Pre-recorded playback of another chimp group(PRU). The pre-recorded videos displayed general activity of a group of chimps of similar composition and location (Zoo compound). It was hypothesised that there would be a greater response to the videos of chimps than the control, and a different response to videos of their own and the unfamiliar group. One video was played to the group per day and the group were played all 4 videos over 4 days. The procedure was repeated one week later using the same videos. Chimp responses were analysed using analysis of variance for factors of video (control; real-time; familiar group; unfamiliar group) and time (1st / 2nd presentation) for each behaviour recorded. Video type influenced the chimps' behaviour. Chimps looked more (mean (s.d.) no. of obs.-C:3.44(.5) RT:8.19(1.8) PRO:13.63(1.2) PRU:8.87(1.2), P<0.001), exhibited more TV directed behaviours (C:3.98(.6) RT:9.13(2.1) PRO-14.19(1.3) PRU:10.18(1.3), P<0.001), less aggression (C:0.48(.2) RT:0.63(.1) PRO:0(0) PRU:0.63(1.1), P=0.02), and more social behaviour (C:1.56(1.7) RT: 4.12(1.2) PRO: 5.93(1.7) PRU:3.43(1.5), P=0.04) when watching the video of their own pre-recorded group than the pre-recorded unfamiliar group or real-time playback. On the second presentation responses to the real-time playback increased whilst responses to the pre-recorded playback decreased (change in response C: +0.6(.1) RT:+4.0(1.1) PRO:-5.75(1.9) PRU:-3.6(1.3), P<0.001). We conclude that visual stimuli presented via TV influences chimp behaviour promoting positive behaviours (e.g. social interactions) and decreasing negative behaviours (aggression, abnormal). Chimps habituated to pre-recorded images however real-time playback sensitized their behaviour and this may offer the potential to influence positively their behaviour over a long period.

Mid-term effects of different handling treatments on heart rate and heart rate variability of dairy cows

Silvia Ivemeyer, Marthe Julie Boll and Ute Knierim
University of Kassel, Farm Animal Behaviour and Husbandry Section, Nordbahnhofstr. 1a, 37213 Witzenhausen, Germany; ivemeyer@uni-kassel.de

While there is evidence that positive handling may improve health and performance, not much is known about the direct and mid-term effects (about one hour later) of the quality of short-term handling on the cows, e.g. in terms of cardiac responses. We therefore exposed six multiparous cows after morning milking to four standardized handling procedures, one per day in random order: positive (POS): TTouch© (PT) and stroking (PS) for each 10 min; negative (NEG): tail lifting (4 times within 10 min, NT) and individual isolation (data from minute 5-10 of isolation, NI), in order to assess their effects on heart rate (HR) and heart rate variability (HRV) during the tests and during the following lying bouts (52±19 min after treatment start). To compare the post-treatment response of POS and NEG with a relaxed, but neutral situation, a lying reference recording from an earlier day before the start of the treatments at the same time of day was used (NEUTRAL). HR and HRV data after POS and NEG were subtracted from NEUTRAL lying data. HR and HRV data were recorded with equipment from Polar Electro™ and analyzed with Kubios 2.2 software. Analyzed sequences were 5 min each. If possible, three successive 5-min sequences were averaged for the lying recordings. General mixed models with repeated measures within cows and treatments as fixed factor (either separately or summarized to POS and NEG) were calculated for HR and HRV (standard deviation of inter-beat intervals (SDNN), root mean square of successive differences (RMSSD), ratio low to high frequency band powers (LF/HF)) as dependent variables. The four different handling treatments had significant effects on HR (means: PT: 66.7, PS: 68.6, NT: 71.8, NI: 74.6 bpm; P=0.002) and HRV (SDNN means: PT: 15.2, PS: 22.2, NI: 47.8, NT: 60.2 ms; P<0.001; LF/HF means: PT: 2.4, PS: 8.2, NI: 11.1, NT: 17.5, P=0.013; RMSSD, n.s.) during treatment. The post-treatment analyses resulted in HR being significantly increased after NEG (0.44 bpm) and decreased after POS (-1.35 bpm; P=0.006) compared to NEUTRAL, with RMSSD being lower after NEG (-0.30 ms) and higher after POS (1.41 ms. P=0.009; for SDNN and LF/HF, n.s.). Despite the limited sample size and the effects not being high and not consistent for all variables, results suggest that (1) short-term handling has slight, but longer lasting effects for at least about one hour (2) POS treatments induce a higher parasympathetic activity compared to negative and even neutral situations. Thus, single positive events may contribute to longer lasting enhanced well-being.

From food-hoarding titmice to stressed poultry: integration of functional and mechanistic approaches to behaviour

Tom V. Smulders

Newcastle University, Institute of Neuroscience, Framlington Place, Newcastle upon Tyne, NE2 4HH, United Kingdom; tom.smulders@ncl.ac.uk

Many species in the family Paridae hoard food for later consumption. How might such a novel behaviour evolve from a non-hoarding ancestor? In a combination of anatomical, field, computational and laboratory studies, I show that the primary function of food-hoarding is as an adaptation to a stochastic and unpredictable food supply. The capacity of the digestive tract of these animals is limited, and although the digestive tract increases in capacity in winter, this is not sufficient to overcome the constraints in exploiting temporarily abundant food sources throughout winter. Hoarding allows them to evenly spread consumption across the few hours of winter daylight and hence to gain more fat reserves and survive the long, cold winter nights. Having understood this primary function of food hoarding, I then investigate the decision rules that govern it. The same factors (both environmental and endocrinological) that lead to an increase in consumption also lead to an increase in hoarding, driven by a common increase in foraging motivation. Once an item has been collected, however, it will be hoarded if it cannot be eaten at that point in time (due to satiation or other factors). The same simple set of rules also explains the long-term, seasonal pattern of hoarding, with a peak in the autumn when food is abundant. Of course, hoarding would not serve any function if the food was not retrieved. For short-term retrieval (within days), the birds use memory for the locations (and possibly contents) of the hoards to have ready access to them. The food hoarded in autumn is not remembered into winter, however. Instead, these forgotten items enrich the birds' winter foraging niche. Again, the rule is simple: memorize every item hidden (this is probably automatic). This memory then makes those items still in memory readily available when no other food is found. We also show that this same memory allows the birds to avoid hiding new items too close to old ones. The hippocampus, the neural structure processing spatial memory, also shows a seasonal plasticity, following the intensity of hoarding and hence of the amount of information stored in memory. More recently, our insights into this hippocampal plasticity in birds has led us along a different, more applied path. The same plasticity mechanisms that respond to increased memory processing are also sensitive to chronic, cumulative experiences of stress. Because hippocampal plasticity can respond to positive and negative stimulation with opposite patterns, we are currently developing the avian hippocampus as a marker of welfare, with the potential to pick up both the intensity and the valence of the animals' experiences.

Feather pecking genotype shows increased levels of impulsive action in a delayed reward task

Patrick Birkl[1], Joergen Kjaer[2] and Alexandra Harlander[1]
[1]UofGuelph, Animal Science, 50 Stone Rd, Guelph, Canada, [2]F-L-Institut, Animal Welfare, Dörnbergstr. 25, Celle, Germany; pbirkl@uoguelph.ca

Impulsivity is a multi-faceted concept. One dimension related to motor control is impulsive action, best characterized in terms of premature responding and the inability to inhibit an initiated response. Impulsive action in animals can be beneficial, for instance when quick responses are required to obtain unexpected food. Increased levels of impulsive action however are symptoms in behavioral disorders such as attention deficit hyper-activity disorder (ADHD). Feather pecking (FP) in laying hens, a motor behaviour which involves pecking at, or pulling out of feathers from conspecifics and thereby damaging the victim, is a major welfare problem in egg production. FP has been proposed as potential model for ADHD. However, whether impulsivity is a key feature of FP and whether it contributes to an individual's presentation of FP is poorly understood. The current experiment aimed to test differences in impulsive action in birds selected for high (HFP) and low (LFP) levels of FP activity and an unselected control genotype (C). To test these differences, 30 laying hens were trained in an operant conditioning chamber (Med Associates, USA) to peck a key using a fixed-ratio 1 schedule of reinforcement. After pecking at the key, the stimulus (illumination) was switched off for 2 seconds and a sliding door opened for food reinforcement (3 sec). Birds were then tested on a 5 second delay schedule. Each trial lasted 4 minutes. The number of rewarded pecks and unrewarded pecks (pecks at non-illuminated key and pecks close to the key; wall-pecks), before and after the stimulus was presented, were recorded. Data were analysed using a GLIMMIX procedure in SAS with number of pecks as the dependent variable and genotype as main effect. Estimated Means and Standard errors (EM±SE) are presented: Overall, rewarded pecks did not differ between genotypes. The total number of unrewarded pecks at the key differed between genotypes; C (3.6±0.2) higher than HFP (2.8±0.2) $P<0.0001$, LFP (3.5±0.2) higher than HFP: $P<0.0001$ but not C vs LFP: $P<0.1$). The number of wall-pecks before the stimulus differed significantly between genotypes; HFP (3.4±0.1) higher than C (3.0±0.1), $P<0.001$, HFP higher than LFP (2.8±0.2): $P<0.002$, but not between C vs LFP, $P<0.34$). The number of wall-pecks after a successful peck at the illuminated key also differed significantly between all three genotypes; HFP (3.6±0.1) higher than C (3.3±0.1), $P<0.01$, C higher than LFP (2.9±0.1), $P<0.02$, HFP higher than LFP, $P<0.0001$. These results provide strong evidence that HFP-birds show increased levels of impulsive action, as expressed by premature responses as well as responses during the delay phase.

Using epidemiology to understand stereotypic behavior of African and Asian elephants in North American zoos

Brian J. Greco[1,2], Cheryl L. Meehan[1], Georgia J. Mason[3] and Joy A. Mench[2]
[1]AWARE Institute, 3212 NW Wilson Street, Portland, OR 97210, USA, [2]University of California, Davis, Department of Animal Science, One Shields Ave, Davis, CA 95616, USA, [3]University of Guelph, Animal Science Department, 50 Stone Road East, Guelph, ON, N1G 2W1, Canada; bjgreco@ucdavis.edu

Stereotypic behavior (SB) is an important indicator of compromised welfare. Zoo elephants have been documented to perform SB, but no previous studies systematically assessed factors contributing to performance rates. We videoed 89 elephants (47 African and 42 Asian at 39 North American zoos) for a median of 24 daytime hours per elephant. A subset of 32 elephants (19 African, 13 Asian) was also observed live for a median of 10.5 nighttime hours per elephant. We used 5-min instantaneous samples to calculate SB rates as percentages of all active behaviors. Stereotypic behavior was the second most common behavior (after feeding), making up 15.5% of daytime and 24.8% of nighttime observations. Whole-body motor movements (e.g. weaving, rocking) made up the majority (89.9%) of SB observations, followed by locomotor movements (e.g. pacing) (8.2%). We used negative binomial regression models fitted with generalized estimating equations to assess the environmental and individual characteristics associated with daytime and nighttime SB rates. The primary statistics produced by the models were risk ratios (RRs) describing the magnitude of effect each independent variable had on increasing or decreasing the risk of elephants performing stereotypic behavior at or above their current rates. Species was a significant risk factor, with Asian elephants at greater risk (daytime: $P<0.001$, RR=4.087; nighttime: $P<0.001$, RR=8.015) than Africans. Spending time housed separately (daytime: $P<0.001$, RR=1.009) and experiencing inter-zoo transfers (daytime: $P<0.001$, RR=1.175; nighttime: $P=0.033$, RR=1.115) increased risk. In contrast, spending more time with juvenile elephants (daytime: $P<0.001$, RR=0.985), in larger social groups (nighttime: $P=0.039$, RR=0.752), in environments with both indoor and outdoor areas (nighttime: $P=0.013$, RR=0.987), and engaging with zoo staff (daytime: $P=0.018$, RR=0.988), reduced this risk. Overall, our results point to the social environment as the most influential factor predicting elephant stereotypic behavior rates, and suggest that reductions in rates can be achieved by maintaining larger multigenerational social groups and reducing social separations. The newly discovered relationships between stereotypic behavior, inter-zoo transfers, staff interaction time, and spending time in spaces that provide both indoor and outdoor access at night are also valuable information for elephant managers and merit follow-up research with elephants and other species kept in zoos.

Neighbour effects confirm that stereotypic behaviours in mink are heterogeneous

Andrea Polanco, María Díez-León and Georgia Mason
University of Guelph, Animal Biosciences, 50 Stone Rd, Guelph, ON N1G 2W1, Canada;
apolanco@uoguelph.ca

Stereotypic behaviours (SBs) are common in farmed mink and other captive animals. In carnivores, SBs typically involve route-tracing and, less often, 'stationary' forms (e.g. head-twirling). Farmed mink may also repeatedly scratch at the cage walls ('scrabbling'), sometimes apparently directing this at neighbouring mink in adjacent cages. We thus hypothesized that scrabbling (but not other SBs) represents frustrated attempts to reach neighbours. We housed 32 young male mink in standard cages between two non-experimental animals of random sex. In Study 1 (conducted when mink were 7-12 months old) and Study 2 (conducted at 10 months old), we compared the proportion of each type of SB performed near each shared cage wall when the relevant neighbour was nearby in the adjacent cage vs when that neighbour was distant. We predicted that scrabbling would be performed close to neighbours. Study 2 additionally compared SB time budgets before and after removing one of each subject's neighbours, to test the prediction that this would reduce scrabbling. In Study 1, a Wilcoxon Signed-Rank Test revealed that at 7 months, scrabbling by subjects with a male neighbour occurred more often on the shared cage wall when that neighbour was nearby (mdn=0.46, IQR=0.68) than when he was distant (mdn=0.02, IQR=0.30, W=17, P<0.05), although this became non-significant at 10-12 months of age. The location of scrabbling was not significantly affected by female neighbours' proximity. Other SBs' locations were also unaffected by neighbour proximity when subjects were 7 months old; but at 10-12 months, stationary SBs occurred significantly less often near the shared cage wall when male neighbours were nearby (mdn=0, IQR=0.21) vs distant (mdn=0.48, IQR=0.67, W=0, P<0.05), and the same held for route-tracing (although a trend) when neighbours of either sex were nearby (mdn=0, IQR=0) vs distant (mdn=0.60, IQR=0.39, W=0, P=0.10). Removing neighbours (Study 2) significantly reduced scrabbling (baseline mdn=0.07, IQR=0.09; treatment mdn=0.03, IQR=0.07, W=40, P<0.05) and stationary SBs (baseline mdn=0.05, IQR=0.06; treatment mdn=0, IQR=0.10; W=8, P<0.05), regardless of neighbour sex (route-tracing being unaffected). Location data for Study 2 are under analysis and will be presented. Thus, scrabbling by male mink is directed towards male neighbours when young (but not when older), and they perform less of this SB if neighbours are removed (regardless of sex). Other SBs were generally performed in locations away from neighbours (although removing neighbours did not increase their performance). Overall, these data suggest that scrabbling may derive from frustrated attempts to access neighbours, and add to growing evidence that SBs are heterogeneous in causation.

Behavioural activity of dairy cows on the day of oestrus v. mid luteal phase

Gemma Charlton, Emma Bleach, Carrie Gauld and Mark Rutter
Harper Adams University, Shropshire, TF10 8NB, United Kingdom; gcharlton@harper-adams.ac.uk

Detecting behavioural signs of oestrus in dairy cows is essential for good reproductive performance, and therefore more producers are using sensors to automatically detect oestrus. The objectives of this study were to compare behavioural activity of dairy cows on the day of oestrus to that in the mid-luteal phase, and to determine whether there is any difference in behaviour, as recorded by accelerometers on the front vs back leg. The study was conducted at Harper Adams University, using 100 post-partum Holstein Friesian cows, fitted with three IceQube® accelerometers (IceRobotics Ltd, UK); one on each back leg (back left (BL) and back right (BR)) and one on the front left leg (FL). The sensors recorded lying duration (min/d), frequency of lying bouts (LB/d), average lying bout duration (LBD; min/bout) and step count (steps/d). Milk samples were collected three times a week (Mon, Wed, Fri) from 14 d post calving until a positive pregnancy diagnosis (PD+), for subsequent progesterone determination. Behavioural data were analysed from the day of oestrus (progesterone levels <2.7±0.13 ng/ml; mean±SEM), confirmed as the day that insemination resulted in a PD+ and the day of highest progesterone concentration (41.2±1.85 ng/ml), 12-16d following oestrus (mid-luteal phase). A one-way ANOVA (Genstat v.17) revealed that cows spent significantly less time lying down on the day of oestrus (570.0 vs 670.6 min/d, respectively; P<0.001), had fewer LB/d (10.9 vs 12.3/d, respectively; P<0.001), which, on average were shorter in duration (55.5 vs 58.0 min/bout, respectively; P=0.026) and steps/d were more than double (2,841.2 vs 1,384.7, respectively; P<0.001) compared to mid-luteal phase. A comparison between accelerometers attached to different legs showed no significant differences between BL, BR and FL in lying duration (621.0±13.51 vs 626.7±13.63 vs 613.9±13.31 min/d, respectively; P=0.734) and steps/d (2004.5±122.13 vs 2059.1±118.67 vs 2265.4±146.98, respectively; P=0.369). However, a Tukey test showed significantly more LB/d (10.8±0.28 vs 11.0±0.29 vs 12.9±0.37/d, for BL vs BR vs FL, respectively; P<0.001), which, on average were significantly shorter in duration (59.9±1.25 vs 59.8±1.27 vs 50.6±1.23 min/LB) recorded from the FL leg compared to BL and BR legs. These results show that dairy cows are significantly more active on the day of oestrus compared to the mid-luteal phase. Differences in LB frequency and average LBD recorded by IceQubes on the rear and front leg may be a result of lying position or more acceleration exerted by the front legs due to more joint flexibility. However, more investigation is required to determine why these differences occur, and which leg gives the most accurate detection of oestrus.

Do laying hens have a motivation to grasp while night time roosting?

Lars Schrader[1], Sabine Dippel[1] and Christine Nicol[2]
[1]Friedrich-Loeffler-Institut, Doernbergstr. 25/27, 29223 Celle, Germany, [2]University of Bristol, Langford House, Bristol, BS40 5DU, United Kingdom; lars.schrader@fli.bund.de

Laying hens have a high motivation to rest on elevated perches and they prefer higher perches compared to lower ones for night time roosting. It has been suggested that hens should be able to grasp perches with their feet as this allows them to rest without muscular effort ('tendon lock mechanism'). However, it has been shown that hens in some contexts will prefer higher grids on which they cannot grasp compared to lower but graspable perches for roosting. In this preliminary study we examined the motivation of hens to grasp by testing the following hypotheses: H(1) At night time hens will show a higher usage of a round and graspable perch than of a flat and non-graspable grid. H(2) At dusk, under conditions when access is denied, hens will show more attempts to reach a graspable round perch compared to a non-graspable grid, and will tak longer to settle elsewehre. Twelve groups of 5 laying hens (F3 of LB × LB) were kept in experimental pens (2.0×3.0 m, height 2.0 m) from an age of 47 weeks. They were accustomed to the pens and the different resting sites for 10 weeks. During experiments each group was offered, at a height of 90 cm and a length of 180 cm, a round metal perch (3.4 cm diameter) and a flat metal grid (12.0 cm deep) in a successive but random order. Six days after installing the respective resting site the number of hens on resting sites at the start of the dark phase was counted from video recordings. On the following day, in each pen access to resting site was denied approximately 2 hours before dark phase by Perspex walls. Attempts of hens to enter the resting site were counted during the 15 min dusk phase (dimmed light) from video. In addition, latencies to rest (no change of resting position for at last 2 min after onset of dusk phase) were recorded. Data were analyzed using Wilcoxon signed rank test. Night time usage did not differ between resting sites (P>0.05; median per pen: 5 hens on grids vs 4 hens on perches). When access to resting sites was denied 24 attempts (total sum in all pens) to reach the grid and 6 attempts to reach the perch were observed during dusk phase (P=0.031). Changes of latencies to rest did not differ between perch and grid when access to resting sites was denied (P>0.05). Results of this preliminary study suggest that motivation of laying hens to roost on a resting site which they can grasp with their feet is not higher than to roost on a non-graspable resting site. Alternatively, other features of the tested resting sites (e.g. surface, depth) may have affected the hens' behaviour.

On-farm risk factors for non-nutritive sucking in group-housed organic Simmental dairy calves

Verena Größbacher, Christoph Winckler and Christine Leeb
University of Natural Resources and Life Sciences, Department of Sustainable Agricultural Systems, Gergor-Mendel-Straße 33, 1180 Vienna, Austria; verena.groessbacher@boku.ac.at

In EU organic dairy farming, group housing of calves is required after the first week. This is perceived as a risk factor for non-nutritive sucking (NNS) leading e.g. to umbilical infections especially in Simmental herds. Therefore, this study aimed at identifying risk factors on-farm to provide farmers with effective preventive measures and to investigate potential associations of NNS with treatment incidences. During one-day visits data were collected by one observer on 31 organic dairy farms with Simmental only. The visits included 90 min of direct continuous behaviour observation after the morning milk meal, semi-structured interviews and analysis of treatment records (available from n=25 farms). The average herd size was 31 (SD=10) cows and 11 (SD=7) calves with a mean of 4 (SD=2) calves per group. Potential risk factors were pre-selected (P<0.2) using univariate analysis and Pearson or Spearman rank correlation. GLM was used for final modelling with backward selection of factors when P<0.2. Associations between behaviour and health data were identified using Spearman rank correlation. NNS was observed on 29 farms (94%) at a median rate of 1.66 (Q1=0.70, Q3=3.00) events/calf×hour. NNS (explained variation 61.3%, Intercept=2.75) decreased when age was similar within group (Estimate=-2.40, P=0.001) and increased when calves were not restrained during the milk meal (as compared with restraint for >30 min; 1.46, P=0.026). NNS was lower when use of nose-clips was not reported as a countermeasure (-2.22, P=0.008); however, nose-clips were not in use for calves during the farm visits. These reports likely reflect farms with a longer-lasting history of NNS. Duration of sucking at teat buckets was negatively correlated with NNS (-0.23, P=0.018). The age at grouping had no significant effect on NNS. There were no significant correlations of NNS and treatment incidences of diarrhoea, respiratory diseases and umbilical infections. This on-farm study comprising Simmental organic dairy herds partly confirms existing knowledge on preventive measures (e.g. homogenous age groups, long duration of sucking) to be applicable on-farm. Furthermore, it provides evidence that grouping after the first week is possible without an increased risk for NNS. The perceived risk of NNS leading to infections could not be shown, as most likely other factors are more relevant.

Affective states and proximate behavioural control mechanisms

Lorenz Gygax
FSVO, Centre for Proper Housing of Ruminants and Pigs, Tänikon, 8356 Ettenhausen, Switzerland;
lorenz.gygax@agroscope.admin.ch

Animals choose a course of action many times a day. To do so, they prioritise behaviour within a set of alternative actions, decide which action to perform, and when its proximate goal is reached. Causation in Tinbergen's sense addresses how animals take these decisions. The study of proximate behavioural control mechanisms is highly relevant for animal welfare because most welfare problems seem to arise if control is disturbed when, e.g. an animal cannot reach the relevant goal. I will set out with a classical conceptual model of behaviour involving the aspects of motivation, behaviour and negative feedback whenever a goal state is reached. I will then extend this model and show how motivational states and feedback can be reflected by affective states and selection of behaviour viewed as decision-making. I therefore integrate aspects of classical models for proximate control (motivation, goal) with ideas on affective states (valence, value, and wanting/liking) and decision making (prioritising options), showing that, to a large extent, these fields study the identical scientific questions and phenomena using different wording. I will end with an extended model integrating the views of these different fields. A conceptual model is only of value if it can be related to an embodied implementation. It can be assumed that the mechanisms of behavioural control are physically implemented in the brain. Indeed, recent research based mainly on fMRI has indicated brain areas that are activated in experimental tasks such as valuating, weighing options, and decision making, all necessary for proximate control. These brain activation patters are therefore sensitive indicators of these processes. However, no such patterns have been identified that are specific for these tasks. Single-neuron networks may need to be identified to study the implementation of proximate control in the brain. Due to the large number of brain cells, the number of their connections and presumed redundancies in the brain, this information may remain inaccessible for a long time to come. In my opinion we should, however, be willing to look at proximate behavioural control mechanisms on a more general level and continue accepting that some brain processes currently remain black boxes. Observing behaviour remains a promising approach then, because behaviour can be viewed as the integrated output of the decision-making process inherent in the behavioural control mechanism. Considering affective states in the approach will lead to novel hypotheses. In addition, behaviour is easily accessible for experimentation. Even such an approach focusing on animal behaviour will increase understanding of the mechanisms of behavioural control and help to tackle welfare issues closer to their roots.

Effects of dietary protein and amino acid supply on damaging behaviours in pigs kept under diverging sanitary conditions

Yvonne Van Der Meer[1,2], Walter J.J. Gerrits[2], Bas Kemp[3] and J. Elizabeth Bolhuis[3]
[1]De Heus Animal Nutrition, Rubenssstraat 175, 6717 VE Ede, the Netherlands, [2]Wageningen University, Animal Nutrition Group, P.O. Box 338, 6700 AH Wageningen, the Netherlands, [3]Wageningen University, Adaptation Physiology Group, P.O. Box 338, 6700 AH Wageningen, the Netherlands; liesbeth.bolhuis@wur.nl

There is a strong incentive to reduce crude protein (CP) levels in pig diets to increase efficiency. Reducing dietary CP without affecting growth seems possible if supplementary essential amino acids (AA) are added to the diet. It has been stated, however, that low dietary CP levels may increase the occurrence of damaging behaviour such as ear and tail biting. This could especially hold for pigs with suboptimal health which may have higher requirements for particular AA. We studied the effect of dietary CP level and AA profile on the behaviour of pigs kept under diverging sanitary conditions. In a 2×2×2 factorial design, 64 groups of 9 tail-docked growing-finishing boars (n=576) were subjected to low (LSC) vs high sanitary conditions (HSC), and fed a normal (NP, NRC, 2012) vs low CP diet (LP, 80% of NP) *ad libitum*, with an AA profile based on the composition of deposited proteins or this profile supplemented with threonine, tryptophan and methionine. HSC pigs were vaccinated (6 diseases) and received antibiotics, after which they were kept in a disinfected part of the farm with a strict hygiene protocol. LSC pigs were kept in non-disinfected pens to which manure from another farm was added fortnightly. At 15, 18, and 24 weeks of age, tail and ear damage was scored. At 20 and 23 weeks, frequencies of biting behaviour and aggression were scored using behaviour sampling, continuous recording for 10×10 min per group per week. LSC pigs showed higher acute phase protein levels and pleuritis scores than HSC pigs, confirming a difference in health status. LSC led to more ear biting (+40%) and more pigs with ear damage (41 vs 33%) compared with HSC (P<0.05). The supplemented AA profile reduced ear damage (33 vs 41%, P<0.05) and, in LSC pigs only, also ear biting (-18%, SCxAA profile, P<0.01). Regardless of AA profile or sanitary status, LP pigs showed more ear biting (+20%, P<0.05), tail biting (+25%, P<0.10), belly nosing (+152%, P<0.01), other oral manipulation directed at pen mates (+13%, P<0.05), chewing toy (+61%, P<0.01) and aggression (+30%, P<0.01) than NP pigs with no effect on ear or tail damage. In conclusion, both low sanitary conditions and a reduction of dietary protein increase the occurrence of damaging behaviours in pigs and therefore may negatively impact pig welfare. Attention should be paid to the impact of dietary composition on pig behaviour and welfare, particularly when pigs are kept under suboptimal (sanitary) conditions.

Explaining daily feeding patterns in pigs: modelling interaction between metabolic processes and circadian rhythms

Iris J.M.M. Boumans[1], Eddie A.M. Bokkers[1], Gert Jan Hofstede[2] and Imke J.M. De Boer[1]
[1]Wageningen University, Animal Production Systems group, P.O. Box 338, 6700 AH Wageningen, the Netherlands, [2]Wageningen University, Information Technology group, P.O. Box 8130, 6700 EW Wageningen, the Netherlands; iris.boumans@wur.nl

Growing pigs housed under conventional conditions typically show two feeding peaks during the day: a smaller peak of feed intake shortly after the onset of light and a larger peak before dark. Feeding behaviour is under control of metabolic processes and affected by circadian rhythms. How circadian rhythms and metabolic processes interact over the day and cause the observed feeding patterns is not understood. Understanding this could provide valuable insights into feeding behaviour and the effect on growth and productivity under imposed conventional housing conditions. The aim of this study was to gain more understanding of the causation of feeding behaviour and to explain the observed daily feeding patterns with two peaks in pigs. We developed an agent-based model in Netlogo in which feeding behaviour over the day emerged from interactions between energy metabolism and circadian rhythms. Energy metabolism included processing of feed in the gastrointestinal tract and signalling in the body about energy absorption and use in the short and in the long term. Circadian rhythms included signals affecting energy expenditure and storage via catabolic and anabolic pathways (e.g. levels of melatonin, leptin and cortisol). These processes affected hunger and satiation of an agent (pig), which caused feeding behaviour via a motivational decision model. After parameterisation, the typical feeding pattern in pigs with a smaller peak in the morning and larger peak in the afternoon emerged from the model. Varying the energy content in the diet affected the size of peaks, while varying the light period affected the duration of peaks (e.g. merging peaks into one peak under 24 h light). Emerging feeding patterns such as daily feed intake, meal frequency and feeding rate were tested by comparing them to those observed in empirical studies. The model explains how metabolic processes and circadian rhythms affect the metabolic energy balance over the day and can cause these daily feeding patterns in pigs. Furthermore, the model shows that feed intake in the late afternoon is most optimal for growth. The developed model gives a better understanding of the pig's motivation to feed and how feeding behaviour can affect growth and productivity of pigs. The model will be further developed to study the role of social interaction in the causation of feeding patterns in group-housed pigs.

Effect of docking length on tail directed behaviour and aggression in finisher pigs

Karen Thodberg[1], Torben Jensen[2] and Karin Hjelholt Jensen[1]
[1]Aarhus University, Animal Science, Blichers Allé 20, 8830 Tjele, Denmark, [2]SEGES Pig Research Centre, Axeltorv 3, 1609 Copenhagen, Denmark; karen.thodberg@anis.au.dk

Tail docking is a widely used preventive treatment to avoid tail biting. Some studies show that docking reduces the risk of tail biting, but the connection between docking length and tail directed behaviour is not clear. Furthermore, it has been speculated that a reduced tail length could influence communication between pigs, due to misinterpretation of the tail posture. The aim of this study was to explore the connection between: (1) tail directed behaviour and tail biting and; (2) the level of aggression in relation to standardised and different docking lengths. We compared the behaviour in pens with finisher pigs that were docked litter wise at day 3 pp (median), leaving 2.9 cm (Quarter), 5.7 cm (Half) or 7.5 cm (Three quarters) of the tail. Pigs from 84 pens in 4 production herds were included and housed according to treatment in groups of 18-25, in pens with different degrees of slatted floor. Two of the herds fed the pigs with dry feed ad. libitum, and the last two herds used restricted liquid feed. Behaviour was registered weekly until 7 weeks after introduction to the finisher section, or until an eventual tail biting outbreak. Data were analysed in either a linear mixed model (activity; scan sampling) or a generalised linear mixed model (tail contacts, aggression; behaviour sampling). Activity level was not affected by docking length but decreased ($F_{6, 358}=17.88$; $P<0.0001$) over weeks and increased with group size ($F_{1,125}=29.98$; $P<0.001$). Likewise, the probability of aggressive events was not affected by docking length but decreased after the first three weeks ($F_{6,362}=5.01$; $P<0.0001$) and was affected by herd ($F_{3, 71}=12.29$; $P<0.0001$). Probability of tail contacts depended on docking length ($F_{2, 74}=5.40$; $P<0.01$) and was higher in Three quarters (0.80±0.05) compared to Quarter pens (0.57±0.05; $P=0.002$) and tended to be higher in Half (0.71±0.06) compared to Quarter pens (0.57±0.05; $P=0.07$). Tail contacts were more likely with increasing group size ($F_{1,362}=8.44$; $P<0.05$) and less likely in the last week ($F_{6,362}=2.13$; $P<0.05$). In the same study we found that pigs in Half pens ($P<0.05$) had, and pigs in Three quarters pens tended ($P=0.06$) to have a higher risk of tail biting compared to pigs in Quarter pens. These findings substantiate the connection between tail directed behaviour and tail length, but question whether reduced tail lengths affect aggressive behaviour. As the activity level did not differ between pens with different docking lengths, the increase in tail interest in pens with longer tailed pigs is most likely attributed to the higher accessibility and visibility of long compared to short tails.

Pigs' fighting ability in the application of game-theory models to address aggression

Irene Camerlink[1], Gareth Arnott[2] and Simon P Turner[1]
[1]Scotland's Rural College (SRUC), Animal Behaviour & Welfare, Animal and Veterinary Sciences Research Group, West Mains Rd., EH9 3JG Edinburgh, United Kingdom, [2]Queen's University Belfast, Institute for Global Food Security, School of Biological Sciences, University Road, BT7 1NN Belfast, Ireland; irene.camerlink@sruc.ac.uk

Aggression between pigs is a longstanding animal welfare issue. Game-theory models from behavioural ecology contribute to understanding aggressive behaviour and have been proposed as a method to address pig aggression. In game-theory models, fighting ability (also called resource-holding potential; RHP) is an important variable. Body weight is commonly used as a proxy of RHP, assuming that the heaviest animal will win. However, in pigs it is not uncommon that the smaller individual wins; suggesting that weight is a poor proxy. We hypothesize that personality traits contribute to RHP, with the more proactive and aggressive contestant being more likely to win. Pigs (n=316) were studied in the backtest (2 wk age); two resident-intruder tests (wk 9), and dyadic contests (C1 wk 10; C2 wk 13). All tests were ethically approved and had end-points in place, preventing injury other than skin lesions and limiting contests to 30 min (average contest time was 5 min). Pigs with more escape attempts in the backtest, which indicates a proactive coping strategy, were more likely to win contest 1 than pigs with no or few attempts (C1: winners 1.4±0.1 attempts; losers 1.1±0.1; P=0.03; C2 P=0.20). Pigs with a more aggressive personality, as assayed through the resident-intruder test, were more likely to initiate aggression in C1 (P=0.03) and more likely to win C2 (P=0.04). Backtest and resident-intruder scores were unrelated (P>0.10). Heavier pigs were more likely to win C1 (P=0.01) but not C2 (P=0.09). Males won most often when staged against a female (both contests P<0.001), even when smaller, and displayed different agonistic behaviour (e.g. foaming P<0.001). Resource holding potential was thus most strongly determined by sex when males fought females, and further depended on aggressiveness, body weight, and coping strategy. These results highlight that RHP is a complex multidimensional trait influenced by a range of factors, knowledge of which will contribute to a more accurate application of game-theory models in animal behaviour science.

Different behavioural responses of stabled horses to delayed feeding and polymorphisms of the dopamine D4 receptor gene

Moyuto Terashima[1] and Shigeru Ninomiya[2]
[1]*Gifu Univeristy, Graduate School of Applied Biological Sciences, Yanagido 1-1 Gifu city, 5011193, Gifu, Japan,* [2]*Gifu Univeristy, Faculty of Applied Biological Science, Yanagido 1-1 Gifu city, 5011193, Gifu, Japan; nino38@gifu-u.ac.jp*

Animal behaviour is a useful indicator of animal welfare, but individual differences in behavioural responses to environmental challenges make animal welfare assessment difficult. Associations have been reported between some horse temperaments (e.g. curiosity and vigilance) or behavioural responses to frustration, and the dopamine D4 receptor gene (DRD4) in stable horses. For this study, it was hypothesised that horses without the A allele in the DRD4 polymorphism display a greater quantity of behaviourally active responses to environmental challenge (delayed feeding time). This study examined 14 horses at two horse riding institutes. As an experimental treatment, their feeding times were delayed for 30 min. Treatments were conducted randomly three times per animal on different days. Behavioural responses during the 30 min delay were observed using continuous behavioural recording. Behavioural categories included eating, resting, standing, moving, investigating (outside their stall or bedding), and frustration behaviours. Time budgets of respective behaviours during the three time delays were averaged for each animal. General linear models incorporated behavioural data as dependent variables, with polymorphism (with or without A allele) as a fixed factor and the riding institute as a random factor. The average time budget of resting during the 30 min delay was significantly lower in horses without the A allele than in those with the A allele (10.2% (S.D.=14.2) vs 28.4% (S.D.=26.2), P<0.05). The average time budget of investigation during the delay was significantly higher in horses without the A allele than in those with the A allele (38.6% (S.D.=30.7) vs 9.4% (S.D.=6.9), P<0.05). No difference was found for other behaviours. These results and those of previous studies demonstrate that horses without the A allele display more behavioural responses to environmental challenges, which must be considered when managing horses and evaluating their welfare.

Feather pecking in laying hens – do control and case flocks differ regarding compliance with recommendations?

Lisa Jung, Christine Brenninkmeyer and Ute Knierim
University of Kassel, Farm Animal Behaviour and Husbandry Section, Nordbahnhofstr. 1a, 37213 Witzenhausen, Germany; lisajung@wiz.uni-kassel.de

Despite extensive research, feather pecking (FP) is still a great behavioural problem on commercial laying hen farms. One challenge in knowledge transfer is the multifactorial nature of FP which leads to largely deviating effects of single factors on individual farms. Possibly, a general farm optimization rather than the fulfillment of single requirements might be an important aspect in the prevention of feather pecking. Therefore, we wanted to know whether problem (FP) and non-problem (non-FP) farms are distinguished by the extent to which they comply with common recommendations for the prevention of FP. Based on 15 existing practice recommendations and results from 84 experimental and 21 epidemiological studies we compiled a list of 37 preventive factors. Using data from three past cross sectional studies with conventional and organic laying hen flocks in eight European countries we selected 60 non-FP flocks (at least 98% hens with very good plumage) and 105 FP-flocks (≥10% hens with highly damaged feathers or featherless areas >5 cm^2). Flocks were frequency matched for age, beak trimming, free-range and egg shell colour. Relating to the recommendation list, data on 18 variables were available which were dichotomized into complying/not complying. Although non-FP flocks complied with a significantly higher number of recommendations (min, median, max: 11.1%, 44.4%, 72.2%) than FP-flocks (min, median, max: 11.1%, 38.9%, 66.7%) P=0.009, U=2,381.0, Mann-Whitney-U Test), the difference in numbers of complied items was very small. Analyses are being continued in order to better grasp differences between problem and non-problem farms.

Confinement before farrowing affects performance of nest building behaviours but not progress of parturition in prolific sows

Christian Fink Hansen[1], Janni Hales[1], Pernille Weber[1], Sandra Edwards[2] and Vivi Aarestrup Moustsen[3]
[1]University of Copenhagen, Department of Large Animal Sciences, Groennegaardsvej 2, 1870 Frederiksberg C, Denmark, [2]University of Newcastle, Newcastle upon Tyne, NE1 7RU, United Kingdom, [3]SEGES Pig Research Centre, Innovation, Vinkelvej 11, 8620 Kjellerup, Denmark; cfh@sund.ku.dk

The effects of confinement prior to farrowing on the performance of nest building behaviour and progress of parturition were investigated using hyper prolific Danbred sows. Forty first parity and 40 second/third parity sows were allocated to one of two treatments: loose housed (L) or confined (C). All sows were housed in a freedom farrowing pen with an option of confinement and free access to a straw rack with long stemmed straw. Loose sows were loose housed throughout the observational period and confined sows were confined from 2 days before expected farrowing until the completion of parturition. Sows were video recorded from 2 days before expected farrowing until birth of the last piglet, and behaviours (biting/rooting pen fittings, straw directed and rooting/pawing floor) and postures (lying sternal, lying lateral, sitting and standing/walking) of the sows during the last 24 hours before farrowing were registered continuously. The time of birth of every piglet was registered from the video recordings, and it was noted if the piglet was alive or stillborn. Treatments were compared by use of linear models with treatment and parity as fixed effects. Results showed that confinement did not influence duration of the nest building period, but affected the performance of nest building behaviour. Loose housed sows tended to perform more nest building behaviour during the nest building period than confined sows (817 (95% CI: 713;929) vs 686 (95% CI: 590;789) s/h/sow, P=0.08). Loose housed sows had fewer bouts of nest building behaviours than confined sows (4.6±0.48 vs 6.1±0.48 bouts/sow/h, P=0.03) but mean duration of bouts was longer (154 (95% CI: 136;173) vs 98 (95% CI: 83;114) s/interval, P<0.001). Loose housed sows tended to spend a greater proportion of time during the nest building period standing/walking (21±1.33 vs 17±1.33 min/h, P=0.05). No differences were found in total born per litter (L: 17.4±0.4; C: 17.6±0.4, P=0.70), the duration of farrowing (L: 283 min (95% CI: 244;324); C: 258 min (95% CI: 222;297), P=0.38), mean birth interval (L: 17.4 min (95% CI: 15.1;19.9); C: 15.7 min (95% CI: 13.5;18.0), P=0.30), or number of stillborn (L: 0.5 (95% CI: 0.3;0.8); C: 0.7 (95% CI:0.5;1.1), P=0.18). In conclusion, confining the sows during nest building decreased the performance of nest building behaviours, but did not prolong progress of parturition compared to loose housed sows.

Behavioural, physiological and scientific impact of different fluid control protocols in the rhesus macaque (*Macaca mulatta*)

Helen Gray[1], Henri Bertrand[2], Claire Mindus[1], Paul Flecknell[2], Candy Rowe[1] and Alexander Thiele[1]
[1]*Institute of Neuroscience, Newcastle University, Newcastle upon Tyne, NE2 4HH, United Kingdom,* [2]*Comparative Biology Centre, Newcastle Medical School, Newcastle upon Tyne, NE2 4HH, United Kingdom; helen.gray1@ncl.ac.uk*

Rhesus macaques are a widely used model throughout behavioural neuroscience due to their advanced cognitive abilities and the similarities in brain structure with humans. To motivate macaques to engage in a neuroscience task, researchers may implement fluid control protocols. These protocols limit the amount of freely available fluid animals receive daily, allowing the monkeys to earn additional fluid rewards for performance of correct trials in the laboratory. The impact of these protocols is controversial, but poory understood, with concerns that animals may become dehydrated or experience poor welfare. We tested two fluid control protocols against a control of free access to water to assess effects on home cage behaviour, physiological measures of hydration status and motivation to perform in a laboratory task. Following a 12-day period of free access to water (when animals were not working in the laboratory), four male rhesus macaques received four-week blocks of a 5-day fluid control protocol (5 days of fluid control followed by 2 days of free water access at the weekend) and a 7-day fluid control protocol (fluid control on every day of the week). The macaques were videoed in their home cages and behavioural measures of welfare were scored blind. Behavioural categories included: abnormal behaviours, self-directed behaviours, consumption, social behaviours and locomotion. Physiological measures of hydration state were also taken throughout the study and compared with a non-restricted control group of similar age. Despite concerns surrounding these protocols, there were no differences between blood measures of hydration taken in any of the three conditions (free acess, 5-day fluid control or 7-day fluid control) and urine became more concentrated during fluid control protocols, indicating well-functioning kidneys. The impacts on behaviour were limited, although levels of foraging decreased on a stricter fluid control protocol and motivation to drink increased, coinciding with higher levels of performance in the laboratory tasks. Overall, fluid control protocols had little measurable impact on the welfare of rhesus macaques whilst ensuring that scientific data of high quality could be obtained.

The role of sexual experience on behavioural responses of rats to natural oestrus odours and odorants

Birte L Nielsen[1], Nathalie Jerôme[1], Audrey Saint-Albin[1], Christian Ouali[1], Sophie Rochut[2], Emilie-Laure Zins[2], Christine Briant[3,4], Elodie Guettier[5], Fabrice Reigner[5], Isabelle Couty[3], Michèle Magistrini[3] and Olivier Rampin[1]

[1]NBO, INRA, Université Paris-Saclay, Jouy-en-Josas, France, [2]UPMC, Paris VI, Paris, France, [3]PRC, INRA, Nouzilly, France, [4]IFCE, R&D, Nouzilly, France, [5]PAO, INRA, Nouzilly, France; birte.nielsen@jouy.inra.fr

Previous results from our lab have shown that sexually experienced male rats are able to detect the smell of oestrus in faeces from females – not only of their own species, but also from mares and vixens, indicating some commonality in oestrus odours across species. Three experiments were conducted to investigate if sexual experience affects the behavioural response of male rats to natural oestrus odours and if it is possible to identify the constituent odorants responsible. These could potentially be used to develop a heat-detection test for mares. In the first experiment, male Brown Norway rats (n=16) were exposed before and after sexual experience to four odours: 1-hexanol (herb odour; control), a ketone mixture (candidate odorants), and faeces from mares and rats in oestrus. These were presented one at a time to individual rats in a Latin square design during a 30-min test. More penile erections were observed during tests when rats were exposed to the mare and the rat faeces as well as the ketone mixture compared to 1-hexanol, both when sexually naïve (1.4, 1.4, 1.5, 0.4 (±0.3), respectively; P=0.032) and experienced (3.2, 3.6, 2.8, 0.9 (±0.5); P=0.002). In the second experiment, male rats (n=44) were tested using a habituation/dishabituation protocol. When the ketone mixture used in experiment 1 was presented first, the rats were able to distinguish between this and another single-molecule ketone (n=23; P=0.028), which has previously been found to elicit erections in male rats. This was not evident when the order of odour presentation was reversed (n=21; P=0.936). Finally in the third experiment, five faeces samples each from male, oestrous and di-oestrous rats and horses (n=30) were analysed using gas chromatography-mass spectrometry (GC-MS). The samples were tested for the presence of three ketones and 1-hexanol using menthol as an added calibrating compound. Although significant intra-species differences in faecal ketone composition were found between males and females in oestrus and di-oestrus, these differences were dissimilar across species. Also, 1-hexanol was present in all six faeces types and thus a proper control odorant. The results indicate that the ketones used here may not be specific oestrus odorants, but may share volatile characteristics with natural oestrus odours. Indeed, ketones as a chemical group may play a significant role in relation to oestrus odours and their potential commonality across mammalian species.

Proximate mechanisms of control of locomotion in horses: corrective input in central pattern generators

Hilde Vervaecke[1], Ella Roelant[1], Sandra Nauwelaerts[2] and Peter Aerts[2]
[1]Odisee University College, Agro-& Biotechnology, Hospitaalstraat 23, 9100 Sint-Niklaas, Belgium, [2]University of Antwerp, Laboratory for Functional Morphology, Universiteitsplein 1, 2610 Antwerp, Belgium; hilde.vervaecke@odisee.be

On a proximate level, rhythmic motor patterns are controlled by central pattern generators (CPGs), neural networks that can endogenously (i.e. without sensory or central input) produce rhythmic patterned outputs. With the corrective input of sensory feedback, these motor patterns can be altered to deal with environmental information. We tested the extent to which horses show evidence of underlying CPG-activity and corrective central input when dealing with visual information on obstacles. Four different tests were performed in an arena on 40 sport horses. (1) When horses were halted after having stepped with the front legs over a short pole, and the pole was subsequently visibly pulled from under the horse, we saw that in 88% of the tests, they lifted the first moved hind foot, futilely. This suggests that a CPG had started and unfolded till the end, without correction by central information on the sudden absence of the obstacle. A minority of the horses did not lift a hind foot, showing the effect of corrective central input. (2) Horses were halted with a short pole just in front of the front legs, with reduced vision by eyepatches. In the majority of the tests they correctly lifted the front legs after a pauze of 10 seconds (73%), 20 seconds (68%) and 30 seconds (74%) and these differences were not significant (P=0.329). The centrally stored information that can regulate the start of a new CPG, can span a time interval that is longer than the reputedly 8 seconds short term memory. (3) When horses were halted, after stepping over a short pole with the front legs, with reduced vision to block sight on the pole under their belly, the horses correctly lifted the back legs after 10 seconds (71%), 20 seconds (76%) and 30 seconds (77%). This time effect was not significant (P=0.356). The CPG that was started by the front legs easily lasted at least 30 seconds. (4) In another test-type, the horses were halted and a pole was slowly but visibly shoved under the belly. The horse had to lift the hind leg in order to avoid the pole. In 85% of the tests, the horses did not lift the hind leg and touched the pole. Possibly, corrective input is more difficult when the motion starts with the hind leg. Simple tests can show how sensory information can modulate centrally generated endogenous patterns in horses. This information on the proximate regulation of locomotory behaviour can help to understand biological limitations of voluntary control of horse locomotion.

May calving site selection of multiparous group housed dairy cows be influenced by the presence of amniotic fluids?

Maria Vilain Roervang, Mette S Herskin and Margit Bak Jensen
Aarhus University, Department of Animal Science, Behaviour and Stressbiology, Blichers Allé 20, 8830 Tjele, Denmark; maria.vilainrorvang@anis.au.dk

Among ungulates, amniotic fluids and placenta contain olfactory cues that stimulate maternal behaviour and facilitate establishment of the mother-offspring bond. On pasture, cow's attraction to amniotic fluids starts hours before calving, and might play a role in calving site selection. This pilot study investigated calving site selection when amniotic fluids were present in the deep bedding (as a result of a previous calving). Ten multiparous Holstein cows moved to one of two group pens 16 days (range 4-29) prior to expected calving were included. Each pen included a group area (9×9 m) connected to 6 secluded areas (4.5×3 m each). Cows could move freely between all areas. Time of calving, location of breaking of the amniotic sac and birth place were recorded. In the first group, all five cows calved in the group area within a radius of approximately one cow length. The first cow calved where her amniotic sac broke. The four subsequent cows calved within a radius of approximately one cow length of where the amniotic sac of the first calving cow broke, and their amniotic sac also broke within this area. The mean interval between calvings was 28.5 hours (range 16-61). In the second group, the first cow calved in the group pen, where her amniotic sac also broke. All bedding in a radius of 1 m of where the amniotic sac broke was afterwards removed for this particular cow. The second cow calved inside a secluded area, where her amniotic sac also broke. For the three subsequent cows their amniotic sac broke, and they also calved, within a radius of approximately one cow length of where the second cow calved. Mean interval between calvings for this group was 88 hours (range 31 to 120). As a result, seven out of 10 cows calved in an environment where amniotic fluids were present in the deep bedding as a result of a previous calving. These cows all calved within a distance of one cow length from where the previous calving took place. To analyze whether these cows calved at randomly chosen locations a 1-sample proportions test with continuity correction was used. The test indicated that cows did not select calving site randomly, but chose to calve where the previous calving had taken place (1-sample proportion test; χ^2=5.14, df=1, P=0.023). Whether this was due to the presence of amniotic fluids in the bedding we cannot conclude from this study. These preliminary observations indicate that choice of calving site might be affected by the presence of amniotic fluids from other cows, and this warrants investigation in a larger study. Such study should aim to clarify when and to what extent this attraction occurs and which underlying mechanisms are involved.

Energetic adaptations of Shetland pony mares to changes in climatic conditions and feed restriction

Lea Brinkmann[1], Martina Gerken[1], Catherine Hambly[2], John R. Speakman[2] and Alexander Riek[1]
[1]University of Goettingen, Department of Animal Sciences, Albrecht-Thaer Weg 3, 37075 Goettingen, Germany, [2]University of Aberdeen, Institute of Biological and Environmental Sciences, Tillydrone Avenue, Aberdeen AB24 2TZ, United Kingdom; lmann@gwdg.de

In cold winter times free-ranging herbivores in the Northern hemisphere can face a twofold challenge: while food quality and quantity is reduced, the energy demand to sustain body temperature is elevated. Studies on wild Northern herbivores suggest that they can reduce their metabolic rate during times of low ambient temperature and food shortage in order to reduce their energetic needs. It is, however, not known whether domesticated animals are also able to modify their energy expenditure with reduced energy supply under cold temperatures. The aim of our study was to determine effects of different seasonal climatic conditions and restricted food availability on energy expenditure and physiological parameters in an extensively kept domesticated horse breed. We exposed 10 Shetland pony mares to summer and winter conditions. In winter, five ponies received 100% of their maintenance energy requirement while the remaining five animals received only 60%. Field metabolic rate (FMR), total water intake (both measured by the doubly labelled water method), locomotor activity, body temperature, resting heart rate and body mass were determined. Ponies showed a considerably higher ($P<0.001$, $F_{1,8}=95.4$) FMR in summer (63.4 ± 15.0 MJ/day) compared to winter (19.3 ± 7.1 MJ/day). Similarly, locomotor activity, resting heart rate and total water turnover increased in summer ($P<0.001$) compared to winter. In winter restrictively fed animals (n=5) compensated for the decreased energy supply by a significant reduction ($P=0.017$, $F_{1,7}=9.6$) of 26% in FMR compared with control animals (n=5). Furthermore, resting heart rate and body mass were lower in restrictively fed animals (29.2 ± 2.7 beats/min and 140 ± 22 kg, respectively) than in control animals (36.8 ± 41 beats/min, 165 ± 31 kg; $P<0.05$). The mean daily body temperature amplitude of restrictively fed animals was with 1.20 ± 0.31 °C higher ($P=0.04$, $F1,8=5.9$) than in control animals (0.66 ± 0.23 °C) indicating elevated nocturnal hypothermia, while locomotor activity and resting time did not differ between both groups. We conclude that ponies acclimatized to changing climatic conditions by variations in their metabolic rate, behaviour and some physiological parameters. When exposed to energy scarcity, ponies, like wild herbivores, exhibited hypometabolism and nocturnal hypothermia. Procedures performed in our study were in accordance with the German animal ethics regulations and approved by the State Office of Lower Saxony for Consumer Protection and Food Safety.

Diurnal and seasonal variation in activity pattern of free-roaming cats in an urban city area in Tokyo

Kana Mitsui, Kiyomi Amano and Yoshie Kakuma
Teikyo University of Science, Department of Animal Sciences, 2-2-1 Senjusakuragi, Adachi-ku, Tokyo, 120-0045, Japan; g554002@st.ntu.ac.jp

In Japan community cats are widely promoted by government and animal protection organisations as a method of managing free-roaming cat populations in urban areas. Community cats are generally fed and controlled by volunteers and TNR (Trap-Neuter-Return) are performed so that the cat populations are able to live alongside human residents. The home range and activity patterns of free-roaming cats in fishing villages are known to be dependent on food availability,but there are few studies examining activity patterns of cats found in urban city areas where actual problems occur. Thus our study aimed to compare the activity of free-roaming cats between time of day and between three months to enable a targeted management of the cats. Individual features,activity and location of each cat were recorded while walking along appx. 2 km route within the study area (12.6 ha) in Tokyo. The census was carried out three mornings (8-12 am), afternoons (0-4 pm), and evenings (4-8 pm) in a month from August to October in 2015. The features of cats recorded were; adult/kitten, coat colour and pattern, sex, ear tipping(only done through TNR), collar, tail length, and other conspicuous body characteristics. The behaviour of each catat the instantthat it was first seen was recorded as one of the following 11 categories; resting, feeding, grooming, eliminating, scratching, urine-spraying, rubbing, nursing, searching, playing, chasing. Individuals were identified by photos afterwards. The number and the behaviour of cats were compared between three time periods of day and months by two-way ANOVA, followed by Scheffe's post-hoc tests. In toatal, 715 cats were found, and of these,159 cats were identified. Only nine cats had their ear tipped (6%), thirty were kittens (16%), and eleven had collars (7%). The number of cats was not different between times (P=0.07) or by months (P=0.66), but more kittens were found in evenings than in mornings (P=0.03). The most common behaviour was 'resting', followed by'searching','grooming', and 'feeding'. More cats groomed in mornings than afternoons (P=0.0007) and evenings (P=0.005), and in afternoons than evenings (P=0.02), and in August than in September and October (P=0.005).Although there is a financial support program available for neutering cats, few cats had ears tipped. Kittens may search for food along with their mothers in evenings when volunteers feed cats. Grooming may be performed in higher temperature as suggested in literature. In conclusion, free-roaming cats in an urban area were more active in evenings,and their activity may be affected by time of day and season. This information should be considered to develop more effcient control of cats in urban city areas.

Impact of aggressiveness on feeding behavior, pubertal development and fat androstenone of entire male pigs

Severine Parois[1], Geoffrey Bissianna[1], Sabine Herlemont[1], Marie-Jose Mercat[2], Benoit Blanchet[3], Catherine Larzul[4] and Armelle Prunier[1]
[1]*INRA, Agrocampus Ouest, UMR 1348 PEGASE, Domaine de la Prise, 35590 Saint-Gilles, France, Metropolitan,* [2]*IFIP, Domaine de la Motte au Vicomte, 35651 Le Rheu, France, Metropolitan,* [3]*INRA, UETP, Domaine de la Motte au Vicomte, 35653 Le Rheu, France, Metropolitan,* [4]*INRA, UMR 1388 GenPhySe, Chemin de Borde-Rouge, 31326 Castanet-Tolosan, France, Metropolitan; severine.parois@rennes.inra.fr*

In boars, social behavior could influence pubertal development and feeding behavior. The objectives of the study were to determine the relationships between aggressiveness, feeding behavior and pubertal development and their influence on fat androstenone. A total of 216 boars of two different crossbred genotypes were used. They were raised in groups of 12 pigs/pen. Animals were observed for 10 hours/day either at the beginning or at the end of fattening. Agonistic behaviors (fighting, hitting, biting, threatening, chasing) and skin lesions (an indicator of aggressiveness) were counted 48 hours after entering the fattening pen, in their pen before the first departure for slaughtering, and on the carcass. Data obtained from the electronic feeders were used to determine the feeding behavior (number and duration of meals, feed intake) over 96-hour periods. At the end of the fattening period, blood was sampled to measure estradiol and testosterone. At slaughter, fat was collected to measure androstenone. The protocol was approved by the local ethical committee. Prior to statistical analysis, variables were normalized if needed. For each genotype, Pearson correlations were calculated on variables corrected for the pen effect. In one genotype, significant positive correlations ($P<0.05$) were demonstrated between the number of skin lesions at slaughter, estradiol ($r=0.42$), testosterone ($r=0.26$) or androstenone ($r=0.38$). These correlations were low ($r<0.04$, $P>0.1$) in the other genotype probably because the measured concentrations of sex hormones were too low (38.6 ± 3.5 vs 55.1 ± 95 pg/ml for estradiol; and 1.43 ± 0.14 vs 2.86 ± 0.31 ng/ml for testosterone). In both genotypes, significant positive correlations were found between the numbers of sexual and agonistic behaviors at the beginning of fattening ($r=0.53$ and $r=0.28$). Finally, boars with the most numerous skin lesions at the beginning of the fattening period were also the biggest eaters in the middle of the fattening period ($r=0.41$ and $r=0.21$).

Nest-related behaviours in pigs before resting: observations on commercial farms with no bedding

Helena Telkanranta and Anna Valros
University of Helsinki, Faculty of Veterinary Medicine, P.O. Box 57, 00014 University of Helsinki, Finland; helena.telkanranta@helsinki.fi

Domestic pigs in semi-natural environments build two types of nests: preparturient sows build farrowing nests, and pigs of both sexes build resting nests for sleeping. Nest-building before farrowing is a well-studied behavioural need. However, little is known of whether nest-related behaviour in connection to resting is also innately motivated in pigs. The aim of this study was to collect descriptive data on pre-rest behaviours in intensively farmed pigs with no bedding on partly slatted floors. We hypothesized that a pre-resting motivation for nesting behaviour would be seen as frequent oral-nasal behaviours when starting to settle to a recumbent position, and as inserting the head or body under an object of plant-based material. The experiments were carried out on three farms in Finland. In Experiment 1 on 167 breeder gilts, aged 4 months, data were collected from video recordings by continuous observation for 2 h. For each event of settling to recumbency, the occurrence (yes/no) and target of an oral-nasal contact were recorded. Of the total 1079 observed events of settling to recumbency, 92% started with an oral-nasal contact. The most prevalent targets were the floor or another pig, in 52 and 31% of the events, respectively. In Experiment 2, pigs were provided with continuous access to objects suspended on pen walls: 128 growing-finishing pigs, aged 4 months, had pieces of recently harvested birch trees, and 51 suckling piglets, aged 2 weeks, had pieces of sisal rope. Behavioural data were collected from video recordings by continuous observation. Each pig was observed once, until immobile in a recumbent position. Data were collected on the frequencies and durations of inserting and moving the head or body under an object, and on remaining under the object when immobile. In growing pigs, 14% of the pigs moved the head or body under an object for durations ranging from 2 to 19 s; 3% of the pigs remained under the object when immobile. In suckling piglets, 16% of the piglets moved the head or body under an object for durations ranging from 3 to 25 s; 14% of the piglets remained under the object when immobile. The results show that in a considerable part of the observed events of settling down to rest, and regardless of the absence of bedding, pigs showed behaviours similar to manipulating nest materials. This suggests nesting behaviours may be internally motivated in pigs, which warrants further research.

Does locomotor behaviour of tethered cattle released to the paddock indicate their motivation during tethering?

Fuko Nakayama[1] and Shigeru Ninomiya[2]
[1]Gifu University, Graduate School of Applied Biological Sciences, 1-1 Yanagido,Gifu, 5011193, Japan, [2]Gifu University, Faculty of Applied Biological Sciences, 1-1 Yanagido,Gifu, 5011193, Japan; u8122027@edu.gifu-u.ac.jp

Tethering is widely used in Japanese beef cattle rearing, but prevents normal behaviour expression. In order to assess the stress caused by tethering, we investigated locomotor behaviour of tethered cattle released to the paddock. It was hypothesised that behaviour reflects their motivation to express normal behaviour built up by tethering and that cattle increase locomotor behaviour after tethering for a longer time. Additionally, tongue-playing and para-tongue-playing were also observed during tethering because of some frustration of being unable to express normal behaviour. We examined 25 Japanese black female cattle in 2014 (n=4 groups of 4, 41.5±21.7 months mean age) and in 2015 (n=3 groups of 3, 38.9±10.7 months mean age). The animals were tethered with a rope in an indoor pen and released to a neighboring outdoor paddock every day 09:30 am-14:30 pm, the tethering duration was 19 h (SHORT). Subsequently by preventing their release for 6 days or 7 days, we observed cattle responses to tethering duration prolonged: 163 h / 187 h tethering duration (LONG). After SHORT or LONG tethering had finished, animals were released to the paddock for 5 h. Then their behaviours were video-recorded for 60 min immediately after release. From the video, the frequencies of jumping and running were observed. Additionally, 8 tethered cattle were filmed from start of 187 h tethering (Day 0) until the day after the treatment finished (Day 8). From the video, we recorded tongue-playing and para-tongue-playing behaviours every day 18:00 pm – 06:00 am using a continuous behaviour recording technique. Tethering durations used in our experiment were possible in Japan because of weather or not having a paddock. In observation when released, the frequency (mean (±SD)) of jumping per cow per hour was 1.1 (±1.3) after SHORT tethering, 22.5 (±25.2) after LONG tethering (Wilcoxon signed rank test, $P<0.05$). The time of running was 4.1 (±3.6) s after SHORT, 37.4 (±32.7) s after LONG (Wilcoxon signed rank test, $P<0.05$). In observation during LONG tethering, the frequency (min / cow / 12 h) of tongue-playing and para-tongue-playing was Day 0: 8.1 (±10.5), Day 1: 12.3 (±13.4), Day 2: 27.2 (±22.7), Day 3: 25.1 (±17.6), Day 4: 25.9 (±21.5), Day 5: 26.8 (±24.9), Day 6: 37.0 (±41.1), Day 7: 26.8 (±21.3), Day 8: 12.8 (±14.2). The data of tongue and para-tongue playing during long tethering tended to be higher than those before and after the long tethering treatment (Day 0, 8). It was suggested that these results indicated cattle's motivation to express normal behaviour built up by tethering.

The potential of transects to reveal the effect of the environment during broiler welfare assessment

Xavier Averós[1], Neila Ben Sassi[1] and Inma Estevez[1,2]
[1]Neiker-Tecnalia, P.O. Box 46, 01080 Vitoria-Gasteiz, Spain, [2]IKERBASQUE, Basque Foundation for Science, Maria Diaz de Haro 3, 48013 Bilbao, Spain; xaveros@neiker.eus

Transects are a simple yet powerful tool to assess on farm broiler welfare, consisting in walks between areas delimited by feeders and drinkers to detect specific welfare problems. When refining the method, a relevant question is its ability to detect any potential effect due to changes in the rearing environment. We monitored 20 commercial flocks. These were classified according to genetics (Ross (R), Cobb (C), mixed (RC)), house type (open (O), closed (Cl)), bedding (rice hulls (RH), wood shavings (WS)), ventilation (longitudinal (L), transversal (Tr)), and light (incandescent (I), fluorescent (F), green/blue light (GB), LED). Numbers of birds observed lame, immobile, small, sick, dirty, with tail, head and back wounds, terminally ill, featherless and dead were collected on weeks 3 (W3), 5 (W5) and 6 (W6) of age using the i-WatchBroiler, a simple smartphone application in which birds with specific welfare problems are noted down during transect walks. Data were expressed relative to the estimated total number of birds/transect (%), assuming a random distribution of birds in the house. Bedding quality was assessed on 3 locations/transect (4-point scale), and stocking density (SD; kg/m^2) was estimated using sampled BW on W3, W5 and W6. House temperature (T) and relative humidity (RH) were continuously monitored, and a Temperature Humidity Index (THI) was calculated. Statistical models tested, on each welfare indicator (binomial distribution), the effects of genetics, SD, house type, bedding material/quality, ventilation, light, week, maximum THI deviation/period (either above (ATHI) or below (BTHI) the optimal range) and interactions SD×ATHI or BTHI, bedding quality×breed, SD×breed, and breed×ATHI or BTHI. All models included a random farm effect. Immobile birds were lowest ($P<0.01$) with I bulbs. Poor bedding had a negative impact on immobile ($P<0.01$), sick, small and dirty birds ($P<0.05$). Dirty birds were also highest with wood shavings ($P<0.05$). Tail wounds were higher in O houses ($P<0.05$). The effect of temperature deviations was modulated by other aspects such as SD, with higher ATHI being particularly detrimental for the occurrence of tail wounds when SD was relatively high ($P<0.01$). For BTHI models, the detrimental effect of increased SD on deaths was only detected for RC flocks ($P<0.01$). Other relevant interactions were found, such as bedding quality×genetics ($P<0.05$) on tail wounds; SD×genetics on tail wounds ($P<0.05$); genetics×ATHI on lame, small and dead birds ($P<0.05$); and genetics×BTHI on tail wounds ($P<0.01$). The rearing environment had a clear impact on the welfare status of broilers, and this was successfully detected using transects.

How to assess cumulative experience in laboratory animals?

Colline Poirier, Alexander Thiele and Melissa Bateson
Newcastle University, Institute of Neuroscience, Framlington Place, NE1 7RU Newcastle-upon-Tyne, United Kingdom; colline.poirier@ncl.ac.uk

Researchers have ethical and legal obligations to optimise the physical and emotional wellbeing of their animals. Furthermore, current European legislation places an emphasis on the animal's lifetime experience. However, current methods for assessing the cumulative experience of animals are poorly validated and suffer from a lack of sensitivity and/or specificity. The general goal of this work was to develop and validate a new method to assess cumulative experience in non-human primates. Neurobiological evidence (both post-mortem histology and *in vivo* imaging) recently accumulated in humans, non-human primates and rodents indicates that the amount of grey matter in the anterior hippocampus decreases with chronic stress and is correlated with frequency and/or intensity of depressive behaviours. Furthermore, the same measure increases with mood-enhancing factors (physical activity, environmental enrichment, anti-depressant drugs). We thus hypothesised that this neurobiological measure could be a sensitive biomarker of cumulative experience. Artificial weaning is a well-established early-life stressor in non-human primates. It is also known to have long-lasting detrimental effects on emotionality, social, sexual and maternal behaviours, as well as growth, immune responses and in some cases survival, inducing a poorer life time experience in individuals weaned earlier. We thus predicted that if the amount of grey matter in the anterior hippocampus is a good biomarker of cumulative severity, it should decrease with earlier weaning age. To test this hypothesis, we measured non-invasively the amount of grey matter in the anterior hippocampus of adult male macaques (n=12) using neuroimaging techniques (MRI) and determined whether this measure is influenced by the weaning age of the individuals. After controlling for covariates including age and brain size, a multiple regression analysis revealed a positive correlation between amount of grey matter in the right anterior hippocampus and weaning age (t=8.8; P=0.02). This result supports the hypothesis that the amount of grey matter in the anterior hippocampus is a sensitive biomarker of cumulative experience in non-human primates. Because the hippocampus is a brain structure extremely well conserved across taxa, a similar approach could be used to measure cumulative experience in many species used in bio-medical research, including rodents.

Characterisation of short and long-term changes in mechanical nociceptive thresholds in pigs following tail injury

Pierpaolo Di Giminiani[1], Mette Herskin[2], Emma Malcolm[1], Matthew Leach[1], Dale Sandercock[3] and Sandra Edwards[1]
[1]Newcastle University, Agriculture Building, NE1 7RU Newcastle upon Tyne, United Kingdom, [2]Aarhus University, Blichers Allé 20, 8830 Tjele, Denmark, [3]Scotland's Rural College, Roslin Institute Building, EH25 9RG Midlothian, United Kingdom; pierpaolo.di-giminiani@newcastle.ac.uk

To date, studies on pain arising from tail docking and tail biting in pigs have mainly focussed on non-evoked behaviours. The development of neuromas and possible abnormal neural activity prompted us to target changes in nociceptive thresholds. We investigated the feasibility of pressure algometry applied to the tail of unrestrained pigs to quantify sensitisation following simulated tail biting and neonatal tail docking. Mechanical nociceptive thresholds (MNT) were determined in 3 tail regions (1: proximal; 2: intermediate; 3: distal) with a pressure application measurement device (PAM). MNT were measured in pigs of 9 (n=41) and 17 weeks of age (n=67) at baseline and one week following resection. To evaluate the effect of time since injury, a subsample of pigs was tested at 8 (n=24), and 16 (n=38) weeks post-resection. Severity of injury was assessed using 3 treatments: 'Intact' (sham-resection); 'short tail' (2/3 of tail removed); 'long tail' (1/3 of tail removed). MNT were also recorded in 17 week-old pigs tail-docked at 3 days of age (n=23) according to the 3 treatments. The PAM induced clear withdrawal responses (response rate 95.2%). Pre-resection MNT were higher in region 1 vs 2 and 3 (P<0.05). Higher pre-resection MNT were recorded in 17 vs 9 week-old pigs (P<0.05). Intra-individual and intra-site variability in MNT was 30% with a fair-to-good level of consistency (Intra-class Correlation Coefficient). One week post-resection, MNT were lower in resected (P<0.05), but not intact tails. Similarly, MNT were lower than intact values at 16 weeks following resections at 9 weeks of age (P<0.05). For resections at 17 weeks of age, MNT were lower in long vs intact at 8 weeks after resection (P<0.05), but no significant difference was observed at 16 weeks following surgery (Two-way Repeated Measures ANOVA for all comparisons). No difference in MNT was recorded across treatments 17 weeks after neonatal docking. The PAM was an effective method for the assessment of MNT in intact and resected tails. Tail injury resulted in acute sensitisation. Long lasting effects were seen up to 16 weeks post-resection performed at 9 weeks, but were not apparent 16 weeks after neonatal docking or resections performed at 17 weeks. The current results provide novel information on the welfare compromise due to tail damage early or later in the life of pigs and inform assessment of the trade-off between tail docking and the risk of tail biting.

Noise around hatching hampers chick communication and reduces hatching synchronisation

Bas Rodenburg[1], Peiyun Li[1], Henry Van Den Brand[2] and Marc Naguib[1]
[1]Wageningen University, Behavioural Ecology Group, P.O. Box 338, 6700 AH Wageningen, the Netherlands, [2]Wageningen University, Adaptation Physiology Group, P.O. Box 338, 6700 AH Wageningen, the Netherlands; bas.rodenburg@wur.nl

Layer chicks hatch in noisy incubators, offering a very different environment from a mother hen. Noise may negatively affect communication among hatching chicks and may reduce hatching synchronisation. Furthermore, the noisy environment may negatively affect physical and behavioural development of the chicks, resulting in smaller and more fearful chicks. Therefore, the aim of this study was to investigate effects of noise during the hatching phase on communication during hatching, hatching synchronisation and physical and behavioural development of layer chicks. Layer eggs were incubated in two separate climate respiration chambers. In one of the chambers, noise from a commercial hatcher was played back, creating a noise level of approximately 90 dB. In the other chamber, no noise was played back resulting in a noise level of approximately 60 dB. From day 19 until day 22 of incubation, every four hours the number of hatched chicks was recorded and birds were marked and weighed. Chick vocalisations were recorded in each chamber by six wireless microphones and analyzed in samples of 10 s, spread across the hatching day. After hatching, chicks were housed in groups of 5 males and 5 females per treatment, with 14 pens per treatment and chicks were weighed at 4 and 6 weeks of age and tested in an open-field test at 5 weeks of age. Data were analysed using mixed models with fixed effects of noise and gender and a random effect of pen nested within treatment. Chicks from the noise treatment hatched later than chicks from the quiet treatment (62% hatched on hatching day 20 compared with 91%; $F_{1,24}=18.02$, $P<0.001$). Fewer vocalisations were recorded in the noise treatment than in the quiet treatment (6.1 vs 18.4 calls per 10 s; $P<0.001$). Chicks from the noise treatment also had a lower birth weight (42.6 vs 43.6 g; $F_{1,24}=10.69$, $P<0.05$), but this difference was no longer found in weeks 4 and 6. No differences between treatments were found in open-field behaviour at five weeks of age. Birds were active in the open-field with on average 57 steps and 182 calls. In conclusion, the negative effects of noise on communication, hatching synchrony and early life development indicates negative effects of loud incubators on chick development, but not on fearfulness. It needs to be investigated whether these negative effects carry-over to later life phases.

Effects of a water spraying system on lying and excreting behaviours of fattening pigs in heat stress

Hieu Nguyen Ba[1], Andre Aarnink[2] and Inonge Reimert[1]
[1]Wageningen University, Adaptation Physiology Group, De Elst 1, 6708 WD Wageningen, the Netherlands, [2]Wageningen UR Livestock Research, De Elst 1, 6708 WD Wageningen, the Netherlands; nbhieu@gmail.com

In intensive pig husbandry systems, pigs normally rest or lie on the solid floor and excrete on the slatted floor. This division changes when pigs are in heat stress. In heat stress, pigs move to the relatively cooler slatted floor for lying and, hence, excrete on the solid floor. Fouling of the solid floor leads to decreased hygiene conditions and increased ammonia emission in the pen, which has negative consequences for pig health and welfare. The aim of this study was to investigate whether a water spraying system on the slatted floor could aid pigs in heat stress to cool down and as a consequence they keep their lying area on the solid floor and their excretion area on the slatted floor. Twelve groups of 12 pigs (initial BW: 63.6±6.3 kg) were housed in 12 pens in a typical Dutch pig housing system. Each pen had a concrete slatted floor at the front of the pen (3.25 m^2), then a solid floor (5.5 m^2) and a metal slatted floor at the back (3.75 m^2). In half of the pens, treatment pens, water was sprayed from a system installed above the slatted floor at the back of the pen. The spraying system was turned on once per hour for 50 seconds from 8:30 to 19:30 on every day. Control pens had no spraying system. Feed and water were accessible *ad libitum*. Percentage of pigs lying and number of excretions per pig per hour in the different areas of the pen were scored from video recordings on eight days in August 2015 (average inside temperature: 26.4±1.5 °C) by 15 min scan sampling and continuous observations, respectively. Behavioural observations of lying and excreting were analysed using restricted maximum likelihood. Pigs in the treatment pens lied on the metal slatted floor less than pigs in the control pens (18.0 vs 21.8% of total number of pigs, P<0.001). However, treatment pigs excreted more on the solid floor than control pigs (1.1 vs 0.8 times per hour per pen, P<0.01). More specifically, in the treatment pens, pigs excreted significantly more on the solid floor during the 30 min after the spraying system was activated compared to the 30 min before (5.5 vs 4.0 times per ten min in all observed pens, P<0.05) which was not seen in control pens. These results show that the spraying system was effective in keeping pigs to lie on the solid floor in times of heat stress, but it was not effective in reducing the number of excretions on the solid floor. It was hypothesized that increased activity on the slatted floor during and directly after spraying caused pigs to move to the solid floor for excretion.

Age differences in exploratory and social behaviour in dairy cows

Alexander J. Thompson, Heather W. Neave, Daniel M. Weary and Marina A. G. Von Keyserlignk
Univeristy of British Columbia, Animal Welfare Program, 2357 Main Mall, Vancouver, BC, V6T
1Z4, Canada; alexithompson42@gmail.com

Dairy cows assess variation in feed availability and quality via exploratory sampling in time and space. This natural behaviour may be influenced by social interactions such as replacements, where one animal displaces another in order to gain access to a preferred feeding space. To date, little is known about the expression of exploratory feeding behaviour in indoor housing systems. In these systems, we expect the number of feeding locations visited to reflect the degree of exploration, but at times of high occupancy (e.g. right after fresh feed delivery) a greater number of locations visited may be due to competitive displacements, especially for younger animals that are likely to be socially subordinate. Our objective was to compare social and exploratory behaviours of primiparous and multiparous cows at the feed bunk at peak (07:00 – 09:00 and 17:00 – 19:00 h) and non-peak feeding times (all other times). Healthy Holstein cows (n=46 primiparous, n=81 multiparous) were housed in groups of 20 with access to 12 electronic feed bins. We recorded the total number of different feed bins visited (i.e. a measure of exploration at the feed bunk) and the total number of competitive replacements at peak and non-peak feeding times for 21 d postpartum. Daily behavioural data were analyzed by week relative to calving; effect of parity was analyzed using a repeated measures model. Parity did not affect the number of bins visited at peak feeding times; however, primiparous cows visited more bins than multiparous cows at non-peak feeding times in wk 1 (mean±SE; 10.7±0.1 vs 10.4±0.09), wk 2 (11.1±0.1 vs 10.7±0.07) and wk 3 (11.2±0.1 vs 10.8±0.08) after calving. Similarly, parity did not affect the number of replacements at peak feeding times during wk 1, 2, or 3; however, primiparous cows were more likely to be replaced at the feed bunk than multiparous cows at non-peak feeding times in wk 2 (17.1±0.7 vs 15.0±0.5), but not wk 1 (13.8±0.6 vs 12.9±0.4) or wk 3 (17.6±0.9 vs 16.5±0.6). During peak feeding times cows may be more likely to move to another feed bin because they were displaced, while at non-peak feeding times the number of bins visited may be more reflective of exploratory behaviour. Furthermore, primiparous cows appear to exhibit greater exploratory behaviour during non-peak feeding times compared to multiparous cows. Our results indicate exploration of the feed bunk is restricted by competition during peak feeding times, and at off-peak feeding times younger cattle are more motivated to explore than older animals.

Citizen science: a next step for applied animal behavior

Julie Hecht
The Graduate Center, City University of New York, Psychology, 365 5th Avenue, 6th Floor, New York, NY 10016, USA; dogspies@gmail.com

'Citizen science' describes public participation in scientific inquiry and research. Citizen science is associated with advances in fields as varied as population genetics, astronomy, ornithology, and entomology. Researchers receive assistance collecting, categorizing, translating, and analyzing data, and public participation projects have both scientific impact and proven researcher benefits. Participant learning outcomes are also possible for members of the general public engaging in these real-world science projects. In recent years, many citizen science projects have investigated questions about the animal kingdom, and participants have provided raw footage (e.g. animal behavior videos), count data (e.g. number of birds), occurrence data (e.g. wildlife sightings), and species identification (e.g. photo documentation), as well as other forms of engagement. Despite the growth and success of citizen science, valid concerns about citizen science remain, often relating to data quality. The disciplines that utilize citizen science do not shy from these concerns and instead address them head on by incorporating data screening and quality control checks. Citizen science projects can take a number of forms, and I present three main citizen science project models for public participation in applied animal behavior research: (1) researchers receive volunteer interpreted data; (2) researchers receive raw data; and (3) volunteers categorize or analyze researcher-provided content. To date, a number of animal behavior and animal cognition studies have successfully utilized these project models in their research design, such as Otter Spotter, NestWatch, Dognition.com, Project: Play with Your Dog, and the Canid Howl Project. The benefits and limitations of each model will be discussed, particularly as they pertain to data quality, bias, and participant engagement and learning outcomes. The citizen science movement continues to grow across the globe with the European Citizen Science Association, the Citizen Science Association (USA-based), and the Australian Citizen Science Association. Given the scope and breadth of the applied animal behavior field, opportunities for applied ethologists to utilize citizen science approaches will be discussed.

Can oral sucrose solution alleviate castration pain in piglets? Results of an evaluation using a behavioural test

Yolande Seddon[1], Krista Davis[1], Megan Bouvier[2], Joseph Stookey[1] and Jennifer Brown[2]
[1]University of Saskatchewan, Western College of Veterinary Medicine, 52 Campus Drive, Saskatoon, Saskatchewan, S7N 5B4, Canada, [2]Prairie Swine Centre, Department of Ethology, Box 21057, 2105 8th Street East, Saskatoon, Saskachewan, S7H 5N9, Canada; yolande.seddon@usask.ca

Oral sucrose is known to reduce pain responses in neonatal humans and rodents. This study evaluated sucrose for its ability to reduce piglet castration pain using a behavioural test previously shown to distinguish pain responses following castration. Male piglets (n=125), 5 days of age, were assigned to one of five treatments (25 piglets/treatment): (1) castrated with 3 ml of water given orally (C); (2) castrated with 3 ml of 30% sucrose orally before castration (SB); (3) castrated with 3 ml of 30% sucrose orally after castration (SA); (4) handled, given 3 ml of water orally and not castrated (SHAM); and (5) handled, given 3 ml of 30% sucrose and not castrated (SUGSHAM). There was no use of anaesthesia or analgesia for any castration. Piglet navigation time (NT) through a chute containing hurdles was tested at 0, 15, 30 and 45 min post-castration as an objective measure of pain. Treatment differences in NT at each time point were compared using Proc Mixed (SAS, 9.3). Mean (± SD) piglet NT across all treatments at each test point was 40±43 s at 0 min, 42±43 s at 15 min, 38±37 s at 30 min, and 42±40 s at 45 min. Between treatments, at 0 min, C piglets had a NT 50 and 47% longer than SHAM and SUGSHAM piglets respectively (P<0.05), with SA and SB being no different. At 15 min, the NT of C piglets was >42% greater than remaining treatments (P<0.05). At 30 min, NT of C piglets remained greater than SA, SHAM and SUGSHAM (P<0.05), with SB no different from any other treatment. By 45 min, NT of C piglets remained >34% greater than SA, SB, SHAM (P<0.05), but SUGSHAM was not different to any group. The longer NT of C pigs to SHAM and SUGSHAM confirm the handling chute is capable of distinguishing piglets experiencing castration pain from those not castrated. The shorter NT in SB piglets at 15 and 45 min and in SA at 15, 30 and 45 min compared to C piglets suggests that sugar provides some level of pain relief following castration.

External validity of single and multi-lab studies: considering reaction norms

Bernhard Voelkl and Hanno Würbel
University of Bern, Animal Welfare Division, Laenggassstr 120, 3012 Bern, Switzerland;
bernhard.voelkl@vetsuisse.unibe.ch

Every year large numbers of animals are used in biomedical research, but increasing evidence indicates poor reproducibility of results and high rates of translational failure. This is partly due to the limitations of the external validity of single studies, an issue that is also relevant to applied ethology. The imminent danger of producing false negative results has led to a testing culture where not only experimental conditions but all aspects of animal husbandry and not least the animals themselves have been highly standardized. While such rigorous standardisation was done in pursuit of lower error variance and higher internal validity of single studies, we argue that increased internal validity, achieved by eliminating biologically meaningful variation, has come at the cost of reducing the external validity of such studies and, hence, an increase of spurious findings that cannot be replicated. In order to escape the standardization fallacy we promote the conduct of multi-lab studies. Here, we show with an extensive simulation study how parallel testing in more than one laboratory can decrease the false discovery rate and increase accuracy and positive predictive value if interactions between treatment and laboratory environment exist. Taking a reaction norm perspective on animal physiology the latter should be expected being ubiquitous. Specifically, we focused on medium scale studies with 12 to 48 animals, a simple treatment/control design and effect sizes of 10 to 30% as such values are common in the literature. Sampling from distributions with specified variances and lab-by-treatment interactions we found that multi-lab designs using the same overall number of animals can increase the positive predictive value from e.g. 0.4 to 0.8 and decrease the false omission rate from 0.4 to <0.1. A higher positive predictive value increases the likelihood that a positive finding is due to a true effect and hence increases reproducibility and reduces translation failure in clinical trials, and a lower false omission rate means that a smaller number of abandoned lines of investigation are due to false negative findings. Both effects can, hence, contribute to reducing the number of animals 'wasted' for inconclusive research. Hence, we argue that increasing the external validity by multi-lab studies will increase reproducibility and decrease translational failure and, in the long-run, reduce the number of animals used for research as advocated by the 3R policy.

Relationship between sow conformation, farrowing floor type and posture change characteristics using accelerometer data

Stephanie M. Matheson[1], Robin Thompson[2], Grant Walling[3], Ploetz Thomas[2], Ilias Kyriazakis[1] and Sandra A. Edwards[1]
[1]Newcastle University, AFRD, NE1 7RU, United Kingdom, [2]Newcastle University, Open Lab, NE1 7RU, United Kingdom, [3]JSR Genetics Limited, Southburn, YO25 9ED, United Kingdom; stephanie.matheson@newcastle.ac.uk

The modern domestic pig has undergone genetic selection to maximise length of back and lean tissue growth rate for meat production. One consequence of this selection has been a change of sow body shape with increased difficulty in the control of posture changes. Sow control during posture changes may have important implications for piglet welfare and mortality due to crushing by the dam. Therefore, the aim of this study was to look at the relationship between sow body length, flank to flank measurement, front and hind leg conformation and sow movement characteristics recorded using rump-mounted accelerometers. Over a 21 week period, pure-bred Landrace sows (n=315) had their movements recorded over the farrowing period using rump-mounted accelerometers. Front and hind leg conformation data relating to leg shape and joint angles were scored. The floor surface of the farrowing crate (FAR-FL), the flooring type and pen size of the gestation housing, and sow parity were recorded. Data extracted from the accelerometers included the mean maximum acceleration (MAX-AC), the mean rate of acceleration change (AC-CHANGE), mean rate of change in movement around the X-axis (ROLL-CHANGE) and the mean rate of change of movement around the Y-axis (PITCH-CHANGE). Significant effects (Proc mixed model with sow-week as a repeated factor) on MAX-AC were straightness of the front legs when viewed from the front (P=0.0040) where X-shaped front legs gave a higher maximum acceleration, straightness of the hind legs when viewed from the back (P=0.051) and a trend for farrowing crate floor to have an effect (P=0.074). For AC-CHANGE, only FAR-FL was significant (P=0.025), where concrete-plastic slat floors gave a lower mean rate. No factors had an effect on ROLL-CHANGE. For PITCH-CHANGE, there was a significant effect of parity (P=0.0003), FAR-FL (P=0.023) where full metal slats gave a lower rate of change of movement around the Y-axis. There was also a trend for sows who placed their hind feet further under the body to have a lower rate of change of movement around the Y-axis (P=0.082). In conclusion, accelerometer data can be used to characterise posture changes and may be used to improve pig welfare, for example by identifying the physical constraints of the environment such as floor type or through selection of sows with better lying characteristics to reduce piglet crushing risk. This research was funded by the EU FP7 Prohealth project (no. 613574).

Porcine facial inference: facial biometrics as noninvasive proxy measures for innate aggression in piglets

Catherine Mcvey

North Carolina State University, Animal Science & Statistics, 2310 Stinson Dr, Raleigh, NC 27695, USA; cgmcvey@ncsu.edu

Ethics and economics both demand that research trials utilize the smallest number of animals necessary to achieve conclusive and reproducible results. Due diligence in the design and execution of experimental protocols can eliminate variance from extraneous environmental factors, but it can be difficult to control for intrinsic variability between individual animals. In small sample studies, failing to control the distribution of personality traits between treatment groups could increase the risk of selection bias and false positive results, but the behavioral tests needed to quantify these traits can be prohibitively time and labor intensive. Horse trainers have casually used facial features as a proxy indicator for numerous personality traits for centuries. Recent scientific studies have revealed correlations between Facial Width-to-Height ratios and aggression in humans and capuchin monkeys, but have not yet been extended to farm animal species. The purpose of this study was to assess the efficacy of facial biometrics in the prediction of individual differences in the aggressive tendencies of swine. To do this, facial photographs were acquired from 120 piglets at 24-48 hours of age. Algorithms developed using the image processing tools in MATLAB for a previous project with horses were adapted to porcine facial structures, and a total of 38 facial biometrics were extracted from each piglet image. At two weeks of age, prior to weaning and mixing, 46 of these piglets were then randomly paired by gender and subjected to an IACUC-approved paired encounter protocol with an unfamiliar conspecific in a novel environment. From behaviors observed during this encounter, piglets were assigned binary classifications for proactive and reactive aggression. Wilcoxon ranked sum tests were used to screen for facial metrics with categorical potential for either response, which were then added to a series of increasingly complex multiple logistic regression models optimized using the R software package. Measures considered by the industry to inform grouping decisions – gender and weight – did not yield statistically significant models. Addition of facial biometrics significantly improved the predictive potential of both models (P<0.001), but only yielded a strong R^2 value (0.50) for the reactive aggression model. Addition of interaction terms with fitness measures and the facial metrics of the opponent were needed to achieve greater accuracy with the proactive aggression model. Final models for both proactive and reactive aggression yielded strong R^2 values of 0.64 and 0.69 respectively, and classification accuracies of 82 and 91%, suggesting facial biometrics could be further developed as a non-invasive proxy measure for aggression.

Welfare assessment of low atmospheric pressure stunning in chickens

J E Martin[1,2], K Christensen[3], Y Vizzier-Thaxton[3], M A Mitchell[1] and D E F Mckeegan[4]
[1]SRUC, Animal Behaviour and Welfare, Animal and Veterinary Science Group, Easter Bush, Edinburgh, EH25 9RG, United Kingdom, [2]University of Edinburgh, R(D)SVS and The Roslin Institute, Easter Bush, Edinburgh, EH25 9RG, United Kingdom, [3]University of Arkansas, Centre of Excellence for Poultry Science, Fayetteville, AR 72701, USA, [4]University of Glasgow, College of Medical, Veterinary & Life Sciences, Bearsden road, Glasgow, G61 1QH, United Kingdom; jessica.martin@ed.ac.uk

A novel approach to pre-slaughter stunning of chickens has been developed: Low Atmospheric Pressure Stunning (LAPS), where birds are rendered unconscious by progressive hypobaric hypoxia through gradual decompression (280 s). We examined behavioural, electroencephalogram (EEG) and electrocardiogram (ECG) responses to LAPS in broilers, and interpreted their welfare impact. Trial 1 characterised the responses of broilers exposed to LAPS in 30 triplets at two temperature settings (TS3 (13-18 °C); TS4 (5-12 °C)). Trial 2 examined the influence of illumination and sham treatment in a 2×2 factorial design (20 pairs per treatment), at TS4 only. In each triplet/pair, one bird was instrumented for recording of EEG and ECG, and the behaviour of all birds was recorded. The trials were authorized by the University of Arkansas Institutional Animal Care and Use Committee (Protocol 15031). Birds consistently exhibited ataxia, loss of posture (LOP), convulsions and became motionless during LAPS. TS4 was associated with shorter LOP latencies than TS3 (TS4=62.3±1.1 s; TS3=57.5±1.2 s (P<0.001)), but in Trial 2 illumination had no effect (dark=54.7±1.3 s; light=55.9±1.2 s (P=0.250)).Durations of consciousness related behaviours (e.g. sitting) were predictably increased in the sham (sitting – LAPS=50.9±5.3 s; sham=216.3±17.4 s (P<0.001)). Head-shaking frequency (suggested discomfort behaviour) was higher in LAPS compared to sham (LAPS=0.8±0.2; sham=0.3±0.1 (P=0.010)), but illumination and TS had no effect. During LAPS, EEG spectral analysis revealed progressive decreases in median frequency and increases in total power (PTOT), followed by decreases in PTOT before the onset of an isoelectric state (brain death). TS had no effect on latency to F50<6.8 Hz (general anaesthetic plane/unconscious), but illumination and sham did (LAPS/dark=39.1±6.3 s; LAPS/light=53.6±11.8 s; sham/dark=12.8±5.2 s; sham/light=88.0±29.5 s (P<0.001)). Illumination increased activity and dark induced sleep, but slow-wave EEG was seen in both. Latency to pronounced bradycardia during LAPS was affected by TS (TS3=52.5±4.5 s; TS4=46.7±2.2 s (P=0.021)). In Trial 2, bradycardia was absent in sham and was not affected by illumination (dark=42.5±1.9 s; light=49.3±4.8 s (P=0.078)). This evidence has recently been presented to EFSA to facilitate approval of the method in the EU regulatory framework.

Accelerometer to quantify inactivity in laying hens with or without keel bone fractures

Teresa Casey-Trott[1,2], Michele Guerin[1,3], Victoria Sandilands[4] and Tina Widowski[1,2]
[1]*Campbell Centre for the Study of Animal Welfare, 50 Stone Rd E, Guelph, ON, N1G 2W1, Canada,* [2]*University of Guelph, Animal Biosciences, 50 Stone Rd E, Guelph, ON, N1G 2W1, Canada,* [3]*University of Guelph, Population Medicine, 50 Stone Rd E, Guelph, ON, N1G 2W1, Canada,* [4]*Monogastric Science Research Centre,SRUC, Auchincruive Campus, Ayr, KA6 5HW, United Kingdom; tcasey@uoguelph.ca*

Accelerometers are used in a variety of species to remotely monitor active and inactive phases for extended periods of time in studies involving pain quantification or mitigation, expression of behavioural patterns, and individual differences in activity. Although commercially available accelerometers are not yet validated in laying hens, behaviour with the potential to be recorded by accelerometers in poultry, e.g. periods of stationary, inactive (SI) behaviours (sitting, standing, and sleeping), are typically altered by pain, sickness, or injury. Our objectives were to validate the Actical accelerometer (Phillips Respironics, USA; 28×27×10 mm; 17.5 g) for quantifying inactivity in laying hens and compare inactivity levels between hens with severely fractured keel bones and hens with minimal to no keel damage. For validation, seven LSL-Lite hens wore Actical accelerometers on necklaces, and were simultaneously observed inside their home furnished cages for a one hour period on two consecutive days for a total of 14 hours of Actical inactivity (AI) counts paired with focal behaviour observation of SI. Actical inactivity was quantified by summing the total number of 15 second intervals with a zero recorded (no acceleration) for the activity count during the observation period. Pearson's correlation was used to assess the relationship between AI and SI. Agreement between the AI and SI was high with an adjusted R^2 value of 0.85 (P<0.0001) with <7% of the differences >±2 SDs from the mean difference. Therefore, the Actical can be used to accurately quantify the amount of time hens spend inactive. Following validation, 61 LSL-Lite hens were equipped with Actical accelerometers to measure inactivity level within their home furnished cages (60 hens/cage; 750 cm^2/hen) for a period of seven days. Hens were selected for inclusion in the study based on palpated keel status selecting only hens with severely fractured keels (F1; n=20) and hens with minimal to no keel damage (F0; n=41). A mixed model analysis of variance assessed the effect of keel status on AI. Statistical analyses were completed in SAS 9.4 with significance reported at P<0.05. Severely fractured hens spent less time motionless (1,280±202 zero counts) compared to F0 hens (1,461±196 zero counts; P=0.0358). Further investigation into inactivity differences related to keel status before and after acquisition of keel fractures or administration of analgesics is warranted.

The effect of pressure vest on the behaviour, salivary cortisol and urine oxytocine of noise phobic dogs

Anne-Maria Pekkin[1], Laura Hänninen[2], Katriina Tiira[2], Aija Koskela[2], Merja Pöytäkangas[2] and Anna Valros[2]
[1]University of Oulu, P.O. Box 800, 90014 University of Oulu, Finland, [2]University of Helsinki, P.O. Box 57, 00014 University of Helsinki, Finland; laura.hanninen@helsinki.fi

Fear of loud noises is a very common welfare problem in pet dogs. The severity of fearful reaction varies from mild anxiety to severe phobia. Commercial treatment vests have been tested on dogs to relief noise phobia and peripheral oxytocin has been suggested to be one of the stress-relieving mediators. The effect of vests has not been tested in a controlled situation. We tested: (1) if pressure vests calm noise phobic dogs in a double-blinded experiment; (2) the effect differs between vests with different pressure level; and (3) if he pressure vest increases oxytocin secreted in urine. A total of 28 noise sensitive dogs (2-11 years) both female (18) and male (10) were recruited. Ethical approved was obtained fro the study. Two commercial vests were individually customized for each dog: a DEEP pressure vest (c. 10-12 mmHg) and a LIGHT pressure vest (c. 2-3 mmHg). Each dog was tested three times in a semi-randomized order, either without vest (CONTROL), or with vests. The dogs' behaviour was filmed during three consecutive 2 minutes periods: PRE-NOISE, NOISE (70-73 dB firework sound) and RECOVERY. Saliva samples were collected four times during each test. Urine samples were collected once when the deep pressure vest was first fitted: before dressing the dog (OXY1) and after 30 min (OXY2), controlling human interaction. We present hear results from the differences between treatments (CONTROL, DEEP and LIGHT) in time spent near owner and saliva cortisol (linear mixed models), activity (Wilcoxon and Friedman), and urine oxytocin samples (Friedman), and correlations between behaviours, cortisol and oxytocin concentrations (Spearman-rank). The DEEP vest reduced median lying time in the dogs during NOISE (P<0.05). Saliva cortisol measured after test correlated positively with the total lying time during NOISE with DEEP and LIGHT vest (Spearman cor. 0.57 and 0.49, P<0.05 for both). Both DEEP and LIGHT vests increased the time the dogs spent near their owners during NOISE (P<0.05). Time spent near the owner when wearing the DEEP vest during the RECOVERY correlated positively with OXY1 and OXY2 (Spearman r: 0.44 and 0.50, P<0.05 for both). These results indicate that increased lying time was a sign of high stress level in the dogs and oxytocin might be related to the dog's tendency to seek owner support and the vests might affect this behaviour positively. We did not find a clear therapeutic effect of using pressure vests in noise phobic dogs in a experimental set-up.

Thermal imaging to monitor development and welfare in broilers

Katherine Herborn[1], Malcolm Mitchell[2] and Lucy Asher[1]
[1]University of Newcastle, Centre for Behaviour and Evolution, Institute of Neuroscience, Henry Wellcome Building, Framlington Place, NE2 4HH, Newcastle, United Kingdom, [2]SRUC, Roslin Institute Building, Easter Bush, EH25 9RG, Midlothian, United Kingdom; katherine.herborn@ncl.ac.uk

Skin temperature changes are a window into biological processes that alter body temperature, such as metabolic rate, stress or infection. Skin temperature can be determined remotely, without the need to handle or instrument animals, by infrared thermography: the measurement of radiation emitted from a surface. Such thermal imaging could allow automated, objective and importantly, non-invasive assessment of a variety of welfare problems on the farm, where links to skin temperature are established. With thermal imaging technology becoming increasingly affordable and available, we demonstrate the potential of this approach in novel applications in broiler (*Gallus gallus domesticus*) welfare. In 50 individuals, we compared whole body surface temperature to feather scores collected at weekly intervals through development to 36 days. Overall body temperature decreased and feather score increased with age, and were negatively correlated together, reflecting the growth of adult plumage that improved the insulation of the body (feather score × age t=-3.15, P=0.002). However, despite this strong correlation, surface temperature did also differ amongst individuals of the same feather score, and this was due to variation in underlying skin temperature. We explain variation in skin temperature using differences amongst individuals in activity level and behavioural markers of anxiety. On its own, feather gain is an energetically expensive process, and feathers provide thermal insulation and protection from injury. As such, using thermal imaging to track the rate of feather acquisition, or quantify within age class variability, could inform husbandry decisions. However, once understood, deviations from the expected relationship of feather score to surface temperature may simultaneously be used to monitor for important changes in underlying skin temperature. Potentially linked to inter-flock or individual variation in energy expenditure due to differences in behaviour or stress levels, these changes in skin temperature may have critical welfare and husbandry implications. As thermal imaging technology becomes more affordable, we anticipate that this objective and non-invasive approach will be adopted widely into applied, real-time animal welfare assessment.

Are rats able to detect stress odours in chicken droppings?

Vincent Bombail[1], Blandine Barret[1], Laurence Guilloteau[2], Christine Leterrier[2] and Birte L Nielsen[1]
[1]*NBO, INRA, Université Paris-Saclay, 78350 Jouy-en-Josas, France,* [2]*PRC, INRA, 37380 Nouzilly, France; vincent.bombail@jouy.inra.fr*

Rats can distinguish between faeces from stressed and non-stressed conspecifics. If their response is the same when excreta from chickens are used, this would indicate a specific odorous stress signature. In the first test, male Brown Norway (BN) rats (n=12) were habituated to be alone in a large T-maze (100×77 cm^2). A perforated metal ball (tea-ball) was placed in each arm of the maze, and each rat was tested 6 times on separate days. Each test lasted 5 minutes, with tests 1 and 6 done with empty tea-balls in both arms. In between, each rat was tested twice with rat faeces and twice with chicken excreta. In each test, the two tea-balls contained stressed (rats: chronic variable stress done as part of another experiment; chickens: 24 h transport stress) and non-stressed excreta, respectively, and always from the same species. The rats and sides were balanced in a Latin square (LS) design. No significant interactions were found between species and stress-state for any of the variables analysed. No significant differences were found on latency, frequency and duration of freezing by the rats when exposed to the different odours. The duration of first sniffing was longer for the rat faeces than the chicken excreta (3.6 vs 2.7 ±0.31 s; P=0.045), as was the total duration of sniffing (17.4 vs 12.2 ±0.92 s; P<0.001). In addition, rats sniffed the control excreta/faeces longer than they sniffed the stressed excreta/faeces (16.6 vs 13.0 ±0.92 s; P<0.001). In the second test, male BN rats (n=16) were habituated to be alone in a test arena (50×25 cm^2). The odour source was presented in a container with a perforated lid, allowing odour to escape but no contact with the source possible. Each test lasted 10 minutes, and all rats were tested 6 times, with tests 1 and 6 done with an empty container (no odour). In between, each rat was tested once with each of the four types of excreta/faeces in a LS design. No differences between odours were found in the latency to investigate the odour (mean=2.1 s). Rats spent longer in contact with (17.4 vs 13.7 s; P=0.020) and longer sniffing the stressed than the control odours (15.5 vs 12.6 s; P=0.036). No differences were found in movements between arena halves (mean=12.1 shifts), and the percentage of time spent in the odour half in contact with the odour container was larger for the stressed odours (10.4 vs 7.9%; P=0.028). The two different testing paradigms yielded different responses to stressed excreta/faeces, but within each test, the same differences were seen between stressed and non-stressed excreta/faeces independent of species of origin. This indicates that both in rats and chickens stress gives rise to specific volatile organic compounds.

Could the UV reflectivity of feathers phenotypically identify hens targeted for feather pecking?

Courtney L. Daigle[1], Patricia Y. Hester[2], Don C. Lay[3] and Heng-Wei Cheng[3]
[1]Texas A&M University, Animal Science, College Station, TX 77843, USA, [2]Purdue University, Animal Science, West Lafayette, IN 47904, USA, [3]USDA-ARS Livestock Behavior Research Unit, West Lafayette, IN 47904, USA; cdaigle@tamu.edu

The UV (UV) reflectivity of feathers in birds may be a useful phenotypic characteristic for identifying behavioural differences in feather pecking (FP). Although humans cannot perceive UV light, birds can. UV reflectivity is commonly used by birds for individual identification and status signaling. However, UV reflectivity is difficult to measure. Testosterone concentrations are positively associated with UV reflectivity of feathers. Since testosterone modulates serotonergic receptor activity with regards to anxiety, fear, and aggression, testosterone concentrations may be indicative of individuals with different physical and behavioural phenotypes associated with FP. As part of a preliminary investigation, twenty W-36 White Leghorn hens (48 wk of age) were placed in 2 floor pens (n=10 hens/pen) with access to a feed trough, water, with wood shavings as litter on d-1. At placement, each hen was given a unique identifier using livestock marker, and all hens were video recorded for 2 consecutive days (d0, d1) at 08:00 – 08:30 and 14:30 – 15:00. The number of gentle feather pecks, severe feather pecks (SFP), and aggressive pecks given and received by each hen was recorded and summed across d0 and d1. On d3, a 3 ml blood sample/bird was collected from the brachial vein after oviposition and analyzed for testosterone. Due to the small sample size, summary statistics and general associations were performed. Based upon SFP (total/hen), hens were assigned a behavioural phenotype. Hens were either feather peckers (20%; gave >1 SFP and received 0 SFP), feather pecker-victims (35%; gave and received >1 SFP), neutrals (10%; gave and received 0 SFP), or victims (35%; gave 0 SFP and received >1 SFP). Testosterone concentrations (pg/ml) were highest for victims (87.1±30.6), whereas neutrals (61.9±1.4), feather pecker-victims (69.6±29.5), and feather peckers (62.5±43.4) had similar testosterone concentrations. Hens who gave more SFP tended to have lower testosterone concentrations (R_{20}=-0.38, P=0.10). As victim hens had higher testosterone concentrations, they may possess brighter UV reflectivity and be more likely to be attractive targets of FP. Further investigation measuring UV reflectivity is needed to quantify the relationship between testosterone, UV reflectivity, and FP in laying hens.

Identification and development of appropriate measures to routinely assess the reactivity of dairy cows towards humans

Asja Ebinghaus, Silvia Ivemeyer, Julia Rupp and Ute Knierim
University of Kassel, Farm Animal Behaviour and Husbandry, Nordbahnhofstr. 1a, 37213 Witzenhausen, Germany; ebinghaus@uni-kassel.de

Behavioural indicators of the human-animal relationship (HAR) are predominantly used in animal welfare science, so far. However, the reactivity of dairy cows may also be of interest in the context of breeding, due to its estimated moderate heritability. Breeding associations routinely use the average milk flow and partly the subjective evaluation of milking temperament by the stockperson as traits to select for improved milkability and manageability. However, it can be questioned whether these traits reflect reliably and accurately the reactivity towards humans. In contrast, the avoidance distance (AD) of cows towards an unfamiliar experimenter is regarded an established behavioural HAR indicator. The aim of the study was to identify and develop measures, which allow reliable conclusions on the HAR under various conditions, and which are suitable to be recorded as breeding traits. On three German dairy farms with loose housing and herd sizes of 45 to 195 cows, AD at the feeding place and AD in the barn as well as innovative HAR measures were applied and tested for inter-observer reliability (IOR) and inter-test associations. Innovative measures were a Qualitative Behaviour Assessment (QBA) of the cow's reactivity in a human-animal interaction, tolerance to standardised tactile interaction (TTI), release behaviour after restraint (RB) and facial hair whorl positions and forms (HW). IOR was assessed using Spearman rank or Kendall W correlation (in case of QBA with three observers), or the PABAK (in case of HW data). Inter-test associations between AD at the feeding place and HW were analysed using a General Linear Model and between all other measures with Spearman rank correlation. IOR was good to very good for all measures: AD feeding place rs=0.79 (n=84, P<0.01); AD barn rs=0.83 (n=36, P<0.01); TTI rs=0.93 (n=55, P<0.01); RB rs=0.90 (n=54, P<0.01); QBA W=0.95 (n=32, P<0.01); HW PABAK=0.77-0.83 (n=58). High inter-test correlations were found between AD feeding place and AD barn (rs=0.77, n=44, P<0.01), between TTI and RB (rs=0.78, n=52, P<0.01) as well as between QBA and RB (rs=0.76, n=18, P<0.01). Moderately correlated were QBA and TTI (rs=0.68, n=18, P<0.01), AD feeding place and TTI (rs=0.50, n=44, P<0.01), and AD feeding place and RB (rs=0.45, n=43, P<0.01). No significant associations were found between HW and AD. The results suggest that TTI, RB and QBA alongside the established AD are appropriate reactivity measures. They partly reflect similar and partly different aspects of the HAR, with an apparent clustering into distance and handling measures.

Assessing lateral bias in dogs using the Kong ball test

Deborah Wells, Peter Hepper, Shanis Barnard and Adam Milligan

Queen's University Belfast, Animal Behaviour Centre, School of Psychology, Queen's University Belfast, Belfast BT7 1NN, United Kingdom; d.wells@qub.ac.uk

Motor bias has the potential to be used as an applied tool for assessing welfare risk in animals. Doing so, however, is very dependent upon categorising an animal correctly as ambilateral, left- or right-limbed. The Kong ball test has been used extensively to assess lateral bias in the domestic dog. Here, the paw used to stabilise a ball filled with food is recorded. Implicit in this challenge is the assumption that dogs use their dominant paw to stabilise the ball. This study examined whether this is the case. A comparative approach was adopted, exploring limb use in dogs and humans. In Experiment 1, the paw preference of 48 dogs was assessed on the Kong ball test. Analysis revealed an even split in the distribution of the dogs' paw preferences, with 16 (33.3%) dogs consistently using their right paw to hold the Kong ball, 16 (33.3%) animals consistently using their left paw, and 16 (33.3%) showing ambilateral paw use. This distribution of paw preference was not significantly different from that expected by chance ($\chi^2=0$, df=2, P=1.00). Significantly more dogs were paw-preferent than ambilateral (P=0.01, binomial test), although there was no significant difference in the number of dogs that were right- vs left-paw preferent (P=1.00, binomial test). Distribution of paw preference was significantly associated with canine sex ($\chi^2=6.18$, df=2, P=0.04). More of the male animals were found to be right-pawed than left-pawed or ambilateral. Female dogs, by contrast, were more likely to be classified as ambilateral than right- or left-pawed. In Experiment 2, 94 adult humans were assessed on their ability to remove a piece of paper from a Kong ball with their mouth, using their left, right or both hands to stabilise the ball. Participants also completed the Edinburgh Handedness Inventory Short Form to assess their handedness quotient (HQ). Analysis revealed a highly significant association between the participants' handedness quotient and the hand used to stabilise the Kong ball ($\chi^2=31.31$, df=1, P<0.001). Most of the left-handed individuals (82%) used their right hand to stabilise the ball, while the majority of the right-handed participants (76%) employed their left hand. The findings suggest that, contrary to the implicit assumption that dogs use their dominant limb (governed by the contra-lateral hemisphere) to stabilise the Kong ball, they may, rather, like humans, be using their non-dominant paw to stabilise the object, using their dominant side instead for postural support. These results have implications for the interpretation of work reporting on paw preferences arising from the Kong test, particularly studies relating directional motor bias to emotional functioning and welfare risk.

Lateral bias and temperament in the domestic cat

Louise Mcdowell, Deborah Wells and Peter Hepper
Queen's University Belfast, Animal Behaviour Centre, School of Psychology, Queen's University Belfast, Belfast BT7 1NN, United Kingdom; lmcdowell17@qub.ac.uk

It has recently been suggested that there may be association between motor bias (e.g. 'handedness') and animal welfare, with ambilateral animals being more vulnerable to stress and reduced well-being than animals showing a right-sided motor preference. If this association were to be proven, it would enable behavioural asymmetry to be used as a simple measure of welfare risk. This study explored the association between paw preferences and emotional functioning, specifically temperament, in a species thus far overlooked in this area, the domestic cat. Thirty left-pawed, 30 right-pawed and 30 ambilateral pet cats were recruited following an assessment of their paw preferences using a food-reaching challenge. Here, the cat had to try to remove a food treat from a puzzle feeder. The paw used (left or right) was recorded on 100 occasions. The animals' temperament was subsequently assessed using the Feline Temperament Profile (FTP). Cats' owners also completed a purpose-designed cat temperament (CAT) scale. Analysis revealed a significant relationship between lateral bias and FTP scores ($F[2,84]=544.89$, $h2=0.66$, $P<0.001$). Cats classified as ambilateral had significantly ($P<0.001$, Bonferroni test) lower FTP scores (mean=3.18, ±95% C.I. 2.06-4.29) than animals classified as left (mean=10.98, ±95% C.I. 9.60-12.36) or right (mean=11.06, ±95% C.I. 9.85-12.26) pawed. Analysis of the 12 scale items on the CAT revealed a significant ($P<0.05$ for all one-way ANOVAs) relationship between the animals' paw preferences and owners' perceptions of their pets' temperament for 5 characteristics. Post-hoc t-tests showed that cats with a right paw preference were rated as significantly ($P<0.05$) more playful (mean=4.33, ±95% C.I. 3.93-4.74) than left-pawed (mean=3.93, ±95% C.I. 3.53-4.34) or ambilateral (mean=3.53, ±95% C.I. 3.13-3.94) cats. Ambilateral cats were also perceived to be significantly ($P<0.05$ for all one-way ANOVAs) less affectionate (mean=3.53, ±95% C.I. 2.89-3.64), obedient (mean=2.97, ±95% C.I. 2.54-3.39), friendly (mean=3.17, ±95% C.I. 2.78-3.56) and more aggressive (mean=2.20, ±95% C.I. 1.93-2.47) than left-pawed (affectionate: mean=4.20, ±95% C.I. 3.82-4.57; obedient: mean=3.70, ±95% C.I. 3.27-4.12; friendly: mean=4.43, ±95% C.I. 4.04-4.82; aggressive: mean=1.10, ±95% C.I. 0.83-1.37) or right-pawed (affectionate: mean=4.30, ±95% C.I. 3.92-4.67; obedient: mean=3.70, ±95% C.I. 3.27-4.12; friendly: mean=4.53, ±95% C.I. 4.14-4.92; aggressive: mean=1.10, ±95% C.I. 0.83-1.37) animals. Results suggest that motor laterality in the cat is strongly related to temperament and that the presence or absence of lateralization has greater implications for the expression of emotion in this species than the direction of the lateralized bias.

Early detection of stress and morbidity in dairy cows using a brush

Roi Mandel[1], Helen R. Whay[2], Christine J. Nicol[2] and Eyal Klement[1]
[1]The Hebrew University, Animal Sciences and Veterinary, P.O. Box 12, Rehovot 76100, Israel,
[2]University of Bristol, Clinical Veterinary Science, Langford House, Langford, Bristol BS18 7DU,
United Kingdom; roi.mandel@mail.huji.ac.il

Keeping animals in an environment that meets their proximate needs (e.g. feeding, drinking, and sleeping) allows them to engage in low-resilience behaviours (also referred to as 'luxury activities'; that is, behaviours that typically decrease when energy resources are limited or when the cost involved in the activity increases). As low-resilience activities are elastic in demand, and therefore can be scheduled flexibly, we hypothesized that engagement in these activities will be inversely correlated with the effort allocated to perform them, particularly at times of internal or external stress. One seemingly low-resilience activity available to cows in an increasing number of dairy farms is scrubbing against an automated brush. Despite the increasing popularity of such brushes, there is hardly any data on factors influencing cows' brush usage, nor any documentation on reduced usage of the device at times of stress and morbidity. This study was aimed at examining the correlation of brush usage with distance from food, heat load, intrusive stressful manipulations in the form of a vaginal examination and insemination, and with lameness. Individualized brush usage data was collected automatically using an RF system built for this purpose. We show that brush usage decreases when food is located distantly from the brush (20.51 m away from the brush, compared to 15.36 m; GEE: 95% CI=0.45-0.88, P<0.01), at high temperature and humidity levels (Pearson correlation: THI; r=-0.79, P<0.01), after stressful manipulation (repeated-measures ANOVA: $F_{2,52}=3.26$, P<0.05), and when cows express signs of lameness (GLMM: $\chi^2_1=5.07$, P<0.05). Cows with early signs of lameness (score: 2-3 on a total of 5) decreased their daily duration of brush usage by 9-33% compared to non lame cows, while those with severe lameness (score: 4-5) decreased their duration of daily brush usage by more than 90%. The results of this study support the idea that changes in low resilience behaviours, such as brush usage, may be utilized to detect morbidity or stressful events. On-farm monitoring of low resilience behaviours, together with existing systems that monitor core behaviours (e.g. activity and rumination), may serve as an improved method for detecting events that compromise the welfare of cows. Further research should be conducted in order to assess the sensitivity and specificity of this tool.

Assessment of human-animal relationship in extensively managed ewes

Carolina A. Munoz[1], Angus J. D. Campbell[2], Paul H. Hemsworth[1] and Rebecca E. Doyle[1]
[1]Animal Welfare Science Centre, The University of Melbourne, Parkville, 3010 VIC, Australia,
[2]Faculty of Veterinary and Agricultural Science, The University of Melbourne, Werribee, 3030 VIC, Australia; cmunoz@student.unimelb.edu.au

The quality of the human-animal relationship is a key factor affecting the welfare of farm animals, however methods for its practical on-farm assessment in sheep are lacking. The current exploratory study investigated some measures to assess the human-animal relationship in extensively managed ewes. The behavioural response of 100 Merino ewes (aged 2-4) to human presence was tested at three time points: late-pregnancy (LP), mid-lactation (ML) and weaning (W). Ewes were managed under extensive conditions, in a year-round outdoor system. The ewes were tested and observed in a pen and a race in four groups of 25. To measure flight distance (FD), a single observer, always the same person, quietly entered a pen (9×8 m holding one of the four groups of ewes), walked around the perimeter and stood opposite the entry point. From here, the observer waited for a ewe to be orientated towards her before approaching the animal in a standardised way (i.e. one-step per second). The test ended whenever the ewe withdrew, defined as stepping away from the observer, and FD was estimated. FD was measured in five ewes, randomly selected from each of the four groups per reproductive stage. At the end of this test, the number of vigilant sheep in the group (defined as head above shoulder height and orientated towards the observer) was recorded. Ewes were then moved to a race and vigilance was again recorded (observer was at 2 m distance and was stationary). A two-way ANOVA (reproductive stage and group) and LSD test revealed that each reproductive stage differed ($P<0.001$) from the other (mean FD were 5.2 m at LP (SE±0.2), 6.2 m at ML (SE±0.3) and 4 m at W (SE±0.2). It was also observed that at all reproductive stages and in both tests, the majority of the ewes were vigilant, however, the number of vigilant animals decreased over reproductive stage. Number of vigilant ewes in the four groups at each stage were 100 (100%), 82 (85%) and 61 (64%) respectively. Vigilant animals in the race were 84 (84%), 77 (79%) and 61 (64%) at each stage respectively. While only one flock at one farm was assessed, the variation observed in FD and vigilance behaviour suggests that these measures may be influenced by the reproductive stage, the presence of lambs at foot, time and/or their familiarity with the test/experimenter. The test performed is consistent with other human-approach tests for livestock; however, our results indicate that the repeatability of the measures needs to be assessed comprehensively and caution may be required when interpreting results at different reproductive stages.

Developing and validating a practical screening tool for chronic stress in livestock

Fabio Gualtieri[1], Elena Armstrong[1], Georgia Longmoor[1], Julia George[2], David Clayton[2], Rick D'eath[3], Victoria Sandilands[3], Tim Boswell[1] and Tom Smulders[1]
[1]Newcastle University, Newcastle upon Tyne, NE17RU, United Kingdom, [2]Queen Mary University of London, Mile End Road, E14NS, United Kingdom, [3]Scotland's Rural College, Midlothian/Ayr, EH259RG/KA65HW, United Kingdom; e.a.armstrong@newcastle.ac.uk

Poor animal welfare can be caused by accumulation of many stressors over a long time period and measurements of acute stress responses, like rises in corticosterone (Cort) levels, are inadequate predictors of welfare state. In mammals, adult hippocampal neurogenesis (AHN) is suppressed by cumulative chronic stress; a process mediated by elevated Cort levels and associated with behavioural markers of depressive-like states. Conversely, positive experiences increase AHN in a process also involving elevated Cort levels. Our study aims were: (1) to test whether AHN in poultry is also sensitive to cumulative chronic stress; (2) to dissociate markers of artificially elevated Cort from those of chronic exposure to stressors; and (3) to develop a quick and reliable way to quantify chronic stress in poultry. 64 HyLine Brown hens (19-26 weeks) were assigned to Control-Water, Control-Cort, Stressed-Water & Stressed-Cort treatments. Stressors were randomized and unpredictable and designated pens received 20 mg/l Cort dissolved in drinking water. We measured previously-used stress markers, markers of AHN and gene expression profiles in the hippocampus. Blood was collected from hens before, during & after Cort treatment. Whilst Cort treatment increased plasma Cort titres ($F_{2,41}=8.5$, P=0.001), chronic stress did not. Cort treated animals also had higher heterophil/lymphocyte ratios ($F_{2,57}=15.5$, P<0.001), but this immunological marker failed to show any effect of the stress treatment. Conversely, stressed animals had more prolonged durations of tonic immobility than the non-stressed group ($\chi^2_1=4.0$, P=0.045), and this behavioural measure did not respond to Cort treatment. We are still quantifying hippocampal AHN in these chickens and will present the results at the meeting. However, RNA sequencing of all genes expressed in the avian hippocampus identified a group of 5, with functions related to vascular inflammation, whose combined expression levels afforded complete discrimination between stressed and control birds. We will quantify these genes' expression levels independently, test their response to Cort treatment and verify their anatomical expression patterns. We conclude that the avian hippocampus is sensitive to chronic stress. Quantification of a small number of hippocampal genes may represent an alternative, reliable way to quantify chronic stress in poultry and potentially other animals; integrating experiences detrimental to welfare without undue influence of current state.

Subclinical inflammation may have multiple welfare implications in pigs

Gina Caplen, Doug Wilson and Suzanne Held
University of Bristol, School of Veterinary Sciences, Langford House, Langford, Bristol, BS40 5DU,
United Kingdom; suzanne.held@bristol.ac.uk

Clinical disease impacts directly upon animal welfare. However, the majority of subclinical disease remains undetected. Sickness behaviour is mediated by a systemic pro-inflammatory response. Although inflammatory markers are upregulated in sub-clinical pig disease, evidence for behavioural alterations remains lacking. We propose that mild sickness behaviour within clinically 'healthy' herds could have negative implications for livestock management. We aimed, therefore, to identify behaviour associated with subclinical inflammation in pigs free from clinical disease. Our sample population comprised pigs (n=120, 60 kg) with/without tail lesions from four UK farms (one without any tail biting history and three with current outbreaks). Baseline behavioural analysis and a novel object test were performed in-situ and a comprehensive immunochemical assessment was conducted on serum collected at a single sampling point. Data were analysed using random-intercept nested multilevel models. Correlations were observed between several inflammatory markers and behaviours associated with sickness. High C-reactive protein (CRP) was associated with low physical activity (P=0.046), social responsiveness (P=0.037), interaction time with a novel object (P=0.045) and feeding (P=0.045). Systemic inflammation was associated with tail biting and being tail bitten. Pigs with low Hb (P=0.032) and high levels of IFN-γ (P=0.046) and CRP (P=0.024) were more frequently tail-bitten, regardless of whether they had a tail lesion. The behavioural profile of 'biters' was suggestive of hyperactivity rather than sickness behaviour. However, tail biting was associated with pro-inflammatory cytokine upregulation (IFN-γ: P=0.028; IL-1β: P<0.001; IL-6: P=0.018). Low Hb (P<0.001), high CRP (P<0.001), and high latency to interact with a novel object (P=0.028), were all observed in pigs with healed tail-lesions, indicative of secondary sub-clinical infection. These results highlight an association between systemic sub-clinical inflammation and altered behaviour which may have wider welfare implications. Reductions in activity and environmental interest, associated with an upregulated inflammatory response may, for example, help to explain the variability in the success of management efforts (e.g. enrichment provision) to prevent/manage detrimental behaviours such as tail-biting. Furthermore, sub-clinical inflammation within our sample appeared to have multiple links with tail-biting (both as a trigger and following lesion infection). Quantification of inflammatory (especially CRP) and behavioural markers of sub-clinical inflammation, therefore, offers potential as an advanced means of identifying welfare risk in asymptomatic pig herds.

Use of a novel object test in the handling race for lambs *in situ* on commercial farms

Jennifer Matthews, Poppy Statham and Sarah Lambton
University of Bristol School of Veterinary Sciences, Animal Behaviour and Welfare, Langford House, Langford, Bristol, BS40 5DU, United Kingdom; j.matthews@bristol.ac.uk

Due to the extensive nature of commercial lamb production, handling for common husbandry practices can be quite stressful, as direct contact with humans is largely infrequent and often aversive for sheep. In addition to their relationship with welfare, sheep fearfulness and behavioural reactivity have been associated with performance, especially with regard to growth. Few novel object tests have been used to measure lamb fearfulness, and those that have used isolation pens, rather than the handling race. Carrying out the test in the handling race means it is easier to set up and perform, and it is more representative of typical handling for lambs on a commercial farm. We present the development of a method to perform a novel object test carried out in the handling race. The novel object is suspended on one side, at least two metres from the race entry, and visible to the lamb immediately upon entering the race. The same novel object is used and placed on the same side of the race each time. Lambs are shepherded into the race individually by the farmer but only encouraged through the race if they haven't moved after 60 seconds. A behavioural reactivity score is allocated to each test using the following scale: (1) initial approach followed by retreat/turn around; (2) walks past with no obvious attention paid to the novel object; (3) walks past with some obvious attention paid to the novel object; (4) approaches and sniffs/investigates the novel object for a number of seconds; (5) approaches and interacts (nose/head-butts) with the novel object for a number of seconds; (6) runs past the novel object. Data collected from two pilot flocks show the full range of behavioural reactions: 5% of lambs = score 1; 37.5% = 2; 25% = 3; 27.5% = 4; 2.5% = 5; 2.5% = 6. Each test is video-recorded for the purpose of subsequent detailed analysis: behavioural reactions, ear positions, and vocalisations, in addition to latency to move and time taken for the lamb to pass the novel object. This novel object test is being developed through use on 5 commercial flocks in the UK and we will present the validation of the method, including correlations with the results of other behavioural reactivity tests such as weigh crate agitation scoring, flight speed testing from the weigh crate and approach testing on the animals' release to the field.

Genetic variation of keel bone damage
Ariane Stratmann, Laura Candelotto, Sabine Gebhardt-Henrich and Michael Toscano
Center for Proper Housing: Poultry and Rabbits (ZTHZ), Division of Animal Welfare,
VPH-Institute, University of Bern, Burgerweg 22, 3052 Zollikofen, Switzerland;
ariane.stratmann@vetsuisse.unibe.ch

Keel bone fractures are a major welfare issue in laying hens due to their high prevalence of 40 to 80% within commercial flocks and their likely association with pain. Besides behaviour, genetics has been identified as one key factor to reduce keel bone fractures. In order to minimize confounding factors and focus on the role of genetics, a keel impact testing protocol which delivered controlled collisions to the keel bone of a deceased bird was used to identify true variation across and within different genetic lines and quantify fracture susceptibility. Birds of five distinct genetic lines (Bovans Brown (BB), Dekalb White (DW), ISA Dual Brown (Dual), experimental brown line (EB), experimental white line (EW)) were assigned to one of 10 genetic line-specific, identical pens with 10 to 30 birds per pen. Feed consumption was recorded at the pen level on a weekly basis throughout the experiment. Daily egg production at pen level was recorded to calculate the hen daily average and ratio of feed consumed per eggs produced (FCR). At 26 weeks of age (WoA), six focal hens per pen were randomly selected and orally administered a dye to link external egg quality measures (i.e. egg mass, breaking strength and shell width) to individual hens. At 28 and 29 WoA all birds underwent keel impact testing with two different energy levels (3.248 and 4.331 J) using a methodology adapted from Toscano et al. Keel bones were excised and inspected for presence of old and experimental fractures, as well as their severity. Data were analyzed using generalized linear mixed effects models. Fracture susceptibility differed between the lines with EB birds having a reduced likelihood of experimental fractures compared to the other lines (Z=-2.76, P=0.0059; fracture (%): BB: 75.5%; DW: 97.9%; Dual: 50.9%; EB: 16.7%; EW: 90%) as well as Dual birds compared to DW birds (Z=-3.29, P=0.007). More old fractures were found in DW birds than in the other lines except in EW birds (Z=3.41, P=0.0006). A higher impact energy resulted in a more severe experimental fracture (t=-2.10, P=0.036). FCR, calculated in the period after each pen reached 50% of egg production, showed that EB birds had a greater value than the other lines (Z=3.22, P=0.032). Focal birds with experimental fractures produced eggs with thinner eggshells (Z=-2.155, P=0.037), particularly EB birds (Z=-3.49, P=0.025). Our results suggest a clear genetic effect on keel fracture susceptibility within a controlled setting that eliminated confounding variation due to behaviour. Identified associations between feed consumption, egg shell quality, and keel bone fractures deserve consideration as well.

The transect method: adding supporting evidence for on-farm broiler welfare assessment

Neila Ben Sassi[1], Xavier Averos[1] and Inma Estevez[1,2]
[1]Neiker-Tecnalia, Campus Agroalimentario de Arkaute, 01080, Vitoria-Gasteiz, Spain, [2]Ikerbasque, Basque foundation for Science, Maria Diaz Haroko, 3, 48013, Bilbao, Spain; nbsassi@neiker.eus

The transect method, previously tested against other assessment approaches, was proposed as a practical method for on-farm animal-based welfare assessment of commercial broiler flocks. It consists on a set of walks (transects) within the areas delimited by feeder and drinker lines. In this study, conducted on 20 commercial broiler flocks, the incidence of validated broilers' welfare indicators were assessed at 3, 5 and 6 weeks of age through randomized transects using the i-watchbroiler app. Welfare indicators included: lame, immobile, sick, small, dirty, terminally ill, tail, back and head wounded, featherless, and dead animals. Data were expressed relative to the estimated total number of birds/transect (%), assuming a random bird distribution within the house. Generalized linear mixed models (with distribution models adjusted for each welfare indicator) were used to determine inter-observer reliability, and to test the sensitivity of the method to detect the effects of observer, bird age, transect location (central or side), genetic breeds (Cobb (C), Ross (R) and mixed (RC)) and their interactions. Pearson correlations were calculated to determine the relationship among welfare indicators. Results indicate a strong inter-observer reliability for all indicators (P>0.05), except for tail wounded in week 6 (observer by age interaction P<0.05; 0.0803±0.049 and 0.0134±0.0087% (mean±SE) for observer 1 and 2, respectively, given a binomial distribution). The method detected a clear increase in the incidence of welfare indicators with age (P<0.05), except for sick birds (P>0.05). Differences across genetics were detected for small and sick (P<0.05), and for genetics by age for tail wounded birds (0.018±0.015 and 0.00046±0.00045% for C and RC, respectively during week 3; P<0.05) and dead birds (0.0108±0.0047 and 0.059±0.0172% for C and RC, respectively during week 5; P<0.05). Pearson correlations also provided some interesting results regarding the relationship among parameters, showing a strong correlation between the incidence of lame and immobile birds for weeks 3 and 5 (r=0.637, P<0.001 and r=0.435, P<0.01, respectively), or between the incidence of small and sick birds (r=0.968; P<0.001), which provide more evidence of the sensitivity of the method. These results demonstrate a good inter-observer reliability as well as sensitivity of the method to study the progress of welfare indicators with age, their relationship with slaughter outcomes, or to detect potential genetic effects. Thus, these results provide additional evidence to support the solid foundation of the transect methodology as a practical tool for on-farm broiler welfare assessment.

Using tri-axial accelerometers to determine energies of hazardous movements for laying hens in aviaries

Nikki Mackie[1], Ariane Stratmann[2], John Tarlton[1] and Michael Toscano[2]
[1]University of Bristol, Matrix Biology, Langford, Bristol, BS40 5DU, United Kingdom, [2]University of Bern, Center for Proper Housing: Poultry and Rabbits, Division of Animal Welfare, Burgerweg 22, Zollikofen, 3052, Switzerland; nm14895@bristol.ac.uk

Keel bone fractures in laying hens are of economic and welfare importance, with a prevalence of approximately 35-90% at the end of lay. Risk factors for keel bone fractures are likely multifactorial including genetics, nutrition and housing design. In terms of housing, aviary systems have become more popular as a result of a global transition away from battery cages. Aviary systems allow birds to use three dimensions and express a wide range of natural behaviours. However, vertical motion (such as possible in aviary systems) can result in falls and collisions, contributing to keel bone fracture. Our study examined the accelerations associated with falls and collisions at the keel and produced quantifiable data for the purpose of indicating hazardous areas of the aviary system. To quantify the energy of various locomotion paths within a commercial Bolegg Terrace aviary, tri-axial accelerometers were used on two different flocks, aged between 21-36 and 49-61 weeks of age. The accelerometer was custom designed to allow the sensor to be placed on the keel; data presented here reflects an acceleration vector (AV) that combined acceleration from the 3 dimensions. Video recordings were matched to a time stamp on the accelerometers linking behaviour, location, and path with accelerometer output. Bird movements were classified as controlled (intentional and successful), falls (non-intentional and/or unsuccessful) and collisions (colliding with a structure, conspecific or the floor). Although data has yet to be statistically analysed, preliminary observations suggest movements that consist of a fall and collision have higher maximum AVs (52.5±45.9 g, n=79) (mean ± standard deviation) compared to falls without a collision (44.5±37.6 g; n=63), collisions without a fall (40.9±26.4 g; n=12) and controlled movements (36.0±25.4 g; n=200). A similar pattern is observed across similar motion paths. When birds navigate from a height of 0.73 m, falls have higher AVs (45.3±34.4 g, n=23) than controlled movements (33.3±20.7 g, n=74). From the greatest height in the aviary (2.7 m) one event with a fall and collision had a maximum AV of 284.9 g, a value much greater than the average AV in the same movement path that was controlled (neither a fall nor a collision) (29.9±10.6 g, n=8). As expected, our data indicates that the AV is proportional to the height of movement and whether a collision occurred. We believe our methods aid in identifying hazardous areas and altering housing design to help reduce the incidence of falls and collisions, and consequently keel bone fractures.

Behaviors associated with high impacts experienced at the keel by hens housed in enriched colony systems

Sydney L. Baker[1], Cara I. Robison[2], Darrin M. Karcher[2], Michael J. Toscano[3] and Maja M. Makagon[1]
[1]*University of California, Davis, Animal Science, 1 Shields Ave, Davis, CA 95616, USA,* [2]*Michigan State University, Animal Science, 220 Trowbridge Rd, East Lansing, MI 48824, USA,* [3]*University of Bern, Hochschulstrasse 4, 3012 Bern, Switzerland, Switzerland; slbaker@ucdavis.edu*

Behaviors resulting in high impacts occurring at the hen's keel bone, such as falls and collisions, are believed to be associated with keel bone damage in aviaries and percheries. Enriched cages (EC) provide hens with less vertical and total space for high impact behaviors to take place, and potential sources of keel bone damage in EC are unclear. Using keel-mounted tri-axial accelerometers, we evaluated the peak combined acceleration of impacts experienced at the keel, and behaviors associated with the highest peak impact category. Hens were housed in 2 types of EC systems differing primarily in perch type and arrangement. Sixty W-36 hens were housed in each of 8 cages (2 cages/room; 4 replicates per EC type). Impacts experienced at the keel by 10 individually marked hens per cage were monitored over a 3-week period when hens were between 52 to 58 weeks of age. Of the 4,356 total impacts recorded (2,884 in cage type A and 1,472 in cage type B), 48.44%, 8.79%, 22.7% and 9.11% had a peak magnitude of 16-20 G, 20-60 G, 60-100 G, and >100 G respectively. Using video footage recorded continuously over each 3-week trial we determined the behavior and location of affected hens when impacts >100 G occurred. Of the 204 high impact readings matched to video, collisions accounted for 65.69%. Hens collided most frequently with a perch (48.5% of collisions), another bird (26.87%), a feeder (12.69%), or the floor (11.94%). Grooming events, aggressive interactions among hens, and wing flapping accounted for an additional 14.22%, 10.29%, and 9.8% of high impacts respectively. A X^2 test showed there was not a difference in the distribution of impacts among behavioral categories between the 2 cage types (P=0.68, χ^2=1.49). This study identifies key behaviors and events that should be further investigated as likely risk factors for keel bone damage sustained by hens housed in EC systems.

An investigation of hair cortisol as an objective measure of castration stress in beef cattle

Kate Creutzinger, John Campbell, Cheryl Waldner, Yolande Seddon and Joseph Stookey
University of Saskatchewan, Western College of Veterinary Medicine, Large Animal Clinical Sciences, 52 Campus Drive, Saskatoon, Saskatchewan S7N 5B4, Canada; kate.creutzinger@usask.ca

Hair cortisol (HC) is a novel biomarker of long-term stress in cattle. An experiment was conducted to investigate the use of HC to detect long-term stress following surgical castration, performed using emasculators. Bull calves located on two farms, site 1: Hereford cross (n=73), 94±16.5 kg (mean±S.D.) at 47±9.6 days of age, site 2: Black Angus (n=85), 88±15.9 kg at 48±11.3 days of age, were equally divided across three treatments: castration with saline (CS, n=52), castration with meloxicam (CM, n=54), and sham castration (S, n=52), balanced for calf age. Hair samples were collected over the left hip prior to castration on d 0, with hair regrowth collected on d 14 from the d 0 sample location. Calf standing time, a validated behavioural measure indicative of castration pain, was recorded on 129 calves (CS=47; CM=42; S=40, balanced by farm site), d0-d7 via accelerometers as a behavioural measure of stress. Additionally, calves were checked for scrotal abscesses on d 14. Data were combined from both farms for analysis. Treatment effects on HC were determined via the MIXED procedure (STATA* 12). Standing time was analyzed using repeated measures MMLR; and again for calves with scrotal abscess present compared to those without via two-sample t-test. On d14 results found, HC was 13.5% higher in CS than S calves (P=0.025) and tended to be higher than CM (P=0.06) with no differences between CM and S. Calves presenting scrotal abscess at d14 stood significantly more than their healthy counterparts on day 0 by 26% (P=0.007), day 2 by 14%(P=0.05) and day 3 by 18%(P=0.014). However, results of standing time across all treatment groups showed that CM tended to stand more than S calves (P=0.052) on d 0-4, with no other treatment differences. Differences detected in HC concentration in calves following castration occurred in a predictable manner. The increased HC concentration in CS calves compared to CM and S on d14, suggests calves castrated with no pain control experienced greater stress. With no differences in HC between CM and S calves, it suggests use of meloxicam may help reduce post-castration stress. Standing time appeared to indicate, calves which developed scrotal abscesses post-castration may be in greater discomfort, as indicated with increased standing. However, the measure did not clearly differentiate a behavioural stress response between treatments. However, the HC results suggest potential as a tool for detecting long-term stress in beef calves.

Sea motions regularity effects on sheep behaviour and physiology

Eduardo Santurtun[1], Valerie Moreau[2], Jeremy Marchant-Forde[3] and Clive Phillips[4]
[1]National Autonomous University of Mexico (UNAM), Ethology and Wildlife, Circuito Exterior S/N, Delegación Coyoacán, 04510 Ciudad de México, D.F., Mexico, [2]LaSalle Beauvais Polytechnic Institute, 19, rue Pierre Waguet, BP 30313, 60026 Beauvais Cedex, France, [3]USDA- ARS, Livestock Behavior Research Unit, 125 South Russell Street, West Lafayette, IN 47907, USA, [4] The University of Queensland, Centre for Animal Welfare and Ethics, School of Veterinary Science, The University of Queensland, Gatton, 4343, QLD, Australia; esanturtun@gmail.com

Roll and pitch sea transport motions have been described to produce stress responses in animals. This study aimed to identify the effects of regular (a standard angle and speed) and irregular (random selections of thirty angles and speeds) sequences of roll (side to side) and pitch (end to end) motions, or a combination of the two on sheep feed and water intake, heart rate variability and body posture. Six merino wethers were restrained in pairs in a crate that was placed on a moveable and programmable platform that generated roll and pitch motions for 60 min in a changeover design over 12 consecutive days. Sheep behavior was recorded continuously in real time by video cameras during exposure to treatment and the data were then analysed using a continuous recording of each animal and Cowlog 2.0 software for coding of behaviors. The duration of time spent in the following mutually exclusive states was continuously recorded: standing, head positions, drinking, eating and licking the bowl feeder. Stepping, pawing, and butting were recorded as events. Feed intake was increased by irregular sequences in the combined Roll and Pitch motion (P=0.04). The two sheep spent more time during Irregular sequences with their heads one above the other (P=0.001), indicating greater affiliative behaviour, and facing down (P=0.001). The combined Roll and Pitch also increased the time sheep spent with their heads down (P=0.007). Sheep spent more time during Irregular sequences standing with their back supported on the crate (P<0.001) or kneeling (P=0.03). Irregular sequences, the combined roll and pitch and the interaction between the two produced more stepping behaviour, indicating loss of balance. Sheep exposed to Irregular sequences of combined Roll and Pitch had increased heart rates (P<0.001). The RMSSD band was significantly reduced during Irregular sequences (P=0.04), and LF/HF ratio increased when Roll and Pitch combination interacted with Irregular sequences (P=0.007) suggesting a reduced activity of the parasympathetic nervous system during stress periods. Therefore, there was both behavioral and physiological evidence that Irregular sequences and the combination of Roll and Pitch caused stress, loss of balance and more affiliative behaviour between sheep.

YouTube to the rescue: factors influencing feather damaging behaviour in parrots

Rutu Acharya and Jean-Loup Rault
Animal Welfare Science Centre, University of Melbourne, Parkville 3010, Australia;
raultj@unimelb.edu.au

Videos posted on social media offer novel opportunities to study animal behaviour. Most attempts in studying the aetiology of behavioural problems involve some form of interference in the animal's routine (e.g. tests, vet clinics), or indirect assessment such as owner/caretaker surveys. Videos posted online offer insight of animals in their environment with minimum disturbance on their behaviour, apart from the owner/observer filming. We analysed videos of feather damaged (n=26) and species-matched control parrots (n=12) posted on YouTube to investigate feather damaging behaviour in pet parrots, an unresolved but salient behavioural problem. We scored plumage condition and recorded parrot's characteristics: genus, sex, age, other behavioural problems; environmental characteristics: owner type, interaction with owner, cage location, presence of other parrots, environmental enrichment; type of feed; and possible interventions against feather damage and its visible effects. We tested the prevalence of feather damage using a contingency table permutation test, and the severity of feather damage using univariate linear regression. The type of owner was associated with the prevalence of feather damage (P=0.002), with more feather damaged parrots owned by single females (38%) than single males (19%), whereas families owned more often control parrots (83% of this group) than feather damaged parrots (31%). The type of feed was also associated with the prevalence of feather damage (P=0.02), with most feather damaged parrots provided seeds only (42%) or feed mix (39%), and none fruits and vegetables only (0%), whereas control parrots most often received feed mix (58%) but also fruits and vegetables only (25%), followed by seeds only (17%). Half of the control parrots, but none of the feather damaged parrots, were provided with foraging devices, which therefore was associated with a lower prevalence of feather damage (P=0.001). None of the variables recorded had significant effects on the severity of feather damaged parrots, in average 4.73±0.09 on the score scale of 10 (intact feathers), but ranging from 0.75 to 8.75. The various interventions attempted by owners (re-homing, environmental enrichment, drug, conspecific, collar or combinations of these) resulted in feather improvement overtime (P=0.01), with re-homing the parrot into a different environment being the most common and effective (+2.80), while parrots that received no intervention worsen overtime (-0.13). Hence, the wealth of data available on video sharing websites can offer significant insight in the aetiology of behavioural problems, especially when interested in the importance of the home environment.

Automated measures of lying behaviour of dairy cows detect welfare problems on farms with automated milking systems

Rebecka Westin[1], Anne Marie De Passille[1], Trevor Devries[2], Edmond Pajor[3], Doris Pellerin[4], Janice Siegford[5], Elsa Vasseur[6] and Jeffrey Rushen[1]
[1]University of British Columbia, Box 220, 6947 Highway 7, Agassiz, BC, V0M 1A0, Canada, [2]University of Guelph, 50 Stone Road, Guelph ON, N1G 2W1, Canada, [3]University of Calgary, University Drive, Calgary AB, T2N 1N4, Canada, [4]Laval University, Rue de l'Universite, Quebec City QC, G1V 0A6, Canada, [5]Michigan State University, 220 Trowbridge Rd., East Lansing, MI 48824, USA, [6]McGill University, 21111 Lakeshore Drive, St. Anne-de-Bellevue QC, H9X 3V9, Canada; passille@mail.ubc.ca

Time constraints during on-farm animal welfare assessments often prevent behavioural measures being taken, which has resulted in interest in automated behavioural recording. The duration of time that cows lie down can be measured automatically and is a behavioural measure related to many threats to cow welfare. Using leg mounted accelerometers (which have been validated as providing accurate measures of lying times), we measured lying time over 4 d for 1378 lactating cows on 38 dairy farms in Canada and the US with automated milking systems (AMS), assessed the prevalence of lameness and leg lesions, and took measures of lying stall dimensions and type and depth of bedding. Average daily lying time varied greatly between farms (median=11.3 h; range=9.5-13.3 h), as did bout frequency (median=9.9/d; range=8.1-13.3/d), and bout duration (median= 74.0 min; range=53.5-93.9min). Lying times on farms with AMS are similar to those of farms with milking parlours. Risk factors for altered lying times were tested using regression analysis. Although lame cows lay down for 36 min longer than non-lame cows (P<0.01), and lameness prevalence varied greatly between farms (median= 12.7%; range=0-45.9%), the prevalence of lameness on a farm did not affect mean farm lying time (P>0.10). Mean farm lying time was reduced by 40 min (P=0.04) on farms where most stalls were too narrow for the cows (based on current Canadian standards), and by 47 min (P=0.05) on farms where most of the stalls had obstructions in the cow's lunge space, and was increased by 61 min (P<0.01) on farms that used sand bedding. Lying time was not related to factors associated with the management of the AMS, such as the numbers of cows per AMS unit. Automated measures of lying time of cows on commercial dairy farms is feasible and provide a novel method of detecting behavioural-based animal welfare problems associated with stall design and stall bedding management.

Using high-definition oscillometry as an indicator of chronic stress in domestic chickens (*Gallus gallus domesticus*)

Anna Davies, Elizabeth Paul, Michael Mendl, Suzanne Held, William Browne, Gina Caplen, Ilana Kelland and Christine Nicol
University of Bristol, School of Clinical Veterinary Science, Langford House, BS40 5DU, United Kingdom; anna.c.davies@bristol.ac.uk

Stress causes activation of the hypothalamic-pituitary-adrenal axis and the autonomic nervous system (ANS) as part of the adaptive allostatic response. Chronic activation of the ANS however, can be maladaptive resulting in the sympathetic nervous system repeatedly overriding the parasympathetic nervous system. This can lead to cardiovascular dysfunction, symptoms of which can include tachycardia and hypertension. The development of a non-invasive technique for determining cardiac function may therefore provide a proxy measure of chronic stress. High-Definition Oscillometry (HDO) has been developed for measuring cardiac activity in animals, but has not yet been used extensively on birds or for detecting long-term differences. We housed 14 domestic chickens (*Gallus gallus domesticus*) in a non-preferred environment (N – containing a wire floor, small low perch, wire nest-box), and 15 chickens in a preferred environment (P – containing shavings, multi-level perches, dust bath, enriched nest-box), for 24 weeks. At the end of the 24-week period, birds underwent a series of behavioural and physiological tests. A specifically designed cuff was applied to the leg of each bird and HDO was used successfully to record pulse rate, systolic, diastolic and mean arterial blood pressure. Each measurement was repeated five times for each chicken and the mean was taken for analyses. All measures of cardiac activity were significantly positively correlated with each other (Pearson's correlation coefficient: all >0.66, P<0.001). Independent samples t-tests revealed no significant differences in the measures of blood pressure between the N and P birds (all P>0.05). The pulse rate however, was significantly higher in N than in P birds (t_{27}=2.46, P=0.021). We also analysed the relationship between cardiac function and other measurements taken within each treatment group. P birds' blood pressure parameters were negatively correlated with their latency to leave a water-box (7 cm deep) (Spearman rank correlation coefficient: all <-0.58, P<0.05), but pulse rate was not correlated (P>0.05). For N birds however, pulse rate (but not measures of blood pressure) was negatively correlated with birds' latency to fly from a 1-metre high perch to receive mealworms (coefficient=-0.61, P=0.20). The difference in the behavioural correlations indicates that the birds responded differently to tasks regarding punishment and reward. The results suggest that pulse rate might be more sensitive for detecting long-term housing stress differences in cardiac function, than measures of blood pressure in birds.

Attention bias: a practical measure of affective state in farm animals

Caroline Lee
CSIRO, Animal Welfare, FD McMaster Laboratory, Armidale, NSW 2350, Australia;
caroline.lee@csiro.au

Measuring affective states is recognised as an important goal for improving our understanding of the welfare of animals. While much recent effort has focused on measuring judgement bias as an indicator of affective state, the methodology requires extensive training of the animal and as a consequence is impractical in applied contexts. We have therefore investigated attention bias as an additional window on affective state in farm animals. Humans and animals show increased attention towards threats, known as attention bias, when they are in increased states of anxiety. The few animal studies in starlings and rhesus macaques which have examined attention bias have applied environmental manipulations to induce anxiety and measured vigilance and attention towards a threat. In an effort to investigate whether or not a similar phenomenon is present in farm animal species, we developed an attentional bias test in sheep. The test involved introducing an individual animal to an arena containing a food source where a threat (a live dog) was visible through a window for 10 s. The dog was then removed and the response to the dog was assesed, including attention towards the direction of the dog, vigilance and latency to eat. Using drugs known to increase and decrease anxiety, we validated the test to show that the predicted effects on attentional biases were demonstrated. Sheep directed increased attention towards a threatening stimulus when they received an anxiogenic drug compared to a control or anxiolytic drug confirming that the test is able to detect differences in attention which is indicative of anxiety states. From here we have validated this test of attention bias in cattle and poultry as well. Measuring attention bias has the potential to provide a practical indicator of anxious states in animals. As this method requires no prior animal training, it presents a potential measure of one aspect of affective state on farm.

Serotonin depletion in a cognitive bias paradigm in pigs

Jenny Stracke[1], Sandra Düpjan[1], Winfried Otten[1], Armin Tuchscherer[2] and Birger Puppe[1,3]
[1]FBN Leibniz Institute for Farm Animal Biology, Instititute of Behavioural Physiology, Wilhelm-Stahl Allee 2, 18196 Dummerstorf, Germany, [2]FBN Leibniz Institute for Farm Animal Biology, Institute of Genetics and Biometry, Wilhelm-Stahl-Allee 2, 18196 Dummerstorf, Germany, [3]University of Rostock, Faculty of Agricultural and Environmental Sciences, Behavioural Sciences, Justus-von-Liebig-Weg 6, 18059 Rostock, Germany; stracke@fbn-dummerstorf.de

The cognitive bias approach provides information on the valence of affective states in non-human animals. In our study, we used a serotonin depletion model using para-Chlorophenylalanine (pCPA) to decrease brain serotonin levels in pigs in order to validate our cognitive bias paradigm and provide insight into its neural underpinnings. All experimental procedures were ethically approved (LALLF M-V/TSD/7221 3-1-066/13). To test the model, 40 female, juvenile pigs received either injections of pCPA or saline on six consecutive days. Tissue samples from seven brain areas presumed to be related to cognitive bias were collected at five time points and serotonin levels were measured. Data were analysed using the MIXED procedure in SAS with treatment, replicate and day as fixed factors. Serotonin levels were reduced significantly in all brain areas (all $F>37$ all $P<0.001$) for at least 13 days after first application of pCPA. This serotonin-depletion model was then applied to measure behaviour in a spatial judgement bias task with a go/nogo paradigm, where the go response was opening a box on one location to get a reward, and the no go response was not opening the box on the opposite location in order to avoid punishment. Testing on three intermediate locations was conducted two weeks pre- and two weeks post-treatment with either pCPA or saline. Relative latency to opening the box was analysed with box location, treatment and time point (before/after treatment) as fixed effects. Pairwise multiple comparisons of the least square means were done by Tukey Kramer t tests. Analyses revealed a significant effect of treatment ($F=6$, $P<0.05$), time point ($F=45.2$, $P<0.001$) as well as an interaction of both ($F=16$, $P<0.001$). Only the pCPA group showed significantly longer latencies to opening the box after treatment (LSM 0.54, SEM 0.01) than before (LSM 0.42, SEM 0.01), resulting in a significant difference between control group (LSM 0.46,SEM 0.01) and pCPA group after treatment ($t=-4.5$, $P<0.001$). These results confirm that serotonin, which plays a key role in cognitive-emotional processing, shifts behavioural responses to a more pessimistic outcome when depleted, and therefore contribute to the validation of our cognitive bias paradigm to make it a reliable tool to measure affective states in pigs.

On how impulsivity can affect responses in cognitive tests of laying hens

Elske De Haas and Bas Rodenburg
Wageningen University, Behavioural Ecology Group, De Elst 1, 6708 WD, Wageningen, the Netherlands; endehaas@gmail.com

Impulsivity is defined as the inability to wait before responding. In cognitive tests, subjects often need to associate a cue with an outcome. When an animal is unable to withhold its response due to high impulsiveness, its choice could be wrongly attributed to its cognitive (in)ability. Moreover, low levels of serotonin (5-HT) seem to underlie impulsivity. We hypothesized that cognitive performance is reduced with high impulsivity and low whole-blood 5-HT levels in chickens. Two experiments were conducted where White Leghorn laying hens needed to associate cue with reward in a two choice set-up and withhold responding. Ten sessions per day with maximum of five training days were performed per test. Hens were socially housed in groups of ten. Cognitive performance and impulsivity of hens with higher and lower than average levels of whole-blood 5-HT (measured once at adult age) were compared with repeated ANOVA and t-tests. Data expressed at means ± SEM. Correlation analysis of response latency and cognitive performance was conducted, expressed as regression coefficient. In the first experiment (n=36), hens cognitive performance was assessed by the number of incorrect choices. Response latencies were related to these. We found that hens with low 5-HT made more errors than hens with high 5-HT (25 to 40 vs less than 18% incorrect choices: $F_{1,96}=5.5$, P=0.03). A high number of errors correlated to short response latencies (r=- 0.3, P=0.005). In the second experiment (n=16, training 5 days), hens reversal learning and impulsivity was measured. In a discount-delay test, hens needed to withhold their response otherwise their choice would be unrewarded. Delays were in progressive ratio, amended to the hens ability to withhold responding. During the reversal test, hens with low 5-HT made more errors than hens with high 5-HT(79 vs 87% average correct choices, t=2.2, P=0.04). Hens with low 5-HT could not wait as long as hens with high 5-HT before switching to the unrewarded feeder (maximum inhibition time: 4.4±0.3 vs 5.8±0.6 s, t=2.12, P=0.05). Our results show that, in laying hens, fast responses could cause errors in cognitive tests, which may derive from impulsivity as affected by low levels of 5-HT. To establish whether these relationships are causal, 5-HT levels should be modified experimentally and effects on impulsivity and cognitive ability should be investigated.

Correlation between cognitive bias and other measures of stress in rats

Timothy H Barker, Alexandra L Whittaker and Gordon S Howarth
The University of Adelaide, School of Animal and Veterinary Sciences, Roseworthy Campus, University of Adelaide, Roseworthy, SA, 5371, Australia; timothy.barker@adelaide.edu.au

Assessment of animal emotion has become an integral component of animal welfare evaluation. One measurement of emotion is through detection of a cognitive bias in response to ambiguity. Despite widespread employment of the cognitive bias test in animals, few studies have correlated bias detection with other traditional measures of emotional state. We sought to determine whether rats trained on a judgment bias paradigm whilst expressing a pessimistic cognitive bias, would respond with a decreased sucrose preference ratio (SPR), increased concentration of corticosterone (CORT) and increased concentration of tyrosine hydroxylase (TH). Pessimistic cognitive biases were encouraged by housing equal numbers of male and female Sprague-Dawley rats in metabolic cages for 6 nights. This housing system is known to compromise welfare and encourage the likelihood of pessimistic biases. Consequently, the maximum time an animal spends in such a cage is strictly legislated. Rats (n=60) were trained to learn the correct response needed to obtain a reward, given the type of stimulus present (rough vs smooth sandpaper). One stimulus was associated with a high-positive reward whilst the other was associated with a low-positive reward. Rats began testing upon learning the discrimination and were introduced to a stimulus intermediate between their learned stimuli (intermediate sandpaper) creating an ambiguous probe, once a day for 5 days. Responses to the probe were regarded as optimistic if the rat responded as if it were positively rewarded, or pessimistic if the response to the probe was if it were negatively rewarded. Rats were separated into either metabolic cages or remained in control and were provided with a bottle of sucrose solution, allowing the SPR to be measured. Faeces were collected, weighed and snap-frozen. A repeated measures 3-way ANOVA and a Pearson's correlation were performed. Animals moved to the metabolic cages responded with a significantly greater number of pessimistic cognitive biases (4.38±0.86) compared to control (0.19±0.4), P<0.001. These results identified the metabolic cage as a housing mode that induces pessimism. Rats expressing pessimistic biases responded with significantly reduced SPR (0.72±0.06) compared to rats expressing optimistic biases (0.89±0.02). There was significant correlation between pessimistic biases and SPR, r=-0.816, n=48, P<0.001. CORT and TH results are pending. Preliminary results demonstrate that detection of cognitive bias is a valid and objective measure of emotion in animals based on reasonable correlation with traditional measures of emotional state.

Dogs with abnormal repetitive behaviours do not show differences in relative cognitive bias than matched unaffected dogs

Bethany Loftus and Rachel Casey

University of Bristol, The School of Veterinary Sciences, BS40 5DU, United Kingdom; bethany.loftus@bristol.ac.uk

Abnormal repetitive behaviours (ARB's) in dogs are thought to indicate compromised welfare, but past studies have found no clear relationships between these behaviours and classical welfare measures. 'Cognitive biases' in animals are a valuable tool in assessing longer term emotional state. The aim of this study was to measure cognitive bias of dogs displaying tail chasing or spinning ARB and matched controls, from both dogs in re-homing centres and the owned population. It was hypothesised that dogs displaying ARB's would judge ambiguous stimuli more 'pessimistically' than their non-ARB counterparts. 70 dogs from the owned population (35 ARB case and 35 matched controls) and 34 from re-homing centres (17 ARB case and 17 matched controls) were included in the study. Tail chasing or spinning type ARB was reported by the carer of the dog, inclusion criteria was for the dog to spin 3 or more times consecutively over 3 or more 'bouts' of the behaviour. All dogs completed a cognitive bias task following a previously published protocol. Wilcoxon signed rank tests were used to explore differences between case and matched controls. Overall, there was no significant difference in latency to reach ambiguous locations between case and control dogs (near positive location Z=-1.138, P=0.255; near negative location Z=-1.376, P=0.169; middle location: Z=-1.079, P=0.281). On dividing the population into owned and re-homing centre dogs, latency to ambiguous locations was not statistically different between case and control, in rehoming centre (near positive Z=-.071, P=0.943; near negative Z=-0.931, P=0.352; middle location Z=-0.876, P=0.381) and owned dogs (near positive Z=-1.507, P=0.132; near negative Z=-1.000, P=0.317; middle location Z=-0.680, P=0.497). Latency to reach the negative location was significantly different in re-homing centre dogs (Z=-2.391, P=0.017), with case dogs running faster than controls. Dogs displaying ARBs did not differ when compared to controls with regards to how relatively 'optimistic' they were in a spatial judgment bias task. This may suggest that display of ARB is not associated with current welfare state, but a 'scar' of previous compromised welfare. However, dogs in re-homing centre's that displayed ARB were apparently faster than their matched control's to reach the negative reference location suggesting that they were more motivated than controls to check for a reward. Differences between dogs with ARBs and controls in this sub-set may be related to differences in other characteristics which co-vary with risk for ARB occurrence, or this result may be an artefact of performing multiple tests.

Crib-biting behavior of horses: stress and learning capacity

Sabrina Briefer Freymond[1], Elodie F. Briefer[2], Sandrine Beuret[3], Klaus Zuberbüler[3], Redouan Bshary[3] and Iris Bachmann[1]
[1]*Agroscope, Swiss National Stud Farm, Les Longs-Prés, 1580 Avenches, Switzerland,* [2]*Institute of Agricultural Sciences, ETHZ, Rämistrasse 101, 8092 Zürich, Switzerland,* [3]*University of Neuchâtel, Rue Emile-Argand 11, 2000 Neuchatel, Switzerland; sabrina.briefer@agroscope.admin.ch*

Crib-biting is a stereotypy in horses that is potentially linked to both chronic stress and genetic predisposition. Chronic stress can cause neurobiological changes such as alteration of the dopaminergic modulation of the basal ganglia, which could alter the learning profile of the horses. We tested 19 crib-biters and 18 non-crib-biting horses (controls) in five challenging spatial tests, in order to test if potential differences in dopaminergic modulation impair learning capacities. The tests were performed in two parts, separated by a break, in a small arena (8×10 m) that was familiar to the horses. For each trial (part 1: 21 trials; part 2: 12 trials), the horses were led to the start zone in front of a solid fence (4 m) and were then left alone in the arena. Their task was then to find a bucket containing food, which was situated at different locations around the fence, depending on the tests (reversal task, detour task and extinction task). The time to reach the food bucket and the trajectory taken by the horse (left or right side of the fence) were recorded. Additionally, salivary cortisol was collected before the tests (baseline), after part 1, and after part 2. We found that crib-biters and controls behaved similarly during the learning tests. However, crib-biters that did crib-bite on the solid fence during the task (group A; 10 horses) behaved differently than crib-biters that did not crib-bite (group B; 9 horses) and controls (group C; 18 horses) for some tests. These differences could be explained by the time taken to crib-bite or by differences in impairment of the dopaminergic system. We also found a difference in salivary cortisol after part 1, between groups A, B and C. Indeed, the crib-biters that did not crib-bite had higher salivary cortisol values than all the other horses (mean±SD: A, 0.51±0.16 ng/ml, B, 0.78±0.17 ng/ml, C, 0.59±0.20 ng/ml; Linear mixed model, $F_{2,17}=5.08$, $P<0.02$). Our results suggest that crib-biting horses that did not crib-bite during the learning tests were more stressed than all other horses. This difference could be due to higher stress sensitivity in crib-biters, which could be reduced by the opportunity to crib-bite. However, the consequence of crib-biting on learning capacity seems to vary between horses.

Observing calves' behaviour during learning in a colour discrimination task throws light on their cognitive processes

Alison Vaughan[1], Anne Marie De Passillé[1], Joseph Stookey[2] and Jeffrey Rushen[1]
[1]University of British Columbia, Land and Food Systems, UBC Dairy Centre, 6947 Highway 7, V0M 1A0 Agassiz, Canada, [2]University of Saskatchewan, Large Animal Clinical Sciences, WCVM, 52 Campus Drive, S7N 5B4 Saskatoon, Canada; alisoncvaughan@gmail.com

Cattle can use visual cues to locate food, and visual cues may help cattle locate new food sources. Examining calves' behaviour during learning may help us understand how cattle learn discrimination tasks. We investigated whether calves generalise prior training with colour cues to a Y maze colour discrimination task and whether their behaviour shows their understanding of the task. 19 female Holstein calves (2-6 weeks of age at beginning of experiment) were either pre-trained to associate colours (red or yellow) with presence or absence of a milk reward or were exposed to randomly presented colour cues (control calves). Calves were trained to approach a teat to receive a milk reward. Unrewarded trials, in which the teat was removed if touched, were gradually introduced. In both training and testing, we recorded latency to touch the teat and behaviours before and after touching the teat in rewarded and unrewarded trials and tested using paired t tests. Trained, but not control, calves showed a shorter latency to touch the teat in rewarded vs unrewarded trials (mean ± SE=5±0.49 vs 9±0.91 s, respectively; $P<0.01$) demonstrating they learned to associate the colour with the reward during training. Calves were then presented with these colours in a Y maze colour discrimination test and their performance recorded. Calves completing >16 correct choices out of 20 within a session for two consecutive sessions were considered to have learnt. Chi squared analysis was used to compare the number of pre-trained and control calves meeting the learning criteria. Differences between pre-trained and control calves were tested using student t tests. Control and pre-trained calves did not differ in proportion of correct choices during the 1st session (56 vs 51%; $P=0.23$) suggesting that they did not immediately generalize the learned association. However, 9 out of 10 pre-trained calves, but none of the control calves, met the learning criteria ($P=0.01$). Pre-trained calves avoided touching the teat when making an incorrect choice ($P=0.01$) and tended to show their tongue when making a correct choice ($P=0.07$). The most common error made by both pre-trained and control calves was to choose the same side as their previous choice (80% and 89%, respectively). 5 out of 7 control calves, but none of the pre-trained calves, displayed a clear side preference (>75% of choices on the same side). Observing behaviour allows us greater insight into calves' understanding of a task than relying on the outcomes alone.

Pavlovian decision-making in a counter-balanced go/no-go judgement bias task

Samantha Jones[1], Elizabeth Paul[1], Emma Robinson[2], Peter Dayan[3] and Michael Mendl[1]
[1]*University of Bristol, Animal Welfare and Behaviour Group, School of Veterinary Sciences, Langford, BS40 5DU, United Kingdom,* [2]*University of Bristol, School of Physiology, Pharmacology and Neuroscience, Biomedical Sciences Building, University Walk, Bristol, BS8 1TD, United Kingdom,* [3]*University College London, Gatsby Computational Neuroscience Unit, 25 Howland Street, London, W1T 4JG, United Kingdom; s.m.jones@bristol.ac.uk*

Information from previous experiences of rewards and punishments has been hypothesized to influence animals' positive and negative affective states and thereby help them predict and act appropriately in the face of uncertainty and ambiguity. This leads to an affect-induced cognitive bias, with positive and negative affective states leading respectively to optimistic and pessimistic judgements and decisions. Many cognitive bias studies use go/no-go tasks, employing a 'go for reward' / 'no-go to avoid punishment' contingency. However, this induces two potential Pavlovian confounds: a choice bias that favours 'go' responses to acquire rewards and 'no-go' responses to avoid punishers; and a vigour bias in which positive or negative affect could enhance or suppress 'go for reward' responses. These would complicate testing of our hypothesis that changes in perceived probability of reward underlie affect-induced 'optimistic' decision-making under ambiguity. To be able to disentangle Pavlovian and perceived probability explanations, we trained male Sprague Dawley rats on a counterbalanced go/no-go task using a shuttle box set-up. Half the animals had to learn to shuttle (go) in response to one tone (e.g. 2 kHz) to get food, and stay (no-go) in response to another tone (e.g. 8 kHz) to avoid an air puff (GO+/NOGO-). The other half learnt the opposite contingency (GO-/NOGO+). The rats readily learnt to go or no-go to get a food reward (Discrimination sessions, mean % correct GO+ (M=95.41%, SE=0.527) vs NOGO+(M=95.91%, SE=1.045): t(10)=-0.431, P=0.676). However, although the rats learnt to stay (no-go) to avoid the airpuff, they did not learn to shuttle (go) to avoid the aversive stimulus (Discrimination sessions, mean % correct NOGO- (Mdn=66.2%) vs GO- (Mdn=6.67%): U=0.000, W_s=21.00, z=-2.882, P=0.002 (exact significance)). The rats therefore seemed to be able to learn to moderate their behaviour appropriately for the valued food rewards, but were unable to do so for the mildly aversive stimulus. The results support the prediction of a Pavlovian predisposition for animals to 'go for reward' and 'no-go to avoid punishment'.

Social learning in ungulates: effect of a human demonstrator on goats in a spatial learning task

Christian Nawroth, Luigi Baciadonna and Alan G. Mcelligott
Queen Mary University of London, Biological and Experimental Psychology, Mile End Road, London E1 4NS, United Kingdom; nawroth.christian@gmail.com

Good animal welfare is not only determined by the standard of husbandry, but also by both physical and mental health of the animal itself. To achieve this, a detailed understanding of the socio-cognitive abilities of livestock is important for applied ethology because it allows husbandry practices to be adjusted to an animals' behaviour and needs. For example, social learning has the potential to spread new and adaptive information quickly through a group of animals. However, previous research has found inconclusive results for social learning in ungulate livestock using conspecifics as demonstrators. In addition, it is not clear if livestock are able to perceive information socially from a human demonstrator. We assessed spatial and social problem-solving abilities of goats using a detour task, in which food was placed behind a V-shaped barrier (height 120 cm, length 400 cm), which either pointed towards or away from the subject, and thus providing the impression of an inward or outward corner from the subjects view. We tested goats' ability to solve the task (i.e. reached the reward) using varying spatial configurations (inward vs outward V-shaped barrier), and the impact of human demonstration on performance of the task (inward V-shaped barrier with, and without human demonstrator). Goats (n=29) were assigned to one of three experimental groups (inward non-social; outward non-social; inward social). We found an effect of experimental group on the time needed for goats to solve the detour (repeated measures ANOVA; $F_{2,26}=12.17$; $P<0.001$). For the spatial component ('inward non-social' vs 'outward non-social'), goats in the outward detour group took significantly less time to solve the task than subjects in the inward detour condition (SNK post hoc test; $P<0.05$). For the social component ('inward non-social' vs 'inward social'), a single presentation of a human solving the detour resulted in a decrease in the latencies of goats to solve the task (SNK post hoc test; $P<0.05$). Furthermore, goats that observed a human demonstration used the same route as the human (binomial test; K=8; n=9; P=0.04). Our results suggest that stimulus and/or local enhancement play a key role in goats decision making, while more sophisticated forms of social learning (e.g. imitation) cannot be ruled out. In general, we demonstrate that livestock such as goats utilise information provided by a human demonstrator in a spatial learning task. These findings provide insights into social learning mechanisms of livestock and have the potential to improve handling practices by using these mechanisms to facilitate habituation processes to new environments.

Effect of stimulation using a stuffed crow and broadcasted crow vocalizations on crow behavior

Naoki Tsukahara[1], Koh Sueda[2] and Fuko Ueda[3]
[1]SOKENDAI, Shonan village, Hayama, Kanagawa, 240-0193, Japan, [2]National University of Singapore, IDM Institute, 21 Heng Mui Keng Terrace, 119613, Singapore, [3]Tokyo University of Agriculture, 1737, Funako, Atsugi, Kanagawa, 243-0034, Japan; tsukahara_naoki@soken.ac.jp

Crows cause multiple problems (like damage to crops) for humans in Japan. However, no effective solutions that overcome these problems have been identified, because crows are one of the most intelligent birds on Earth. Nevertheless, the development of techniques to moderate crow behavior will be useful. Crows can recognize members of their own group and other individuals via the contact call ('Contact'). In addition, pairs of crows, which have defined territories, are known to threaten intruders in defense of their territory. In this study, we aimed to stimulate some threatening behaviors of crow pairs using a stuffed crow and broadcasted crow vocalizations in their own territories. We hypothesized that broadcasts in the presence of a stuffed crow would stimulate a stronger reaction from the crows in the territory, which could include pairs of crows attacking the stuffed crow, than in tests without the stuffed crow. The tests were conducted within the territories of 10 pairs of crows in Kanagawa prefecture, Japan. A Contact of a crow recorded in Tochigi prefecture was broadcast using a speaker. The behaviors and vocalizations of birds were recorded using a video camera. Moreover, we counted the incidences of threatening behavior (moving close to the speaker, vocalizing threatening calls, etc.) in during the test periods and in the recordings of the tests. Each test was run for 20 min, and five tests were conducted overall. We observed the behavior of five pairs of crows as a control during the first 5 min of the test. Next, the Contact was broadcast for 5 min. Thereafter, the observations were recorded. Finally, the Contact was broadcast again. The same test was repeated using a stuffed crow in five different crow territories. Following observation for the first 5 min, the stuffed crow was placed near the speaker and kept there until the end of the test. Threatening behavior by the pairs without the stuffed crow occurred at a significantly higher incidence than by pairs with the stuffed crow ($P<0.05$). For example, we observed 10.2 ± 3.4 incidences of threatening behavior during the first broadcasts without the stuffed crow and 2.6 ± 4.2 incidences with the stuffed crow. The results of the present study suggested that the stuffed crow suppressed territorial defense behavior. The crow pairs may perceive the stuffed crow as a safe object rather than a stranger because it is absolutely still. This work was supported by Academist Inc. in Japan.

Higher-order cognition in horses: Shetland ponies show quantity discrimination

Vivian Gabor
University of Goettingen, Department of Anmial Sciences, Albrecht-Thaer Weg 3, 37075, Germany;
vgabor@gwdg.de

To date only limited information is available on the higher-order cognitive capacity of the horse including skills such as perception and discrimination of quantity. Previous results concerning the numerosity judgment in horses remain controversial. In the present study we wanted to show whether Shetland ponies are able to transfer a previously learned concept of sameness to a numerosity judgment. Seven ponies first had to solve a 'matching to sample' task, where they learned to relate abstract symbols to another which were presented on a LCD screen. Three of them reached the learning criterion and were subsequently tested in two experiments. In experiment 1, the ponies had to relate two similar quantities to another, paired in contrasts (1 vs 2, 3 vs 4 and 4 vs 5) of the same stimulus (dot). When the learning criterion for 1 vs 2 was reached they entered the next step: 3 vs 4 and so on. To exclude discrimination due to the shape of the stimuli, the dots were varied in size and arrangement. The stimuli were presented in a triangular arrangement on the LCD screen; the sample stimulus was presented in the middle above and the discrimination stimuli in the two lower corners. The pony received a food reward by choosing the correct stimulus. Each learning session consisted of 20 decision trials. The overall learning criterion was set at 80% correct responses in two consecutive sessions. With P_0=0.5, the binominal probability of selecting 16/20 in two consecutive sessions is P=0.000035. Subsequently in experiment 2 specific pairs of quantities (all differing by one) of up to five different geometrical symbols were presented to investigate whether the performance was transferable (2 vs 3, 3 vs 4 and 4 vs 5). All of the three Shetland ponies met the learning criterion in experiment 1 within 2-8 sessions. In experiment 2 one pony could transfer all judgments to the mixed symbols (4 vs 5), another pony at a level of 3 vs 4 and the third at a level of 2 vs 3. These are the first reported findings that ponies are able to show quantity discrimination. Task solving seemed to be easier for the ponies when homogenous objects were presented, than in the case of heterogeneous symbols. The responses occurred within few seconds, suggesting that the ponies used subitizing for their numerosity judgment.

Emotional contagion of distress in young pigs is potentiated by previous exposure to the same stressor

Sébastien Goumon and Marek Špinka
Institute of Animal Science, Ethology, Pratelstvi 815, 10400 Prague, Czech Republic;
sebastien.goumon@gmail.com

Empathic responses to conspecific distress are relevant to the welfare of group living animals as an individual may be affected by the emotional state of others. This is particularly the case in farm animals which, under commercial conditions, are exposed to negative (aversive) situations like routine handling (e.g. vaccination, restraint, tail docking, castration, teeth clipping), transport or slaughter. This study tested whether emotional contagion occurs when piglets directly observe a pen mate in distress (restraint) and whether there is an effect of previous experience on the response to subsequent restraint or exposure to conspecific distress. Piglets (49.7 ± 0.7 days) were exposed in pairs to two stress phases (SP1 and SP2) in an arena divided into two pens by a wire mesh wall. During SP1, one of the pigs of a pair was either restrained (Stress treatment) or sham-restrained (Control treatment), while the other pig was considered observer. During SP2, the previous observer was restrained, while its pen mate took the observer role. Heart rate variability, locomotion, proximity, vocalisations, body/head/ear and tail postures were monitored. Data were analysed using GLM (PROC MIXED) or GLMM (PROC GLIMMIX) models and Holm-Bonferroni adjustments were used. During SP1, observer pigs responded to conspecific distress with increased indicators of attention (looking at: 68.2 vs 31.5%; SEM: 3.7; $P<0.001$; proximity to: 85.9 vs 70.3%; SEM: 3.9; $P<0.001$; and snout contacts with the distressed pigs: 3.9 vs 1.3; SEM: 0.3; $P<0.001$) and increased indicators of fear (reduced locomotion: 2.5 vs 5 squares crossed; SEM: 0.7; $P<0.01$; increased freezing: 3.0 vs 0.6; SEM: 0.2; $P<0.001$). During SP2, the observer pigs that had been restrained previously reacted more strongly (through higher proximity: 92.9 vs 78.9%; SEM: 2.1; $P<0.001$; decreased locomotion: 3.5 vs 6.2 squares crossed; SEM: 0.6; $P<0.05$; and increased freezing: 2.4 vs 1.0; SEM: 0.3; $P<0.01$) to observing the pen mate in restraint than pigs without the previous negative experience. This study suggests that young pigs are susceptible to emotional contagion and that this contagion is potentiated by previous exposure to the same stressor. These findings enlarge the list of species in which emotional contagion was proven and have important practical and ethical implications for aversive procedures that are being routinely applied to group housed domestic pigs.

Development of two methods for investigating cognitive bias in laying hens

Amanda Deakin[1], William J Browne[2], James J L Hodge[3], Elizabeth S Paul[1] and Michael Mendl[1]
[1]University of Bristol, School of Veterinary Science, Langford, BS40 5DU, United Kingdom, [2]University of Bristol, Centre for Multilevel Modelling, 35 Berkeley Square, Bristol, BS8 1JA, United Kingdom, [3]University of Bristol, School of Physiology, Pharmacology and Neuroscience, Biomedical Sciences, University Walk, Bristol, BS8 1TD, United Kingdom; amanda.deakin@bristol.ac.uk

Affect-induced cognitive judgement biases mean that animals in a more negative affective state tend to interpret ambiguous cues more negatively than animals in a more positive affective state and vice versa. Investigating animals' responses to ambiguous cues can therefore be used as a proxy measure of affective state. We investigated laying hens' responses to ambiguous stimuli using two novel cognitive bias tasks – the hole-board task and the screen-peck task. In both tasks, hens were trained to associate one colour with a food reward (positive cue – P) and a second colour with either punishment or lack of reward (negative cue – N). Ambiguous cues were gradations of colour between the P and N cue (near-positive – NP; middle – M; near-negative – NN), and were unrewarded. In the hole-board task, hens were trained to peck through P coloured paper but not N coloured paper in a 12-hole board to gain access to a reward (P) or nothing (N). When hens had learnt this discrimination (33 sessions over 33 days), ambiguous cues were presented alongside P and N cues. Pecking of ambiguous cues showed a clear generalisation curve from P through NP, M, NN to N, suggesting that hens were able to associate colours with reward/lack of reward and could discriminate between colours more or less similar to learnt cues. In the screen-peck task, hens learnt to discriminate between P and N colours on a computer screen (29 sessions over 16 days). Pecking the P cue led to a food reward whilst pecking the N cue led to an aversive air puff. Again, ambiguous cue pecks produced a clear generalisation curve. Using a repeated measures design, we aimed to manipulate affect by changing temperature during testing to either c.20 °C or c.29 °C. Hens have been shown to prefer higher temperatures in this range hence we assumed that exposure to the higher temperature would induce relatively positive affect. Hens tested under warmer conditions were significantly faster to peck the M probe (t_4=3.86, P=0.018) but not the NP (Wilcoxon Z=-1.10, P=0.273) or NN (Wilcoxon Z=-1.63, P=0.102) probes than those tested at cooler temperatures, providing provisional evidence that increased temperature may induce a positive cognitive bias in hens. The screen-peck task provides an effective method for investigating cognitive bias that is faster to train than most tasks thus far trialled in chickens. The hole-board task has potential for further use with a few adjustments.

'Liking' and drinking: a microstructural analysis of hen drinking behaviour as a possible indicator of reward valuation

Elsa Mendl, Gina Caplen, Ilana Kelland, Carly Betts and Elizabeth Paul
University of Bristol, School of Veterinary Sciences, Langford House, Langford, Bristol, BS40 5DU, United Kingdom; emendl422@gmail.com

In rodents, 'Liking' tests based on the microstructure of drinking behaviour have been developed to measure the nature and magnitude of animals' immediate responses to rewarding stimuli. The present study was conducted with the aim of developing similar measures in domestic chickens (*Gallus gallus domesticus*). Seven laying hens were filmed for two three-minute intervals drinking water and four concentrations of monosodium glutamate solution (MSG 0.15, 0.3, 0.6, 1.2% w/vol), a liquid that was hypothesised to be preferred by hens because of its protein-based 'umami' flavour (this preference is currently being testing in a separate experiment). Microstructural analyses were used to investigate whether MSG flavoured water would elicit different patterns of drinking behaviour than water alone. It is already known that when chickens drink from an open source, they show a stereotyped pattern of behaviour in which the beak is 'clapped' several times in a head-down position in the water/solution and then several times again with the head in an upward position, out of the liquid. We found that the number of head-up beak claps was almost twice the number directed to the water/ solution itself (head up/down ratio for de-ionised water 1.96). The mean number of beak claps per drinking bout shown to the different concentrations of MSG were: Water 67.72; MSG 0.15% 80.17; MSG 0.3% 128.52; MSG 0.6% 100.55; MSG 1.2% 41.34. Concentration of MSG was significantly related to number of head-up beak claps shown per drinking bout ($F=6.42$, $P<0.05$), with concentrations of 0.3% MSG producing the highest mean number of beak claps (i.e. a non-linear relationship). We conclude that microstructural analysis of drinking behaviour may prove useful in future studies of 'liking' in hens if an association is found with measures of preference for MSG solutions of varying concentrations.

Optimism in an active-choice judgement bias task is independent from cognitive abilities in pigs

Eimear Murphy, Sanne Roelofs, Elise Gieling, Rebecca Nordquist and F. Josef Van Der Staay
Utrecht University, Emotion & Cognition Group, Department of Farm Animal Health, Yalelaan
7, 3584 CL Utrecht, the Netherlands; eimear.murphy@vetsuisse.unibe.ch

Biases in judgement of ambiguous stimuli in a judgment bias task (JBT) are increasingly being used as proxy measures of emotional states in animals. Optimistic and pessimistic judgements are said to reflect positively- and negatively-valenced states respectively. Such biases have often been investigated with respect to treatment effects, but little work has been done to validate the paradigm in other ways. We recently proposed criteria for behavioural tests of emotion, one being that responses on the task should not be confounded by, among others, differences in learning capacity. Three independent studies using different cohorts of pigs (S1: n=15; S2: n=37; S3: n=18) were conducted. Pigs were first trained in a spatial holeboard task (HBT), a square arena containing 4×4 matrix of 'holes'. For each pig, the same subset of 4 holes were always baited and in each trial pigs could freely explore the holeboard. The slopes of working memory (WM) and reference memory (RM) performance were calculated across successive blocks of trials in the acquisition phase. Pigs were then trained in a two-choice apparatus on a conditional discrimination task (CD), where tone-cues signalled the availability of a large or small reward in an associated goal-box. Ambiguous tone-cues were then interspersed among the training cues and optimism was measured as responses to the large-reward goal-box (mean area under the curve; AUC). As shown in rodent studies, we expected that WM and RM in the HBT represent different memory domains. Further, if judgement bias is truly independent of learning ability, we predict no relationship between AUC and CD learning, WM, or RM. Data were analysed using Pearson product moment correlations or Spearman's rank correlations, where appropriate. As expected, acquisition of RM (Mean±SEM; S1: 0.04±0.00; S2: 0.05±0.00; S3: 0.05±0.00) and WM (S1: 0.01±0.00; S2: 0.02±0.00; S3: 0.02±0.00) in the HBT were uncorrelated in all three studies (S1: r=0.25, P=0.34; S2: r=0.26, P=0.12; S3: r=-0.07, P=0.77). Secondly, AUC in the JBT (S1: 42.13±5.04; S2: 70.75±2.69; S3: 51.98±4.45) was unrelated to CD (S1: 10.60±0.81; S2: 16.32±0.59; S3: 22.22±1.39) learning (S1: r=0.31, P=0.27; S2: r_s=0.08, P=0.63; S3: r=0.35, P=0.16), WM (S1: r=0.28, P=0.31; S2: r_s=0.06, P=0.72; S3: r=-0.42, P=0.08) or RM (S1: r=0.45, P=0.09; S2: r_s=0.02, P=0.91; S3: r=0.20, P=0.43). We therefore conclude that optimism as measured in our active choice JBT is independent from the cognitive abilities assessed using the HBT and CD tasks. These findings go one step further towards validating the use of judgment bias as a gauge of emotional valence in animals.

Behavioural responses of pet dogs to visual stimuli with phonetic sounds

Megumi Fukuzawa, Arisa Tanaka, Kai Tamura and Osamu Kai
Nihon University, College of Bioresource Sciences, Kameino 1866, 252-0880, Japan;
fukuzawa.megumi@nihon-u.ac.jp

The number of dogs kept in houses is increasing, and there are many owners reporting problematic behaviours such as destruction, elimination and barking. The purpose of this study was to estimate the influence of a picture with phonetic sound on the behaviour of dogs. Seven healthy pet dogs (four female, three male; aged 8 to 117 months, weight 7.2 to 30.5 kg) participated in this study. There were no dogs showing a fear response to every picture with sound or separation anxiety from the owner. Female and male experimenters familiar to the dogs performed the three different situations to them with exaggerated gestures and expressions respectively (invitations to 'play', 'scold' loudly, and 'ignore'). These stimuli were recorded beforehand, and showed on a monitor. The pitch or emotional context of each stimuli differed, so the volume was controlled during playback; the maximum/ minimum volume 75.3/ 49.1dB ('play'), 77.9/ 47.4dB ('scold'), and 52.5/ 42.6dB ('ignore'). Each picture stimulus consisted of total of 9 minutes of a blank screen for 3 minutes, a stimulus screen for 3 minutes and a blank screen for 3 minutes again. The looking time, the direction of the body and the attitude to the monitor, behaviour (eight categories) and posture (four categories) of a dog were observed continuously. An ANOVA and a paired t test were used for statistical analysis. The effect of the gender of the stimulus presentation person was not significant, but dogs gazed at the monitor significantly longer during stimulus presentation compared to a blank screen (Tukey, $P<0.05$). The direction of the body and the attitude to the monitor were different in a picture stimulus. Both 'approach' and 'face towards' the monitor duration increased in 'play' and 'scold' picture stimulus presentation, while the reaction to the monitor was weak and time to turn the back on a monitor was long in 'ignore' picture stimulus. The posture of the dog showed a similar tendency in all the observation periods, but there was an increase in lying time after the 'scold' stimulus presentation. These results suggest that dogs can distinguish the features of the picture with the phonetics sound.

Do piglets (*Sus scrofa domestica*) use human visual and/or vocal signals to find hidden feeding reward?

Sandy Bensoussan, Maude Cornil, Marie-Christine Meunier-Salaün and Céline Tallet
INRA, PHASE, PEGASE, Agrocampus Ouest, 35590 Saint-Gilles, France;
sandy.bensoussan@rennes.inra.fr

Though animals rarely use only one sense to communicate, few studies have been done on the use of combinations of signals between animals and humans. Piglets' ability to use human visual signals (i.e. pointing gestures) is not clearly demonstrated. They mainly use auditory signals to communicate with conspecifics and might be sensitive to human voice. This study was carried out on 28 weaned piglets. Among them, 16 piglets were submitted to a pre-test with object-choice task including four conditions to find the reward ('experienced' piglets): the experimenter statically pointed at the reward with (1) or without (2) directing a loudspeaker broadcasting vocal instructions to it; (3) the loudspeaker alone was directed to the reward or the experimenter stood motionless. All piglets were submitted to three successive individual object-choice tasks related to combinations of human signals: (1) the experimenter statically pointed at the reward and directed the loudspeaker to it; (2) the experimenter used a dynamic pointing gesture and the loudspeaker (3) the experimenter performed again the first static combination. During the pre-test, piglets did not found the reward in any case (50.0±0.01% of mean success, P=0.42) and results suggest that previous experience did not influence success rates (mixed model analysis, P=0.58). Piglets found the reward in test 2 and 3 (respectively 63.4±2.1 and 61.6±2.1% of mean success, binomial tests P<0.05; test 1: 51.9±2.1%). Unexperienced piglets do not use vocal and visual signals but they can learn to use combinations of vocal and visual signals. The experience acquired on successive tests – a change from static to dynamic signal or a second presentation of a signal – could be more efficient than a previous stimuli exposure. Further investigation is needed to analyse individual variations in responses.

Hens judge from appearance: changing phenotypes in stable groups of different size

Guiomar Liste[1,2], Irene Campderrich[2] and Inma Estevez[2,3]
[1]Royal Veterinary College, Hatfield, Hertfordshire. AL97TA, United Kingdom, [2]Neiker-Tecnalia, Arkaute, 01080 Vitoria-Gasteiz, Spain, [3]IKERBASQUE, Basque Foundation for Science, Bilbao 48013, Spain; iestevez@neiker.eus

Higher aggression is a concern in alternative poultry production, where larger than conventional-cage groups are housed with potential negative effects on social dynamics. Accidental phenotype alterations compromising the ability to recognize conspecifics could also play a role. We aimed to analyse the impact of changing phenotypes and varying group sizes on the behaviour of hens. Birds from stable groups with artificially altered phenotype (100, 70, 50, 30 or 0% of altered birds/group) were altered further by sequentially marking (M; black mark on back of head) or unmarking (UM) a proportion of the group at a time to observe behavioural responses. 1050 Hy-line brown layers were housed in 45 experimental pens housing 10, 20 or 40 birds (15 pens/treatment at constant density). Behaviour was observed in at least 6 focal individuals/pen, from 27-46 weeks of age, with The Chickitizer software. The experiment was approved by local and regional ethic committees and it conforms to the ISAE ethical guidelines. Time budgets for each focal bird and behaviour were calculated and averaged by time and pen (statistical unit). GLMM assuming a gamma distribution were built, with phenotype and group size as fixed effects and pen as random effect. The occurrence of some behaviours was low and Kruskall-Wallis tests were used to assess the same effects. Initial observations before the changes showed no significant differences in behaviour (P>0.05), suggesting stable social groups. After the 1st change, aggression given (30M: 0.21±0.21a, 70UM: 2.10±0.94c; 30UM: 0.68±0.30b, 70M: 2.89±0.77c) and received (30M: 6.21±0.81b, 70UM: 0.00±0.00a; 30UM: 5.18±1.06b, 70M: 0.00±0.00a) by birds in the recently altered groups was severely affected by phenotype (P<0.001, both cases). Eat, rest and social preen were similarly affected (P<0.05, all cases). The 2nd change showed similar effects on aggression given and received (P=0.013 and P<0.001, respectively), but overall levels of aggression lowered. Differences in levels of aggression within recently changed groups could still be detected after the 3rd change (P=0.034 and P=0.013 respectively) but levels of aggression dropped to values similar to those in control groups. Group size did not seem to affect the behaviour of hens (P>0.05). Phenotype alteration had a greater impact on hen behaviour, especially aggression, than group size. The originality of the phenotype alteration wore off with repetition over time. These results suggest that, recognizable phenotypes, novelty and familiarity play a relevant role on social dynamics and the plasticity of behavioural adaptation.

Effects of two different cognitive challenges on zebrafish social behaviour

Becca Franks, Courtney Graham, Christine Sumner and Daniel M. Weary
University of British Columbia, Animal Welfare Program, 2357 Main Mall, Vancouver BC, V6T
1Z4, Canada; beccafranks@gmail.com

Cognitive enrichment – exploration opportunities, learning, and cognitive challenge – is increasingly acknowledged as a potential contributor to good welfare. Thus far, the majority of cognitive enrichment research has focused on mammalian or avian species, with little study on how such enrichment may affect the well-being of other cognitively sophisticated taxa such as cephalopods and fish. To investigate the effect of two different types of cognitive challenge, we exposed six tanks of zebrafish (10 fish/tank) to two feeding challenges: Rigid and Dynamic. All tanks were exposed to both treatments and were furnished with sloping gravel substrate, plants, rocks and hiding spaces. In the Rigid feeding regime, we changed the food location at the start of treatment and maintained the new, mid-tank location for six weeks. In the Dynamic feeding regime, we switched the location daily: over six weeks, food-delivery alternated each day between the ends of the long, rectangular tanks. We hypothesized that improved welfare would be reflected in improved social dynamics: lowered aggression and enhanced sociopositive interactions. Before feeding the fish, we recorded aggressive displays (chases and charges), coordinated swimming (the degree of synchrony in swimming behaviour), and the number of fish in each area of the tank. After feeding the fish, we noted the time it took for the majority of fish to find the food. In both conditions, fish showed signs of learning: latency for the majority of the fish to find food decreased from 8.79 (SE=3.64) seconds to 0.77 (SE=0.39) seconds in Rigid treatment (generalized linear mixed models [GLMM], P<0.0001) and from 44.77 (SE=6.34) seconds to 8.48 (SE=1.06) seconds in Dynamic treatment (GLMM: P<0.0001). However, what the fish learned differed: by week 6 in the Rigid treatment, fish learned to anticipate food delivery (on average 7.14 SE=0.90 of the 10 fish were in the correct location pre-feed), but in the Dynamic treatment fish had not learned to anticipate the correct location and instead learned to go to one end of the tank or the other (on average 7.83 SE=0.43 fish at either end of tank pre-feed). These differences in learning content – predictable vs probabilistic food-location – corresponded to differential effects on social behaviour. Across the Rigid treatment, aggression decreased (GLMM: P<0.05) and coordinated swimming increased (GLMM: P<0.01), but in the Dynamic treatment aggressive displays and coordinated swimming did not change. These results suggest that cognitive challenges influence zebrafish social dynamics and that beyond learning, the content of what is learned may also matter.

A measure of fear of humans in commercial group-housed sows

Lauren Hemsworth, Candice Powell, Maxine Rice and Paul Hemsworth
Animal Welfare Science Centre, The Faculty of Veterinary and Agricultural Sciences, The University of Melbourne, Parkville, VIC 3010, Australia; lauren.hemsworth@unimelb.edu.au

There is evidence that the human-animal relationship (HAR) affects the welfare and productivity of commercial farm animals; consequently an animal-based measure of HAR, which assesses a pig's behavioural response to a human, would be a valuable addition to on-farm welfare assessment in group-housed sows. The human-approach test (HAT), in which the behavioural response of a pig to a stationary experimenter is measured individually in a test arena, has been used extensively in research to assess fear of humans in pigs. The HAT has consistently shown that handling treatments designed to differentially affect the pigs' fear of humans produced the expected variations in pig approach behaviour. However, the HAT is impractical for regular on-farm assessment of fear, and there is currently no well-validated practical test suitable for use in commercial group-housed sows. The aim of this study was to (a) develop a behavioural test to assess fear of humans in commercial group-housed sows, and (b) validate the developed test (HT_A) against the previously validated HAT. The validity of the HT_A was investigated using 232 crossbred pregnant gilts and sows to examine the relationship between the HAT, behavioural response of sows individually held in an arena to a stationary experimenter; and HT_A, behavioural response of sows in a group-pen to an approaching experimenter. A principal component (PCA) analysis was conducted on each data set for the HAT and HT_A to identify a set of factors for each test that represents the underlying relationships between the variables in each test. For the HAT and HT_A, a single factor was identified that provided an appropriate summary of the data (69.9 and 66.2 variance accounted for, respectively). Spearman rank correlations were used to investigate the relationships between the factors determined in the PCA. The effect of parity on the factor scores was also studied using one-way analysis of variance. A significant moderate correlation was found between the HAT and HT_A scores (r=0.390, P<0.01); the reduced proximity of the sow to a stationary experimenter in the HAT was correlated with a greater avoidance of the approaching experimenter by the sow in the HT_A. Parity affected the HT_A score (P=0.02) but not the HAT score, and post-hoc comparisons (Bonferroni) indicated that gilts in general showed less avoidance of the approaching experimenter in the HT_A than sows. The moderate association between the two tests indicates that, with further refinement to improve validity whilst maintaining feasibility, the HT_A may provide a practical and valid on-farm assessment of fear of humans in group-housed sows in a commercial setting.

Effect of group size and health status on social and feeding behaviour of transition dairy cows

Margit Bak Jensen[1] and Kathryn L Proudfoot[2]
[1]Aahus University, Department of Animal Science, Blichers Allé 20, 8830 Tjele, Denmark, [2]The Ohio State University, Veterinary Preventive Medicine, 1920 Coffey Road, Columbus, OH 43214, USA; margitbak.jensen@anis.au.dk

Dairy cows are often moved to a large group after calving, where they may have to compete for access to resources. These cows are also at high risk of disease, which may impair their ability to compete. Thus, housing in small groups with minimal competition for the first days or weeks after calving may be beneficial for cow welfare. The aim was to investigate the effect of group size and health on social and feeding behaviour of cows during the first 24 h after introduction to a new group. Fifty-six multiparous Holstein-Frisian cows were moved from an individual pen and individually joined an existing group pen for 6 (N6) or 24 cows (N24) on d4 after calving. Cows were considered sick if they were diagnosed with and treated for milk fever or mastitis, or diagnosed with subclinical ketosis within 3d of calving (n=24; balanced across treatments). Stocking density of both pens was the same. Behavioural data were collected from video and electronic feedbins. Variables were analysed using PROC MIXED (SAS) including fixed effects of treatment, parity, health status, and treatment×health status; month was considered a random effect. No interactions between health status and treatment were discovered. N6-cows displaced other cows from feed less frequently than N24-cows (4.7 vs 9.2 ($\pm$2.8) times/24 h; $F_{1,44}$=22.9; P<0.001), were less likely to access feed after a displacement (1.5 vs 2.7 ($\pm$0.9) times/24 h; $F_{1,44}$=8.32; P<0.01), and were less frequently butted by another cow (square-root transformed (back-transformed estimates in brackets) 0.65 (0.4) vs 1.18 (1.4) ($\pm$0.22) times/24 h; $F_{1,44}$=5.85; P<0.05). There was no effect of treatment on DMI (P=0.6), but sick cows ate less than healthy cows (14.0 vs 15.4 ($\pm$0.6) kgDM/24 h; $F_{1,45}$=4.5; P<0.05). However, cows in N6 visited the feeder less often (40.6 vs 55.0 ($\pm$5.2) no./24 h; $F_{1,45}$=7.6; P<0.01), and tended to consume more feed per visit (0.4 vs 0.3 ($\pm$0.04) kgDM/visit; $F_{1,45}$=4.0; P=0.05). Results suggest that cows experience less competition when moved into a smaller group after calving regardless of health status. Thus, minimising competition by housing dairy cows in a small group for the first days or weeks after calving may improve cow welfare under commercial conditions.

Mixing sows into groups during lactation: presence of piglets reduces aggression

Emma Catharine Greenwood[1], Jessica R. Rayner[2], Kate J. Plush[2], William H.E.J. Van Wettere[1] and Paul E. Hughes[2]
[1]*The University of Adelaide, School of Animal and Veterinary Science, Roseworthy campus, 5371, South Australia, Australia,* [2]*South Australian Research and Development Institute, Roseworthy, 5371, South Australia, Australia; emma.greenwood@adelaide.edu.au*

In most commercial settings the mixing of sows into groups occurs after weaning, insemination or pregnancy detection. Altering the timing of mixing, by mixing sows in lactation, may reduce the affects of mixing stress on reproduction and the levels of aggression that accompany mixing. This study aimed to determine the effect of the presence or absence of piglets on sows grouped in lactation. We hypothesised that sows in a multisuckle pen would display reduced aggression at mixing than those that were separated from their piglets and mixed in lactation. The study utilized 60 multiparous (parity 3.7±0.8) Large White × Landrace sows. Sows were mixed into groups of six, with sows either group-housed with litters from d21 lactation (calc. from day 28 wean) (MS) or removed from individual crates and litters and mixed into groups for 7 hrs daily, and then returned to their piglets and crates (seperation and mixing from 06:45 to 13:45 hrs) from d21 lactation (SEP). Behaviour was observed through video records from 07:00 to 13:00 hrs on M0 (mixing), M1 and M6 and W0 (weaning), W1 and W6. Data were analysed using a linear mixed model in SPSS. Data were not normally distributed so transformed mean ± SEM are presented with untransformed means in brackets. There was consistently higher aggression in the SEP group compared to the MS group ($P<0.05$) with higher fight number per sow per hr [0.7±0.1 (0.2) vs 0.2±0.1 (0.1)], higher bite [0.1±0.2 (2.1), -0.5±0.2 (0.4)] and knock number per sow per hr [0.1±0.1 (1.5), -0.6±0.1 (0.4)], and consequently higher total injury number [5.8±0.4 (34.3), 3.9±0.4 (11.8)]. Sows mixed into groups during lactation in the absence of their piglets exhibit higher levels of aggression and injuries than those housed with their piglets. This could be related to several factors, the absence of piglets, the reduction in lactation stimulation or the repeated reintroduction of sows to each other. Therefore, multisuckle systems rather than sow separation methods should be used when mixing lactating sows.

Sows with low piglet mortality are not more careful towards their piglets than sows with high piglet mortality

Janni Hales[1], Danielle K.F. Pedersen[1], Grith K. Guldbech[1], Vivi Aa. Moustsen[2], Mai Britt F. Nielsen[2] and Christian F. Hansen[1]
[1]University of Copenhagen, Department of Large Animal Sciences, Groennegaardsvej 2, 1870, Denmark, [2]SEGES Pig Research Centre, Innovation, Axeltorv 3, 1609 Copenhagen, Denmark; hales@sund.ku.dk

Early piglet mortality is a concern in pig production, especially when sows are loose housed since this is one of the factors that explain a higher mortality in loose housed systems. High levels of maternal care are in general expected to be related to low levels of piglet mortality. The aim of this study was to compare sow pre-lying behaviour and piglet behaviour before litter equalization for sows with high and sows with low level of piglet mortality. Sows (parity 1-2) housed in SWAP farrowing pens were observed from the end of farrowing until litter equalization took place within 12-24 h after farrowing. Video recordings of sows with low live born mortality (L-sows; 0-1 dead piglet; n=16) or high live born mortality (H-sows; 2 or more dead piglets; n=12) in the observational period were used for continuous registrations of sow postures and pre-lying behaviour (rooting, pawing, contact to piglets) throughout the observational period. Piglet location (sow, creep, other) and piglet behaviour (walking/playing, nursing/lying at the udder, lying away from sow) was registered by use of scan sampling every hour of the observational period. Data was analyzed using linear models. Litter size was similar in both groups (L-sows: 17.3±0.57, H-sows: 17.4±0.67; P=0.91) but live born mortality percentage differed (L-sows: 2.2±1.1, H-sows: 16.1±1.4; P<0.001). The number of times per hour sows lied down did not differ between L-sows and H-sows (1.0±0.09 vs 0.8±0.11; P=0.14). Sows rolled similar number of times per hour (L-sows: 0.5±0.11; H-sows: 0.3±0.13; P=0.12). The sows moved to lateral position and ventral position at similar speed (P=0.41 and P=0.76, respectively). Rooting/pawing before lying down occurred 0.37±0.06 times/hour for L-sows and 0.39±0.07 times/hour for H-sows (P=0.846) and snout contact with piglets before lying down occurred 0.43±0.14 times/hour for L-sows and 0.56±0.17 times/hour for H-sows (P=0.58). Piglets were seen in the creep in 29% of observations for both L-sows and H-sows (P=0.32). Piglet activity in L-sow litters and H-sow litters was observed to be nursing/lying at udder in 35 and 33% of observations (P=0.51), walking/playing in 32 and 30% of observations (P=0.29) and lying away from the sow in 33% and 37% of observations (P=0.19). The results indicate that L-sows did not display more careful behaviour towards their piglets than H-sows. Also, piglet activity and location in the pen was very similar for all sows, suggesting that these parameters did not indicate risk of high piglet mortality.

Behavioural reactions of horned and dehorned dairy cows to herd mates whose horn status was manipulated

Janika Lutz[1], Beat Wechsler[1], Hanno Würbel[2], Joan-Bryce Burla[1], Lorenz Gygax[1] and Katharina Friedli[1]
[1]Federal Food Safety and Veterinary Office FSVO, Institute for Livestock Science ILS, Agroscope, Centre for Proper Housing of Ruminants and Pigs, Tänikon, 8356 Ettenhausen, Switzerland, [2]Veterinary Public Health Institute, University of Bern, Division of Animal Welfare, Länggassstrasse 120, 3012 Bern, Switzerland; janika.lutz@agroscope.admin.ch

Horns play an important role in the social behaviour of cows. Nonetheless, in order to prevent injuries caused by horns, disbudding of calves is a common practice in dairy farming. This raises the question of how different aspects of social behaviour are influenced by the presence or absence of horns. In the present study, we tested the interactions of horned (n=15) and dehorned cows (n=17) towards familiar dehorned cows (n=32) by manipulating the horn status of the latter. The experiment was conducted on four farms. All herds were kept in loose housing systems and composed of horned and dehorned cows. During experimental sessions, pairs of cows were matched randomly and separated from the herd for 15 minutes in a pen measuring approximately between 25 and 35 m^2. On three days, each pair was tested in three different conditions. The familiar dehorned cow was wearing either a 'halter with horns', a 'halter with rabbit ears' (a nonsense alternative) or only a 'halter' (control), whereas the other animal (a horned or dehorned cow) remained unmanipulated. During testing, the frequency of agonistic and affiliative behaviours was recorded by direct observation. In addition, the position of the cows' heads (low, middle, high) and the distance between the cows were registered every two minutes. Data were analysed using linear-mixed effects models and stepwise backward elimination. P-values were generated by parametric bootstrap. The horn status of the unmanipulated cow but not the halter type of the manipulated cow had an influence on behaviour. Compared to dehorned cows (median = 0, range= 0-17 interactions) horned cows (median = 2, range = 0-14; as actor or receiver) were more often involved in agonistic interactions with the manipulated cow (horn status: P=0.003). No significant effects were found for affiliative behaviour (horn status: P=0.166, halter type: P=0.257). Further, the probability of carrying the head in a higher position was larger in non-manipulated horned than dehorned cows (horn status: P=0.01). In conclusion, the temporary manipulation of the horn status of a familiar cow did not specifically alter the dyadic interaction of the manipulated and unmanipulated cow. However, dyadic interactions were influenced by the horn status of the unmanipulated cow indicating that the long-term experience with a partner's horn status is of greater relevance than the temporary manipulation of the horn status.

Potential influencing factors on social behaviour and animal welfare state in large danish dairy cow herds

Marlene K. Kirchner, Stephanie Højgaard Mahrt and Nina Dam Otten
University of Copenhagen, Large Animal Sciences, Groennegaardsvej 8, 1870 Frederiksberg C, Denmark; nio@sund.ku.dk

Social behaviour as such was seen as a substantial part of welfare assessment schemes in recent years. However, for dairy cows the welfare quality assessment protocol (WQ) is evaluating agonistic behaviour only. Earlier studies with the WQ in Danish Dairy herds reported low frequencies of agonistic behaviour. This was raising the questions if Danish cows express social behaviour on a very low level in general and or which farm factors are associated. The aim of the study was therefore to investigate socio-positive behaviour additionally to the WQ, revealing more data on social behaviour of Danish Dairy herds. Further we were analysing effects of certain farm factors on social behaviour. The study was performed in 30 Danish dairy cow farms with an average milking herd size >100 cows in loose house systems. Recording of data and computing of the criterion score for social behaviour were performed according to the WQ for cattle. To integrate the positive behaviour in the welfare assessment, we calculated the WQ-criterion score for 'social behaviour with the WQ-procedure for fattening cattle (S_{social}), combining incidences of agonistic and socio-positive behaviour. S_{social} includes an expert-weighting and can range from 0 (poor welfare) to 100 (excellent welfare) points. Potential risk factors such as herd size (NO_COWS), *ad libitum* feeding (ADLIB_FEEDING), the average yearly milk yield/cow (MILKYIELD), the presence of a feeding automate (FEEDAUTOMATE), if cows were kept in one or more groups (GROUPING) and the type of feed rack (FEEDRACK) were assessed on-farm and analysed in a linear model with S_{social} as the outcome variable. The incidence of socio-positive and agonistic behaviours was 0.67±0.42 and 1.08±0.61 on average per cow and hour, respectively. The Danish herds achieved an average WQ S_{social} score of 55±14 points, stating a good welfare state. The final model for the potential influencing farm factors included NO_COWS, ADLIB_FEEDING, MILKYIELD, FEEDAUTOMATE, GROUPING, FEEDRACK and explained 30% of the variance (adjusted r^2=0.30; P=0.04). S_{social} scores, indicating the welfare state, were significantly lower in herds with cows kept in several groups (GROUPING) according to milk yield (P<0.0001) or parity (P=0.02), most likely explained by accompanied, frequent re-grouping. We conclude that the incidences of 'social behaviour' on large Danish dairy cow herds in our study were quite varying and on average welfare state calculated with the combined S_{social} was good. We were able to identify influencing factors on S_{social} in a linear model and significant higher scores, meaning better welfare state, in herds without grouping of cows.

Milker behaviour affects the social behaviour of dairy cows

Stephanie Lürzel[1], Denise Reiter[1], Kerstin Barth[2], Andreas Futschik[3] and Susanne Waiblinger[1]
[1]University of Veterinary Medicine, Vienna, Institute of Animal Husbandry and Animal Welfare, Veterinärplatz 1, 1210 Vienna, Austria, [2]Johann Heinrich von Thünen Institute, Institute of Organic Farming, Trenthorst 32, 23847 Westerau, Germany, [3]JK University Linz, Department of Applied Statistics, Altenberger Str. 69, 4040 Linz, Austria; stephanie.luerzel@vetmeduni.ac.at

Milker behaviour was associated with cows' social behaviour in on-farm studies but experimental evidence is lacking. We exchanged the usual milker for a team of two experimenters. During the positive phase, they showed predictable behaviour including gentle tactile and vocal contact with the cows (n=89); during the negative phase, they behaved unpredictably and produced noise. The first replicate consisted of the pre-experimental 'normal' phase (usual milker, Nor1), a positive phase (10 d, Pos1) and a negative phase (2 d, Neg1); the second replicate consisted of a positive phase (12 d, Pos2), a negative phase (2 d, Neg2) and the post-experimental 'normal' phase (Nor2). We observed the cows' behaviour during milking and their social behaviour afterwards twice at the end of each phase during both replicates. Observers of social behaviour were blinded to the treatment. During milking, cows had a shorter duration of 'high ear' positions in Pos1 than in Neg1 (n=20, Friedman test, $P<0.05$; Wilcoxon tests, $P=0.01$) and of 'head upwards' positions in Pos1 than in Nor1 (Friedman test, $P<0.05$; Wilcoxon tests, $P=0.004$). In Pos2, they showed more low head positions than during Neg2 (Wilcoxon test, $P=0.03$; data loss for Nor2, so no comparison possible). The cows showed more sociopositive behaviour after negative milkings (0.04 behaviours/cow (Neg) vs 0.02 (Nor) and 0.03 (Pos), GLMM, $P=0.02$) and less agonistic behaviour after Pos1 than after Neg1 and Nor1 (0.25 vs 0.29 and 0.26, interaction phase×replicate, $P<0.001$). Lower levels of alert behaviour ('high ear' and 'head upwards' positions) during positive milkings may be a sign of a more relaxed emotional state. A positive milking style reduces agonistic behaviour after milking, which may decrease the risk of injuries and indicate a lower occurrence of negative emotional states. The high level of sociopositive behaviour after negative milkings confirms the suggestion that it may reduce social tension. Altogether, milker behaviour has a direct effect on dairy cow social behaviour and welfare.

Social structure stability in farmed female capybaras (Mammalia, Rodentia)

Sergio Nogueira-Filho, Pauliene Lopes, Djalma Ferreira and Selene Nogueira
Universidade Estadual de Santa Cruz, Ciências Agrárias e Ambientais, Rod. Jorge Amado km 16, 45662-900, Ilheus, Bahia, Brazil; slgnogue@uesc.br

Free-ranging capybara (Hydrochoerus hydrochaeris) live in stable groups composed of adult males and females with their young. A strictly linear dominance hierarchy characterizes the capybara male's social structure, while there is no consensus on the dominance relationships among capybara females. Thus, the aim of this study was to describe the social structure of farmed female capybaras. We also intended to evaluate the effect of the group size on the stability of the females' social structure. We analyzed the occurrences of agonistic behavioral patterns at the feeding time of five capybara groups, comprising one to two adult males and four to 13 adult females, kept in outdoor paddocks with an area of 400 to 4,500 m^2. The paddock size did not affect the results as the individual space allowance for each animal during feeding was maintained as the group size increased. For each group we determined the Landau's corrected linearity index (h'), the 'directional consistency index' (DCI), and the steepness of females' hierarchy among groups. We ran a Pearson matrix correlation to evaluate the relationship between h', steepness, and DCI with the number of females. Females' linearity index ranged from h'=0.73 (P=0.0001), in the largest group, to h'=1.0 (P=0.02) in the smallest one. The linearity index h' ($r_{Pearson}$=-0.97, P=0.006) and the dominance hierarchy steepness ($r_{Pearson}$=-0.97, P=0.005) were inversely correlated with the number of females. There was also a tendency to decrease the females' DCI when the number of females increased ($r_{Pearson}$=-0.85, P=0.06). Our results suggest that farmed female capybara exhibit a linear social dominance hierarchy that prevents the occurrence of physical contact aggressions, which are limited to ritualized patterns – mainly to the threat and push patterns to maintain the social dominance rank. There was, however, less stability of the females' dominance hierarchy with increasing group size. Further studies will be conducted to confirm our findings using a larger sample size and to investigate the effect of females' rank in their reproductive parameters.

Spontaneous, intensive shoaling in laboratory zebrafish: quantifying a previously undescribed social behaviour

Courtney Graham, Marina A. G. Von Keyserlingk and Becca Franks
University of British Columbia, Applied Animal Biology, 2357 Main Mall, Vancouver BC, V6T
1Z4, Canada; courtney3graham@gmail.com

Zebrafish, a highly-social, shoaling species, are widely studied across many fields of scientific research. Nevertheless, as they are typically housed in barren environments, little is known about their social behavioural repertoire in more naturalistic laboratory settings. In particular, spontaneously occurring, socio-positive behaviours have rarely been reported. Housing adult zebrafish (n=60) in large semi-natural tanks (n=6) with sloping gravel substrate, artificial plants and rocks, we observed a previously undescribed social behaviour: episodes of spontaneous, intensive shoaling. The aim of this project was to quantify the behavioural characteristics of these distinctive episodes and compare parameters of sociality to other periods of commonly observed behaviour, including periods of heightened aggression, food anticipatory activity (FAA) and normal/baseline behaviour (i.e. behaviour during the remainder of the day). Continuously scanning 6 days of video footage, from 08:00 to 18:00 h, we identified a total of 17 discrete shoaling episodes lasting on average 8.60±8.00 min (Mean±SD) and a total of 10 episodes of heightened aggression. FAA occurs twice daily, before each feeding, and we randomly selected 35 FAA episodes to analyze. 55 periods of time were randomly sampled throughout the day as baseline behaviour. Within each of these episodes (17 shoaling, 10 aggression, 35 FAA and 55 baseline), we randomly selected 100 second clips to analyze for various dimensions of social behaviour: cohesion (inter-fish distances), aggression (chasing and charging) and spatial organization (the degree to which fish adhered to similar swimming patterns). Averaging across the 100 second clips, we found that spontaneous shoaling episodes were characterized by high cohesion, with average inter-fish distances of 1.13±0.18 fish-lengths (Mean±SE) compared to 3.19±0.09 fish-lengths during all other periods (multilevel models accounted for repeated sampling of each tank; P<0.001), and low aggression, with aggressive behaviour occurring 20% of the time during shoaling episodes vs 45% during baseline (P<0.001). We also found evidence of a high degree of spatial organization during shoaling episodes, with nearly all fish swimming along the same axis compared to all other periods when fish were swimming in a variety of different directions (P<0.01). As the first description of spontaneous, intensive shoaling behaviour, this research begins to fill the gap in our knowledge regarding the range of social dynamics in laboratory zebrafish and contributes to the growing body of literature aimed at enhancing fish welfare.

Gangs of piglets: would they have fought less?

Nikolina Mesarec, Maja Prevolnik Povše, Dejan Škorjanc and Janko Skok
University of Maribor, Faculty of Agriculture and Life sciences, Pivola 10, 2311 Hoče, Slovenia;
janko.skok@um.si

Post-weaning stress accompanied by outbursts of aggression is a big issue in contemporary pig production, since it compromise piglets' welfare and cause health problems and poorer growth performance. Therefore, it is of great importance to prevent piglets from excessive aggressive confrontations right after weaning. The aim of this study was to reduce post-weaning aggression with the experience driven 'identification learning' during lactation period. During lactation, simple 2D shapes, either white circle on the black background (WC) or black cross on the white background (BC), were installed on the walls of the farrowing pens (8 signs of the same 2D shape per pen) – thus, from birth piglets were constantly exposed to the same visual stimuli. Experiment involved 120 cross-breed piglets from 12 litters divided into three repetitions. One experimental repetition involved four litters by 10 piglets of the same age (40 piglets), which were than weaned into two weaning groups, each consisted of 20 piglets from two litters. At weaning test (TG) and control (CG) group were formed; in both one WC and one BC litter was combined into one weaning group. TG pen was virtually separated on WC and BC part – i.e. half of the pen and feeder was equipped with WC and the other half with BC, while there were no WC/BC signs in the CG pen. We hypothesised more distinctive territoriality of WC/BC piglets in the TG, since piglets will have tendency to stay in the part of the pen equipped with 'familiar' 2D shape. Therefore, reduced aggression and better growth performance was expected in TG compared to the CG. However, preliminary results showed no significant differences (GLM; general linear model; $P \geq 0.05$), neither in the frequency of fights nor in the weight gain or body weight. It is important to note that lack of differences in the measured/observed parameters might be due to a high density in the weaning pens (2.0×2.0 m; 0.2 m^2/piglet), thereby an intensive confrontation between piglets was likely inevitable. Therefore, we will re-test the potential efficiency of this approach also on the smaller weaning groups.

The impact of familiar and unfamiliar conspecific pairings on heart rate variability of female white leghorn chickens

Allison M. J. Reisbig[1], Sheila Purdum[2] and Marisa A. Erasmus[3]
[1]University of Nebraska-Lincoln, Child, Youth and Family Studies, 105 Home Economics Building, Lincoln, NE 68583-0801, USA, [2]University of Nebraska-Lincoln, Animal Science, C206j Animal Science Building, Lincoln, NE 68583-0908, USA, [3]Purdue University, Animal Sciences, 125 S. Russell St., West Lafayette, IN 47907-2042, USA; areisbig2@unl.edu

Heart rate variability (HRV), the variation in the interval between heartbeats, is reflective of both human and animal emotional responses and overall health and well-being. Specifically, higher HRV is associated with better health and well-being. Studies of HRV as a measure of poultry well-being are limited and the relationship between HRV and poultry well-being is poorly understood. This proof of concept study, following appropriate institutional approval, investigated HRV in mature white leghorn hens (n=4; two hens each from an aviary and conventional cage) using implanted electrocardiogram (ECG) telemetry. HRV was measured when hens were paired with a familiar (cage/pen mate; three recordings/hen) and unfamiliar (three recordings/hen) conspecific in a novel environment (wire-partitioned floor pen). A 2×4 ANOVA compared the effects of familiarity/unfamiliarity and bird on HRV. The mean inter-beat (RR) interval was higher for birds paired with a familiar (211.83 ms; $F_{1,16}$=13.32, P=0.002) vs unfamiliar (189.23 ms) hen, suggesting that pairings with a familiar conspecific may buffer against stress when placed in a novel environment. Mean RR intervals for only the two aviary birds were significantly different. The R-wave amplitude for one aviary bird was consistently twice that of the other three birds' and may be due to electrode placement or heart abnormalities. This points to the importance of visual inspection of ECG traces in HRV research for measurement artifacts as well as deviations from normative parameters. Furthermore, cardiac activity is greatly affected by behavior, and behavior is affected by conspecifics. Because hens may behave differently in the presence of a familiar vs an unfamiliar conspecific, further research is needed to examine how cage/pen mate status impacts individual hen behavior and HRV. Overall, these preliminary results indicate that HRV may be an objective method for assessing hen well-being. Future research may benefit from examining the role of conspecific relationships in positive welfare.

Social hierarchy and feed supplementation of dry cows

Gabriela Schenato Bica, Ana Beatriz Torres Almeida, Karoline Tenfen, Jessica Rocha Medeiros, Alexandre Bernardi, Dennis Craesmeyer and Luiz Carlos Pinheiro Machado F°
UFSC, Universidade Federal de Santa Catarina, LETA, Laboratório de Etologia Aplicada, Rodovia Admar Gonzaga, 1346 Bairro Itacorubi, Florianópolis 88034-100, Brazil; gsbica@gmail.com

Cows on pasture may have unequal access to grain supplement due to the effect of social dominance. Subordinate cows are known to have less access to resources when a competition exists. Therefore this trial was designed to test if: (1) distributing the grain along the fence in individual piles would grant better access to cows than in continuous line; (2) when entering a new paddock with supplement, subordinate cows will graze new grass instead of competing for the grain supplement, in order to achieve similar level of nutrition. Four groups of nine cows each were tested in a cross-over design with two treatments: individual piles (P) and continuous line (L), along the fence supplementation. Animals were managed under Voisin's rotational grazing system, being moved every morning to a new paddock provided with grain along the fence (1 kg/cow), water and mineral salt. Each period had 3 days for habituation followed by 7 days for data collection. Animals were observed for one hour from the moment they entered the new paddock (8 am to 9 am). A one-minute interval instantaneous scan sampling (based on Altman, 1974) of each cow was taken, and their behavior registered as: grazing, eating grain supplement, ruminating, contesting or other. All agonistic interactions were recorded (aggressor-victim), and a sociometric matrix was estimated for each group. The cows of each group were then divided in high, medium and low rank. An ANOVA was performed to test differences between treatments (t-test) and the interaction treatment×rank. Pearson's correlation test of cow rank and feeding behavior was performed. The procedures were approved by the UFSC's Ethical Committee on the Use of Animals. Treatment did not affect animal behavior, however, there was an interaction rank×treatment on feeding behavior. In P, subordinated cows ate less supplement than dominant ones (P=0.04). Regardless of treatment, high rank cows ate more supplement (r=0.45; P=0.006) than low rank cows, which had more grazing events (r=-0.60; P=0.000). Distribution of grain in piles or lines didn't affect cow's access to the supplement as a whole. However, in P subordinate cows had less access. Based on these results we can assume that offering grain supplement under rotational grazing must be done when cattle enter a new paddock, to allow subordinated cows to strategically graze high quality forage while dominant cows eat supplement.

Assessment of the fear of human in pigs of the Kemerovo breed

Dmitriy Orlov, Elena Rubtsova, Marina Kochneva, Konstantin Zhuchaev and Olga Bogdanova
Novosibirsk State Agrarian University, Biotechnological faculty, Dobrolubova str. 160, 630039
Novosibirsk, Russian Federation; dmitriy_orlov88@mail.ru

In conditions of intensive pig farming the fear of humans experienced by pigs can be an important issue. Fear is connected with stress which can result in suppression of reproductive and immune functions, change in food intake, feed conversion, daily weight gain and meat quality. Animals, which are scared of humans, can be injured trying to escape the contact. Prolonged experience of fear can lead to apathy, high level of anxiety and the expression of harmful behaviors. The object of the research was the fear of human in the Kemerovo pig breed in the conditions of intensive pig farming. Pigs of this breed are exclusively kept in the study site farm. The research was focused on 94 sows and 12 boars. Amongst them, 14 sows farrowed once, and 80 farrowed twice and more. The boars were aged from 1.6 to 3 years. The evaluation of the reaction of pigs to an unfamiliar human was carried out by using methods presented in Welfare assessment protocol for pigs (Welfare Quality® Assessment for pigs, 2009). Score '0' conformed to absence of the avoidance reaction in animals, score '2' to absolutely fearful animals, and score '1', respectively, to an intermediate reaction. The acquired experimental data was subjected to statistical analysis. Reliability of the differences between the samples was assessed by φ-conversion of Fisher criterion. The analysis of the distribution of the reaction to humans revealed that animals without fear were predominant both among sows and boars. The total percentage of fearful sows was 33%, boars 50%. The analysis of sows' fear of humans depending on a pregnancy period revealed that the number of anxious animals increased respectively with the period of pregnancy. Up to the third month of pregnancy the share of fearful sows increased by 30% ($P<0.01$). Distribution of sows by reaction to humans depending on the number of farrows was also assessed. The share of animals with a score '0' of fear of human among first-farrowing sows was by 22% higher than among animals with more than two farrowing, but the difference was not statistically significant. The share of animals with scores '1' and '2' of fear of humans increased by 22% ($P<0.01$). To conclude, sows and boars with a score '0' of fear of humans were predominant in the Kemerovo pig breed sample. Results indicated the sufficient adaptability of pigs of the breed to the intensive pig farm conditions. The share of fearful animals increased in relation to the period of pregnancy. It can be assumed that the tendency of increased fearfulness is related to animals being inclined to escape the stressful situations in order to preserve the embryos.

Aggression in male mice: why can't we all just get along?

Elin M Weber[1], Brianna N. Gaskill[2,3], Kathleen R Pritchett-Corning[2,4], Lilian K. Wong[1], Jerome Geronimo[1], Jamie A. Dallaire[1] and Joseph P Garner[1]

[1]*Stanford University, Department of Comparative Medicine, Stanford, CA, USA,* [2]*Purdue University, Department of Comparative Pathobiology, West Lafayette, IN, USA,* [3]*Charles River, Wilmington, MA, USA,* [4]*Harvard University, Faculty of Arts and Sciences, Office of Animal Resources, Cambridge, MA, USA; elweber@stanford.edu*

Aggression is a serious welfare issue in laboratory mouse husbandry that can lead to injury or death. Mice are commonly kept in same-sex groups of four or more in small transparent cages. This housing is fundamentally unnatural and aggression is particularly common when male mice are regrouped. Many researchers require individuals of the same weight, and mice are therefore regularly regrouped, particularly before shipping. In piglets, mixing individuals of differing weights has been found to decrease aggression. Piglets of the same weight are hypothesized to contest dominance more strongly, and therefore fight more intensely. To investigate if housing male mice in groups of mixed weights similarly decreases aggression, C57BL/6NCrl and BALB/cAnCrl mice were housed in groups of four in either same- or mixed-weight groups. Mice began the experiment at 15, 17, 19 or 22 g body weight. They were conspicuously marked, video recorded for behavioural observations, and checked daily for wounds. Data was analyzed using a GLM with Poission distribution. We found no effect of treatment on the percentage of days on which mice were found to have new wounds in BALB/C (P=0.272) and more wounding in C57 housed in mixed-weight groups (P=0.021), which was the opposite of our prediction. Mouse growth rates were rapid enough that mice changed weight during the experiment, making differences between groups less pronounced. Based on weights near the midpoint of the experiment, at days 5-8, variation in weight within a cage was associated with significantly more wounding among mixed-weight groups (P=0.031), but with significantly less wounding among same-weight groups (P=0.011). During behavioural observations we could not see clear dominance or territorial behaviour; submission without observable threat was often noted and some animals ignored threats from conspecifics. This might be a result of animals failing to cope with the highly unnatural environment they are housed in; unable to escape from threats, they cannot properly respond to signals from conspecifics. We could not support the finding in pigs that animals in mixed weights groups fight less and suggest that it might be misleading to view intra-cage aggression in mice solely as a way for animals to establish social dominance.

Breed-specific stress-coping characteristics and welfare concerns on pig production in China

Ruqian Zhao
Nanjing Agricultural University, Key Laboratory of Animal Physiology & Biochemistry, Ministry of Agriculture, No. 1 Weigang, 210095, Nanjing, China, P.R.; zhao.ruqian@gmail.com

Animal welfare assessment is based on the perception of animals. Different pig breeds show significant differences in growth rate, meat quality, as well as stress-coping characteristics. We compared a Chinese local pig breed, Erhualian, with Western pig breeds, Pietrain or Large White, and found significant differences in back-test and stress-coping strategy during transport. We observed breed-specific regulation in the limbic-hypothalamus-pituitary-adrenal axis (LHPA) that plays critical role in stress sensitivity and welfare. We identified glucocorticoid receptor (GR) as an important player that integrates different traits (behavior, metabolism, meat quality and obesity) in difference tissues including brain (hippocampus), liver, muscle and fat. GR is particularly essential for maintaining homeostasis under stressful situations. Maternal nutrition and stress during gestation and lactation may program offspring GR expression through epigenetic mechanisms and thus determine the stress-coping style and welfare status of the offspring. Experiments were conducted on Chinese local breed Meishan and Western crossbred Landrace×Yorkshire sows to investigate the effects of maternal dietary protein deficiency and betaine supplementation on offspring performances and GR-mediated epigenetic mechanisms were found to be responsible for the fetal programming of hepatic, muscle and hippocampal expression of genes involved in metabolic and brain functions. China is the largest pig producer in the world with almost half of all pigs being raised in China. Nevertheless, the level of industrialization is still low, about 60% of all pigs being raised in farms with more than 500 sows, and the rest being produced by small family farms in rural backyards. Due to highly diverse housing and rearing systems and mixed pig breeds (local, cultivated & imported), it is difficult to implement a universal standard in pig production. In China, some of the welfare concerns arise from a mismatch between animal's genetic background and the environment provided. Imported western breeds are raised under housing and feeding conditions of lower quality than generally found in Europe and the US due to lack of resources, leading to poorer performance for these western breeds. On the other hand, native Chinese breeds are often raised according to NRC standards established in the US, and their genetic selection is based on the criteria developed for fast-growing western breeds. Such mismatch between genotype and environment may lead to the irreversible change of the breed characteristics of the native Chinese breeds with eventual loss of genetic resources through fetal programming and epigenetic modifications over several generations.

Can animal welfare science have a role in creating a sustainable future for animal agriculture?

Janice Swanson[1] and Joy Mench[2]
[1]Michigan State University, Dept. Animal Science, East Lansing, MI 48824, USA, [2]University of California, Dept. Animal Science, Davis, CA 95616, USA; swansoj@anr.msu.edu

Farm animal welfare is increasingly viewed as an important element in ensuring that animal production systems are economically, socially and environmentally sustainable. Assessing animal welfare in the context of sustainability calls for new approaches to conducting research that incorporate broad multidisciplinary expertise as well as a high degree of communication with food system stakeholders to ensure that the research contributes to informed decision-making. We provide an overview of a unique public-private partnership, the coalition for sustainable egg supply (CSES), formed to fund and support research evaluating the sustainability of different laying hen housing systems. Because of concerns about hen welfare the egg industry globally is transitioning away from conventional cages. The goal of the CSES project was to help inform this transition by conducting holistic research involving simultaneous assessment of hen health and welfare, egg safety and quality, worker health and safety, production economics, and environmental impacts on a commercial farm in the USA that had three different housing systems. The CSES members represented a variety of stakeholders, including food retailers and distributors, egg producers, trade groups, universities, and governmental and non-governmental organizations. The CSES was facilitated by a not-for-profit intermediary responsible for communicating the research results to a range of food system stakeholders. We will describe the structural aspects of the CSES responsible for the successful completion and dissemination of the research, as well as the insights that were gained regarding multidisciplinary/multi-institutional collaboration, conducting commercial-scale research, fostering and maintaining stakeholder interaction, and communicating research results. In addition, the challenges associated with information integration to elucidate trade-offs and arrive at meaningful sustainability metrics will be discussed. Discerning how animal welfare interacts with other elements of animal production systems, in consideration of social expectations for resource sufficiency and maintaining or enhancing the functional integrity of those systems, is the next trail to blaze for animal welfare science.

How to make sure cows age well: incorporating economics, environment and welfare for truly sustainable animal farming

E Vasseur[1], J Rushen[2], AM De Passillé[2] and D Pellerin[3]
[1]McGill, Montréal, QC, Canada, [2]UBC, Vancouver, BC, Canada, [3]Université Laval, Québec, QC, Canada; elsa.vasseur@mcgill.ca

Sustainable agriculture is 'ecologically sound, economically viable, and socially just'. A main concern in incorporating all components of sustainability is that efforts to reduce environmental impact and improve animal welfare may come into conflict (e.g. animals are kept in confinement to manage manure and facilities constructed for utility over comfort). An avenue to reduce environmental footprint while improving welfare would be to increase the lifespan of dairy cows and, in turn, reduce the number of animals reared. Often cows leave the herd too early due to poor welfare; this low longevity induces lost income and puts the industry under scrutiny for its attitude towards cow welfare and the environment. This presentation will summarize our research results on understanding how better welfare could lead to improved cow longevity, present research avenues to improve both welfare and longevity, and explore the interaction between welfare, environmental and economic issues. We developed a cow welfare assessment tool integrating behaviours as indicators of how animals cope with their environment (e.g. 40 cows/herd fitted with activity logger to measure herd resting time), along with outcome measures of discomfort (e.g. injuries, lameness) and the identification of environmental risk factors for poor welfare. Longevity was evaluated on % of cows in lactation ≥3 and turnover. On 240 herds assessed, we found most farmers were not following all recommendations, leading to poor outcomes (e.g. >50% of cows were housed in stalls too short and had ≥1 severe injury), and that some recommendations would need further validation to maximize welfare. We found a lack of perception of the problems often explains why producers do not adopt improved practices (e.g. farmer estimates of lameness in tie-stalls was 4.1 lower than researcher estimates) and that new knowledge transfer methods would need to be tested to encourage broader adoption. We identified risk factors for inadequate comfort that had implications for poor longevity (e.g. involuntary culling of young cows increased 15% for each 10 cm decrease in stall width) and that reliable predictors for longevity would need to be identified to help farmers make informed decisions to keep profitable cows. Canadian farmers are earning their social licence by building public trust in implementing an industry-driven framework to demonstrate how they responsibly produce milk. Our current research focus on supporting them in providing a suitable environment for future compliance needs, by better understanding how to keep comfortable and healthier cows longer in the herds, while balancing perceived societal demands with efficient management systems.

An assessment method for the management of farm animal welfare in global food companies

Heleen Van De Weerd[1], Nicky Amos[2] and Rory Sullivan[3]
[1]Cerebrus Associates Ltd, The White House, 2 Meadrow, Godalming, Surrey GU7 3HN, United Kingdom, [2]Nicky Amos CSR Services Ltd, Old Broyle Road, Chichester, West Sussex, PO19 3PR, United Kingdom, [3]University of Leeds, University of Leeds, Leeds LS2 9JT, United Kingdom; heleen@cerebrus.org

The welfare of farm animals can be improved through various pathways, such as legislation, consumer pressure expressed via NGOs or legal routes. Another pathway is to incorporate animal welfare in the Corporate Social Responsibility policies of food companies. The views that these companies hold about the welfare of animals and the management practices that they adopt are of critical importance in determining the behaviour and welfare of the 70 billion land animals annually farmed for food. The Business Benchmark on Farm Animal Welfare (BBFAW) is a structured, annual evaluation of farm animal welfare-related practices, reporting and performance of food companies. The evaluation framework focusses on: (1) Management Commitment: policies for managing farm animal welfare (FAW); (2) Governance & Performance: systems and processes for managing FAW; (3) Leadership & Innovation: efforts to advance FAW. Each category contains questions with scores reflecting how close a company is to best practice (set at 100%). In 2012 and 2013, the evaluation framework was applied to published information (e.g. corporate websites, annual and corporate responsibility reports, press releases and consumer brochures) of 70 of the world's largest food companies. Around 70% of the companies acknowledged FAW as a business issue. The overall mean BBFAW score increased significantly from 23% (2012) to 28% (2013) (Wilcoxon Signed-Rank test, P<0.001, n=68 companies assessed in both years, excluding 1 company that was demerged into 2 companies in 2013). Despite this positive trend, only 34% (2012) and 46% (2013) of companies had comprehensive FAW policies, with a further 12% (2012) and 10% (2013) having only a basic policy statement. In both years many of the policies were quite limited in scope and provided limited information on how the policy was to be implemented. While a small group of companies have integrated FAW into their business strategy, the majority have yet to provide a comprehensive account to their stakeholders and to wider society of their approach to FAW. There is evidence that the BBFAW influences company policies as it provides clear (published) expectations and guidance, while also enabling companies to benchmark themselves against their industry peers. It is difficult to disentangle the Benchmark's effects from wider pressures on companies to take action on FAW, but the increased overall scores and the feedback suggest that companies take note of their results in the annual ranking. This can ultimately drive positive change to animal's lives.

A decade of progress in the United States working to eliminate intensive confinement of hens and gestating sows

Sara Shields
Humane Society International, Farm Animals, 2100 L St., NW, Washington, DC 20037, USA; sshields@hsi.org

The field of ethology has clearly demonstrated that animals have behavioral needs, and that intensive farm animal confinement systems – cages and crates – thwart these needs. In the last ten years efforts to end the use of these systems have gained momentum in the United States. In 2006 an Arizona ballot measure passed making it a crime to confine a pig such that she cannot lie down, extend her limbs or turn around freely. In 2008 The Humane Society of the United States helped launch Proposition 2 in California, a citizen initiative that, for the first time in U.S. history, prevented conventionally used battery cages for egg-laying hens. The proposition passed by 63% of votes in favor, the largest margin ever in the state's history. Polling from 2010 indicated that 69% of consumers were willing to pay more for ethically produced products, a category that includes the humane treatment of animals, and producers and retailers are responding to changing consumer sentiment in a trend that continues today. Smithfield Foods, the world's largest hog producer, announced a transition away from gestation crates, and now over 80% of their company owned sows are in group housing, although their contract growers are transitioning more slowly, toward a 2022 deadline. Several more states now have bans on battery cages, gestation crates or both. In 2013 Federal legislation that would have banned battery cages failed to progress in Congress, but in that same year over 50 U.S. companies had policies in place to reduce the use of gestation crates in their supply chains, including Heinz, Costco, and Target. In 2015 an announcement by McDonald's was the tipping point for laying hens, as now over 100 companies have pledged to eliminate battery cage eggs from their supply chains. Many companies have set time-lines for battery-cage phase-outs, which vary, but usually aim between 2020 and 2025. In 2015 U.S. cage-free egg production expanded by 37%. A measure has qualified for the 2016 ballot in Massachusetts addressing cage and crate confinement. An evolving social climate is helping build momentum, but at the heart of the advancing change is progress in the field of ethology demonstrating management, technology and housing that makes alternative systems possible. To house sows in groups, aggression must be controlled, and this is being addressed by feeding behavior interventions, as competition during meal presentation is a common trigger for agonistic interactions. For laying hens, research on improved aviary designs is addressing the challenge of keeping birds safe while permitting perching, dustbathing and nesting behavior. The movement away from cages and crates reflects a true case of application of the animal behavior science.

Comfort amongst crisis: is comfort of working animals a luxury within the international development context?

Rebecca Sommerville and Kimberly Wells

The Brooke, 5ᵗʰ Floor Friars Bridge Court, 41-45 Blackfriars Road, London SE1 8NZ, United Kingdom; rebecca.sommerville@thebrooke.org

Comfort is recognised as a central component of animal welfare by scientists and non-government organisations. Comfort needs combine the avoidance of pain and physical disorders with positive aspects of existence involving what animals want. Whilst more reachable within the developed world, the reality for animals can be vastly different in the Global South. In low income contexts where equids are key workers in agriculture, construction, tourism and labour or transport for individuals and families, local risk factors (shocks, trends and seasonal fluctuations) make livelihoods vulnerable. The consequence of peoples' survival-based decisions results in animal welfare being less of a priority. The Brooke is an international organisation which aims to make sustainable improvements to the welfare of working horses, donkeys and mules across Africa, Asia and Latin America. Standardised welfare monitoring shows that issues considered severe in developed contexts are often the working equine population norm. Data collected in Nepali brick kilns in 2015 showed almost 90% of animals unable to walk normally, 50% with at least one body wound, 90% in poor body condition and 20% displaying fearful behaviour towards a human assessor. Healing, and thus loss of comfort, can be prolonged as animals continue to work. Essential comfort can be promoted amongst challenged human livelihoods. Examples include training local service providers and owners in handling and behavioural modification to improve comfort during treatment at Indian equine fairs. A 'pleasant life' is even more challenging at times of crisis. Following the major earthquake in Nepal in 2015, basic comfort needs were provided through feed, veterinary care and resting areas, alongside supporting human livelihoods. In Nicaragua, novel approaches are being trialled to provide sufficient feed for comfort during challenging climatic conditions, including drought resistant crops. Recent droughts in Ethiopia and Senegal risked animals' survival, but comfort remained a consideration. The lack of a 'comfortable' life is the trade-off and reality for a working animal in the developing world; however improvements to comfort are possible if addressed within welfare impact strategies involving the people responsible for their welfare. Research is needed to investigate how strong livelihoods can be achieved in these contexts, without risking animal welfare. Based on evidence from welfare assessment, experience in field conditions has shown that achieving win-win scenarios for animals and people must take into account humans' survival-based decisions.

The big picture: synergies and trade-offs between animal welfare and other sustainability issues

Eddie A.M. Bokkers

Wageningen University, Animal Production Systems group, P.O. Box 338, 6700 AH Wageningen, the Netherlands; eddie.bokkers@wur.nl

Sustainable development of livestock systems involves innovations and the adaptation of these systems in a manner that not only meets the needs of current human generation, but also the potential needs and aspirations of future generations. Innovations can be for example novel feeding practices, breeding strategies or farm technologies and designs. To prevent unforeseen consequences, innovations should be studied in a broad context with an integrated research approach that must combine three key themes for livestock systems: animal welfare, environmental impact (e.g. use of resources and emissions to air, water or soil) and profitability (e.g. farm income). With such an integrated analysis synergies and trade-offs can be identified. For about ten years now, our team has worked on projects that aim to improve animal welfare and reduce the environmental impact of livestock systems while ensuring economic viability. We studied, for example, trade-offs between animal welfare and the environment in organic and conventional pig production and found that land use and contribution to climate change related to feed showed to be higher (83% resp. 36%) in organic pig production, while welfare indicators gave a variable picture. Different production systems for layer hens were compared for various sustainability issues and we found that systems with a potential better welfare performed economically better but had higher greenhouse gas emissions, land occupation and acidification potentials. We also did an integrated analysis of veal diets to improve calf welfare, while taken into account environmental and economic consequences and found that a diet containing more roughage improved welfare, was economically attractive and neutral for the environment. A study in dairy cows showed that diseases after calving increase the contribution to climate change with 3.1% at cow level and the economic costs with €111 cow-1 y-1. These examples show that for a sustainable development of livestock systems it is not enough to focus on one theme at a time. One challenge to this type of study is to collect data for animal welfare, environmental and economic indicators from the same farm at the same point in time. Nevertheless, this is the best way for identifying synergies and trade-offs and is therefore essential if research is to significantly contribute to sustainable development of livestock systems in the future. Describing and analysing livestock systems based on more than one theme of sustainability creates complex, but more complete pictures. It is critical that researchers in animal welfare also consider the impact of innovations on environmental and economic performance if these innovations are to be long-lasting.

Organic pig husbandry in Europe: do welfare and environmental impact go hand in hand?

Christine Leeb[1], Gwendolyn Rudolph[1], Sabine Dippel[2], Davide Bochicchio[3], Gillian Butler[4], Sandra Edwards[4], Barbara Früh[5], Mirjam Holinger[5], Gudrun Illmann[6], Armelle Prunier[7], Jean-Yves Dourmad[7], Tine Rousing[8] and Christoph Winckler[1]
[1]*Univ. of Natural Resources & Life Sciences, Gregor-Mendelstraße 33, 1180 Vienna, Austria,* [2]*FLI Celle, Dörnbergstraße 25/27, D-29223 Celle, Germany,* [3]*CRA-SUI, 345 Via Beccastecca, 41018 San Cesario sul Panaro, Italy,* [4]*Newcastle University, Agriculture Building, NE1 7RU Newcastle upon Tyne, United Kingdom,* [5]*FiBL, Ackerstrasse 113, 5070 Frick, Switzerland,* [6]*Institute of Animal Science, Přátelství 815, 10400 Prague, Czech Republic,* [7]*INRA, Domaine de la Prise, 35590 Saint-Gilles, France,* [8]*Aarhus University, Blichers Alle 20, 8830 Tjele, Denmark; christine.leeb@boku.ac.at*

Organic pig systems are diverse, ranging from indoor pens with concrete outside run to outdoor systems all year round. It is often discussed, that husbandry systems common in organic farming enhance animal health and welfare (AHW) but impair environmental impact (ENV). However, the level of AHW and ENV has never been quantified for those systems using on-farm data. In total 74 pig farms in 8 countries were visited in 3 different systems: indoor (IN; n=34), partly outdoor (POUT; n=28) and outdoor (OUT; n=12). Animal based parameters including exploratory behaviour of pregnant sows (SO), weaners (WE) and fattening pigs (FA) were assessed by trained observers using an adapted Welfare Quality® Protocol. Life cycle assessment (LCA) was conducted for 64 production chains (PC), combining piglet production and finishing stage to quantify ENV for the criteria greenhouse gas emissions (GHGE), acidification (AP) and eutrophication potential (EP). Across systems, prevalences of most AHW areas were low; exceptions were respiratory problems (IN, POUT), diarrhoea (IN), short tails (IN, POUT) and total suckling piglet losses (IN, POUT, OUT). Regarding GHGE, no differences were found between systems. POUT showed lower AP than IN and lower EP than OUT. Hierarchical cluster analysis revealed three clusters: a 'low ENV' cluster; an 'intermediate ENV' cluster and a 'high ENV' cluster. No significant association was found between AHW and ENV when comparing the ENV clusters with regard to an overall AHW summary score (GOOD%), summary scores per animal category (GOOD%_SO, GOOD%_WE or GOOD%_ FA) and single animal-based parameters or correlations between GHGE, AP, EP and GOOD%. The lack of association between AHW and ENV found in this study is promising, but does not necessarily mean that no associations exist. The main reasons for a lack of associations may be the fact that LCA includes impact areas (e.g. manure storage), which do not relate to AHW. Still, this study generated a starting point to explore associations between AHW and ENV to be tested either on a larger number of PCs or between specific AHW and ENV areas.

Know thy neighbour – does loose housing at farrowing and lactation positively influence sow welfare post-weaning?

Sarah H. Ison[1,2], Maria Camila Corredor[3,4] and Emma M. Baxter[2]
[1]Michigan State University, Animal Science, 474 S. Shaw Lane, East Lansing, MI 48824, USA,
[2]Scotland's Rural College (SRUC), Animal Behaviour and Welfare, Roslin Institute Building,
Easter Bush, Roslin, EH25 9RG, United Kingdom, [3]La Salle University, Cra. 7 No. 179, Bogotá,
Colombia, [4]University of Edinburgh, Royal (Dick) School of Veterinary Studies, Easter Bush
Veterinary Centre, Roslin, EH25 9RG, United Kingdom; shison@msu.edu

Legislation and consumer pressure is leading to a global shift in sow housing. Welfare concerns have led to a reduction in confinement during gestation and more recently farrowing and lactation. Crate-free systems must maintain a high level of productivity in order to feed an ever-increasing global population. This study characterized behaviour, lesions and condition of sows mixed into familiar groups of up to six after weaning from a loose-housed farrowing system (PigSAFE pens, PS, n=22) or conventional crates (C, n=24). PS allowed tactile contact between neighbouring pens. Eight groups housed together during gestation were returned to the same group at weaning 31 days later. Sow weight and back-fat were measured on entry to and exit from the farrowing system. Focal observations 11 hours post-mixing included posture (stand, sit, lie) and aggressive interactions (fight, bully, bite). Fresh lesions were counted before mixing, and one, four and 24 hours post mixing. Time spent lying next to, and aggressive interactions between PS or C neighbours (N) and non-neighbours (NN) was calculated on a per sow basis. Data were analyzed using generalized linear models. C sows did not show significant differences in aggressive interactions or duration spent lying with sows that were neighbours or not (P>0.05). However, PS sows spent more time lying next to (N=55.0±11.1, NN=20.3±4.9), and had fewer fights (N=0.07±0.04, NN=0.3±0.08), bites (N=1.0±0.04, NN=2.0±0.7), and bullies (N=0.2±0.1; NN=0.8±0.3) per sow towards N than NN (all P<0.05). Over lactation, PS sows lost less weight (PS=-22.5±2.8 kg, C=-36.6±3.8; P=0.04) and tended to lose less back-fat (PS=-4.9±0.8 mm, C=-7.3±1.1; P=0.09) than C sows. For post-mixing lesions, PS sows received more lesions to the front of the body (PS=15.6±3.0, C=7.5±0.8), whereas C sows had more at the rear (PS=2.7±0.4, C=7.6±1.5), leading to a housing × lesion location interaction (P<0.001). For the frequency of wounding aggression, there was a behaviour × system interaction (P=0.009), as PS sows had fewer fights (PS=0.9±0.20, C=2.3±0.4) and delivered more bites than C sows (PS=7.4±2.4, C=4.8±0.9), but had a similar frequency of bullies (PS=0.9±1.3, C=0.9±0.8). Better body condition at weaning and greater familiarity between individuals could benefit the production potential of loose-housed sows.

Dog and human behaviour in a mass street dog sterilization project in Jamshedpur, India

Tamara Kartal, Joy Lee, Amit Chaudhuri and Andrew N. Rowan
Humane Society International, 2100 L St., NW, Washington, DC 20037, USA; tkartal@hsi.org

A mass street dog sterilization program was launched in Jamshedpur, India, in 2013. The sterilization project focused on the major part of Greater Jamshedpur (covering 1,337,131 people) that included the communities of Jugsalai, Mango and Aidityapur (total human population of 944,006). A baseline survey of these communities estimated a total of 24,774 street dogs (2.62 dogs per 100 people) using mark/recapture methods. In India, street dogs used to be caught using metal tongs or wire nooses but those inhumane methods gave way to the use of nets. Because the dogs rapidly learn to run from the net carrier, likely because it often includes negative human-canine interactions (e.g. chasing the dogs), the Jamshedpur project developed a new catching technique focused on catching as many dogs as possible by hand. Staff were recruited and trained in dog behaviour and humane handling techniques and were then deployed initially to the Hindu areas of Mango, where staff alerted the community to the project. Every encountered dog was assessed for behavior and scored as follows: B1: Defensive aggression; B2: Afraid and runs away; B3: Timid but accepts food interaction; B4: Comfortable being petted but not picked up; B5: Friendly, can be petted and carried. Scoring was done on smartphones using an android app that also recorded the GPS coordinates when a photo of the dog was taken while being assessed, its body condition score, whether it was male or female and, if female, whether it was lactating. This data was then imported to a database and used to inform the CNVR (Capture-Neuter-Vaccinate-Return) strategy for the community. Of the 15,845 dogs that were caught, approximately 70% could be hand caught with variations between communities. The engagement with the community and the catching process was observed to have a positive impact on how the local community interacted with the dogs. On average, 50% of the dogs scored B4 or B5 and over 90% had a body condition score of 3 (scores 1-5) or higher at the time of assessment, indicating that street dogs in Jamshedpur have regular access to food. Before, during and after the sterilization process, the animal welfare officers ensured that the dogs are comfortable and that stress is kept to a minimum. As a result dogs did not show defensive or fear aggression towards staff in the facility and were very tolerant of post-op assessments both in the clinic and when back on the streets. This Project demonstrates that high street dog welfare can be maintained during CNVR sterilization. In addition, it appears that hand catching facilitates access to the dog during post-operative monitoring by project staff as well as indirectly promoting gentler human-dog interactions in the community.

Poultry housing systems and gas emmissions; panacea for good animal welfare in a developing country

Olufemi Mobolaji Alabi[1] and Oyebola Sabainah Akinoso[2]
[1]Bowen University, Department of Animal Science and Fisheries Management, Iwo, Osun State, 230001, Nigeria, [2]University of Ibadan, Mathematics Education, Ibadan, Ibadan, Oyo State, 232001, Nigeria; femiatom@yahoo.com

Green house gas emissions are inevitable in animal agriculture and are of topmost interest in the assessment of general well being of both ruminant and non-ruminant animals. However, the issue is not being given the deserved attention most especially in developing countries where little or no emphasis is being placed on animal welfare hence the need to sensitize researchers and animal handlers in this region of the world on the subject matter. Accumulation of nitrogenous gases in poultry houses depends on the housing system adopted and the management therefore coupled with the production purpose. Ammonia gas emission from the decomposing faeces of laying chickens housed inside battery cages extensively is very common. This may lead to eyes infections among the chickens and drop in their production performances with high mortality rate. The roofing of the houses are also prone to corrosion and damages and the farm workers are also exposed to the risk of gas inhalation and subsequent effects on human health. Moreover, chickens housed on deep litter system may experience the same thing from stale and wet litters. This will definitely affect their behavioral patterns in term of feeding, alertness, physiological responses as may be reflected in the growth pattern, feed concersion efficiency, egg production, egg fertility and hatchability and their immune system. Also, chickens under this housing conditions may exhibit agonistic behavior towards flock members and the handlers. However, prompt disposal of the wastes and open sidedness of the poultry pen are ways of alleviating the challenge in battery cage system while good litter management and provision for good ventilation is necessary for good deep litter housing. Meanwhile, organic free range system would have been the best option to minimize the green house gas in poultry production subject to availability of land and fund as may be applicable to developing countries where there is need to promote issues of behavior as indicator of animal welfare.

Who cares about the sick cow in Brazil? Dairy health consultants, the farmers or do they fall between the cracks?

G Olmos[1,2], MAG Von Keyserlingk[3] and MJ Hötzel[2]
[1]Swedish University of Agricultural Sciences, Uppsala, 750 07, Sweden, [2]Universidade Federal de Santa Catarina, Lab. Etologia Aplicada e Bem-Estar Animal, Florianópolis, 88034-001, Brazil, [3]University of British Columbia, Animal Welfare Program, Vancouver, V6T 1Z4, Canada; gabriela.olmosantillon@gmail.com

Animal disease has undesirable effects for animal welfare and farm profitability; reducing the number of sick cows must be a priority. Brazil deals with animal health in a descending linear hierarchy. Although the federal government agenda is set around international trade, it recognizes state authorities and farmers as being semi-autonomous and thus is an arms length stakeholder in farm level decisions. However, given this policy decision the question remains as to whether or not this approach works, particularly from the perspective of the 'sick cows'. Our aim was to determine the current strategies in place for treating the sick cow on small grazing based dairy farms in southwest Brazil. To provide insights into this complex topic we audio recorded face-to-face semi-structured conversations with farmers (n=21) and associated dairy health consultants (including veterinarians) (n=13). Our findings indicate that state authorities believe that farmers should be responsible for all animal health issues, with the exception of those controlled by federal directives (e.g. foot and mouth disease). Although farmers expressed a desire to improve health and, thus, productivity only 1/3 of those interviewed had a defined strategy. Yet, no farmer kept detailed health records, which are a necessary tool to evaluate current status and effectiveness of management changes on animal health. A common practice when dealing with sick cows was to call for a health consultant for diagnosis but this was frequently limited to a telephone conversation. Farmers justified this behaviour by their desire to reduce veterinary costs and the reluctance to hear the sales pitch that was often linked with visits. Although the consultants disapproved of this indirect method of diagnosis, they argued that they were overburdened by the number of farms in their area, stating ratios of consultant to farms from 1:30 to 1:1000. Conversations also highlighted that not all clinical cases were dealt by veterinary professionals; additionally, farmers were often unable to distinguish between a trained veterinarian or another type of health provider. It was also clear that farmers' rationale for choosing whom to contact for advice was based primarily on trust achieved through previous experiences, and not on professional credentials. This work shows that there is no clear structure in place for dealing with a sick cow on dairy farms in this region, which is likely further complicated by the perceived lack of access to professional advice on animal health.

Behavioural indicators of cattle welfare in silvopastoral systems in the tropics of México

Francisco Galindo[1], Lucía Améndola[1], Javier Solorio[2], Juan Ku-Vera[2], Ricardo Améndola-Massioti[3] and Heliot Zarza[4]
[1]*Universidad Nacional Autónoma de México, Facultad de Medicina Veterinaria y Zootecnia, Av. Universidad 3000, 04510 Mexico City, Mexico,* [2]*Universidad Autónoma de Yucatán, Facultad de Medicina Veterinaria y Zootecnia, Itzimná s/n, 97100 Merida, Yucatán, Mexico,* [3]*Universidad Autónoma de Chapingo, Departamento de Zootecnia, Texcoco, 56230, Chapingo, Mexico,* [4]*Universidad Autónoma Metropolitana-Unidad Lerma, Departamento de Ciencias Ambientales, Lerma, 52005, Lerma de Villada, Mexico; galindof@unam.mx*

Silvopastoral systems are a good alternative for sustainable livestock production as they provide ecosystem services and can improve animal welfare. The aim of the study was to compare the welfare of cattle in an intensive silvopastoral (ISS) system based on high densities of *Leucaena leucocephala* combined with grass (*Cynodon nlemfuensis*), and in a monoculture system (MS) based on C. nlemfuensis, both systems in the Yucatan peninsula. In total we observed eight heifers per system. In each system, the eight cows were observed from 7:20 – 15:30 during four days in each of three paddcoks per system, both during the dry and the rainy seasons. GLMM models were carried out to compare time budgets (foraging, ruminating, idling, lying) in both systems. The temperature-humidity index (THI) was higher in the MS than in the ISS ($F=37.61$, $P>0.001$). The temperature-humidity index (THI) was higher in the MS than in the ISS ($F=37.61$, $P>0.001$). Daily foraging times were significantly longer in the MS than in the ISS (t.test: $t=9.50$, $df=13.99$, $P<0.001$) while daily ruminating times were significantly longer in the ISS than in the MS during both seasons (dry: $t=-13.09$, $df=13.81$, $P<0.01$; rainy: $t=-11.30$, $df=9.51$, $P<0.01$). In the ISS heifers spent significantly more time idling than heifers in the MS during the dry season ($t=-7.7$, $df=13.62$, $P<0.01$), but not during the rainy season ($t=-1.43$, $df=13.35$, $P=0.17$). Heifers in the ISS spent significant more time lying than heifers of the MS ($t=-2.69$, $df=11.22$, $P<0.05$) during the dry season but not in the rainy season ($t=0.79$, $df=12.43$, $P=0.45$ during the rainy season). Within the ISS, heifers spent significantly more time lying during the dry than the rainy season ($t=3.89$, $df=7$, $P<0.01$); differences between seasons in the MS were not significant ($t=0.09$, $df=07$, $P=0.10$). Our results indicate that heifers in ISS reduce foraging time and may benefit from better quality lying bouts and be less influenced by heat and humidity.

Litter investment and sibling competition – high cost of modern selection for sow and piglet welfare

Marko Ocepek and Inger Lise Andersen
Norwegian University of Life Sciences, Department of Animal and Aquacultural Sciences, P.O. Box 5003, 1432 Aas, Norway; marko.ocepek@nmbu.no

Genetic selection is likely to have produced a shift in maternal investment towards early litters. Sows of maternal lines have a greater number of, and higher fitness, piglets than the number of functional teats. The aim of this project is to study the effect of high litter investment on sow welfare in terms of loss in body condition score and development of shoulder lesions, and the consequence of increased sibling competition at the udder for maternal lines compared to a breed that is not selected for high productivity and maternal traits. Differences in total litter investment (litter weight at weaning, weight of mummified/stillborn piglets, weight of piglets that died after farrowing but before weaning), in the proportion of disused teats, in teat constraints (teat pair distance, teat length, teat diameter), and in sibling competition (teats per piglet, piglets with two/without a teat) between breeds that are subject to different selection pressures (Landrace (L: selected maternal line, n=12), Landrace and Yorkshire (LY: crossbreed selected line, n=15), Duroc (D: non-select line, n=12)), were analysed using a GLM. Differences in physical characteristics (body condition (1-9), shoulder lesions (0-4)) were determined using GENMOD procedure (SAS). The maternal sow line (L) had higher total litter investment in comparison to LY and D breed (146.8, 138.9, 73.7 kg, respectively; P<0.05). Such a large litter investment and higher milk production in the maternal line imposed higher costs for sows through larger losses in body condition score (2.2, 1.4, 1.5, respectively; P<0.01) and a higher prevalence of shoulder lesions during lactation (1.0, 0.7, 0.3, respectively; P≤0.001). Irrespective of breed, around 22% of teats are not accessible shortly after birth (P=0.392), and a large teat pair distance is the main constraint for teat use (due to excessive height above ground in the upper row and poor exposure of the lower row). The non-selected line (D) had more functional teats per piglet (D: 1.7, L: 1.2, LY: 1.1; P<0.001), had a higher proportion of piglets using two teats (D: 17.3%, L: 8.6%, LY: 3.7%; P<0.001) and a lower proportion of piglets failing to access teats shortly after birth (D: 3.6%, L: 11.5%, LY: 12.4%; P<0.001). The results underline the importance of understanding how selection impairs sow welfare through increased maternal investment (without improving sow physical characteristics) and influences piglet welfare due to increased sibling competition at the udder (without improving important teat constraints).

Animal welfare consequences of broiler transport: comparison between a Brazilian abattoir and a United Kingdom abattoir

Ricardo Lacava Bailone[1,2], Ricardo Borra[3], Roberto De Oliveira Roca[2] and Moira Harris[4]
[1]Ministry of Agriculture, Livestock and Supply, Brasilia, DF 70.043-900, Brazil, [2]Universidade Estadual Paulista Júlio de Mesquita Filho (UNESP), Department of Animal Health, Veterinary Public Health and Food Safety, Botucatu, Sao Paulo 18.618-970, Brazil, [3]Federal University of São Carlos (UFSCAR), Department of Genetics and Evolution, São Carlos, Sao Paulo 13.565-905, Brazil, [4]Harper Adams University, Department of Animal Production, Welfare and Veterinary Sciences, Edgmond, Newport TF10 8NB, United Kingdom; ricardo.bailone@agricultura.gov.br

Transport is a stressful process that can result in mortality, injuries and increased physiological stress with associated effects on meat quality. Data were collected from two large abattoirs, one in Brazil and one in the UK. Numbers of broiler birds dead on arrival (DOA) at the abattoir, numbers of injuries on post-mortem inspection and corticosterone levels in blood serum (measured by ELISA test) were examined over two seasons (summer and winter) and three transport distances from farm to abattoir (short, 0-50 km, medium, 51-150 km and long, 151-300 km). Percentages of DOA were higher in Brazil than in the UK (0.72 vs 0.13% overall). In Brazil, in winter there was no relationship (P=0.33) between DOA and distance while in summer there was a positive correlation (P=0.035). In both seasons there was a positive correlation between injures and distance (P=0.004 (winter) and P=0.003 (summer)). In the UK, in summer there was no relationship between DOA and distance while in winter DOA percentages were higher after long distance transport (P<0.01). UK injuries data are currently being analysed. Regarding corticosterone levels, UK data showed significantly lower levels for long transport compared to short and medium which did not differ from each other: mean (±SE) = 1,137.1 (69.4) pg/ml, 999.2 (69.4) pg/ml and 712.8 (70.4) pg/ml in short, medium and long distances respectively (P<0.001). Brazilian corticosterone data will be collected in June 2016. To summarise, Brazil had higher levels of DOA than the UK. In Brazil, DOA was higher in summer than winter and in the UK the opposite was true. Long distance transport in the UK was less stressful than short or medium distance. In conclusion, in Brazil, during summer, more birds arrived dead at the abattoir after long distance transport and in the UK, long distance transport during winter resulted in higher DOA, suggesting that mortality in transport is related to heat stress in Brazil and cold stress in the UK.

Authors index

A

B